Diversity in the Knowledge Economy and Society

THE GWU/NIFU STEP SERIES ON SCIENCE, INNOVATION, TECHNOLOGY AND ENTREPRENEURSHIP

Series editors: Elias G. Carayannis, *Professor of Science, Technology, Innovation and Entrepreneurship, School of Business, George Washington University, USA and* Aris Kaloudis, *NIFU STEP Studies in Innovation, Research and Education, Oslo, Norway*

There is ample and growing evidence that intangible resources such as knowledge, know-how and social capital will prove to be the coal, oil, and diamonds of the twenty-first century for developed, developing, and emerging economies alike. Moreover, there are strong indications and emerging trends that there are qualitative and quantitative differences between the drivers of economic growth in the twentieth and twenty-first centuries.

This new era is punctuated by:

- Development of a service-based economy, with activities demanding intellectual content becoming more pervasive and decisive.
- Increased emphasis on higher education and life-long learning to make effective use of the rapidly expanding knowledge base.
- Massive investments in research and development, training, education, software, branding, marketing, logistics, and similar services.
- Intensification of competition between enterprises and nations based on new product design, marketing methods and organizational forms.
- Continual restructuring of economies to cope with constant change.

This valuable new series concentrates on these important areas by focusing on the key pillars of science, technology, innovation, and entrepreneurship.

Diversity in the Knowledge Economy and Society

Heterogeneity, Innovation and Entrepreneurship

Edited by

Elias G. Carayannis

Professor of Science, Technology, Innovation and Entrepreneurship, School of Business, George Washington University, USA

Aris Kaloudis and Åge Mariussen

NIFU STEP Studies in Innovation, Research and Education, Oslo, Norway

THE GWU/NIFU STEP SERIES ON SCIENCE, INNOVATION, TECHNOLOGY AND ENTREPRENEURSHIP

Edward Elgar

Cheltenham, UK • Northampton, MA, USA

Published by
Edward Elgar Publishing Limited
Glensanda House
Montpellier Parade
Cheltenham
Glos GL50 1UA
UK

Edward Elgar Publishing, Inc.
William Pratt House
9 Dewey Court
Northampton
Massachusetts 01060
USA

A catalogue record for this book
is available from the British Library

ISBN 978 1 84720 211 6

Printed and bound in Great Britain by MPG Books Ltd, Bodmin, Cornwall

Contents

Contributors

Elias G. Carayannis is Full Professor of Science, Technology, Innovation and Entrepreneurship as well as co-Founder and co-Director of the Global and Entrepreneurial Finance Research Institute (GEFRI) and Director of Research on Science, Technology, Innovation and Entrepreneurship, European Union Research Center (EURC) at the School of Business of the George Washington University in Washington, DC. Dr Carayannis has ten books published and several others under contract with CRC Press, Praeger, Macmillan, Edward Elgar, and other publishers, as well as more than 50 publications in both academic, peer-reviewed and practitioner, US and European journals such as *IEEE Transactions in Engineering Management*, *Research Policy*, *Journal of R&D Management*, *Journal of Engineering and Technology Management*, *International Journal of Technology Management*, *Technovation*, *Journal of Technology Transfer*, *Engineering Management Journal*, *Journal of Growth and Change*, *The Review of Regional Studies*, *International Journal of Global Energy Issues*, *International Journal of Environment and Pollution*, *Le Progres Technique*, and *Focus on Change Management*. He has consulted for several large as well as small technology-driven government and private organizations, such as the World Bank, the European Commission, the Inter-American Development Bank, the US Agency for International Development, the National Science Foundation Small Business Innovation Research Program, the National Institute of Standards and Technology Advanced Technology Program, the National Coalition for Advanced Manufacturing (NACFAM), the USN CNO Office, Sandia National Laboratories' New Technological Ventures Initiative, and others.

Helge Godø is a senior researcher and Program Director at NIFU STEP. He is also a professor (adjunct) in telematics and computer technology at UniK – University Graduate Center, Kjeller, affiliated with the University of Oslo. His main research interest is technological development and innovations, in particular ICT and new energy technologies, and how this interacts with societal development, explaining how and why technologies are created and how social and cultural factors are incorporated in technological solutions. In this, the role of R&D and its relationship to innovation processes are essential – and how these are influenced and constrained by different types of boundary conditions, networks and socioeconomic environments.

Eric J. Iversen has been a researcher at NIFU STEP in Oslo, Norway, since 1994. He is currently working on a PhD at the Australian Innovation Research Centre at the University of Tasmania in Australia. His work in the field of innovation and the economics of technological change centers on the framework conditions of technological change with special focus on quantitative and qualitative analyses of standardization, IPRs, and their interrelationship.

Aris Kaloudis is Program Director and a senior researcher at NIFU STEP – Studies in Innovation, Research and Education – in Oslo, Norway. He is an economist specializing in research and innovation policy analysis and impacts. He is participating in a number of large EU projects, such as INNO Policy TrendChart, ERAWATCH and Sectoral Innovation Watch. Currently his interests are in formal and informal training, skills, and human capital formation and specialization.

Mark Knell is a senior researcher at NIFU STEP. Mark has published widely on various aspects of technology and innovation, and economic transformation, as well as in the history of economic thought, and has edited two books for Edward Elgar on the transformation of Eastern Europe. He has taught these subjects at several universities in the United States and Europe, and has been an economic affairs officer in the United Nations system.

Åge Mariussen is Senior Research Fellow at the NIFU STEP institute. He is a sociologist with a background in studies of regional development, national systems and innovation in Norway and other Nordic countries. He has worked at the University of Tromsø, Nordland Research Institute, Bodø, Nordregio, Stockholm and in the Step Group in Oslo.

Svein Olav Nås is educated in economics, political science and history. He is presently a senior researcher at NIFU STEP. From 1990 to 1993 he worked as a researcher at the Norwegian Computing Centre, after which he took part in establishing the independent research institute STEP. Research topics include studies of innovation systems and, in particular, issues related to measurement of innovation indicators. A particular interest has been the utilization of administrative data sources to investigate issues related to human resources and business demography, and their relation to innovation activity. He has taken an active part in OECD work on indicators through the NESTI group, and has been involved in the development and revisions of the Oslo Manual on innovation over several years. He has been involved in national commissions on research and education and benchmarking, both as a member and a secretary.

Finn Orstavik is a senior researcher at NIFU STEP in Oslo, Norway, with a doctorate in sociology from 1996. His key research interest is the overlapping field of innovation studies and sociological theory. His core competence is in systems theory and in innovation process research. In addition, he is an expert on the developments in the policy system in Norway during the post-World War II period. Orstavik is currently leading a large project studying innovation in the construction industry.

Tore Sandven is a sociologist. His empirical research has mainly centred on quantitative studies of innovation and on business demography. He has also worked on issues relating to the philosophy of the social sciences, with a special focus on the implications of different conceptions of human rationality for the understanding of economic systems and processes.

Olav R. Spilling was educated as an engineer at the Norwegian University of Technology, Trondheim, Norway, 1970 and is a senior researcher at NIFU STEP. He held a professor position at the Norwegian School of Management BI (1995–2005) and has been Adjunct Professor at the Lillehammer University College since 2006. Spilling has a broad background in the fields of regional development, small business and entrepreneurship and dynamic processes related to this. He has also done research on the industrial impacts of mega-events (e.g. the Winter Olympics hosted in Lillehammer, Norway, in 1994) and on the evolution of high-technology clusters and gender relations in entrepreneurship and small business management. His main interests are the evolution of high-technology clusters (2000–02, at the Norwegian School of Management), theoretical and empirical studies of gazelles, and gender relations in the Norwegian industrial structure.

Heidi Wiig Aslesen is a senior researcher and Programme Director in the field of Innovation Studies at NIFU STEP and has a PhD from Lunds University addressing factors affecting innovation and economic growth in city regions, focusing on the role of knowledge intensive business services in heterogeneous agglomerations. Her core research competence is related to regional industrial development, with special emphasis on issues concerning regional growth and regional inter-firm linkages and systems of interactions. In the last few years her research focus has been on cities and knowledge-intensive business services. She has published several book chapters, articles and papers and has been part of projects funded by the Norwegian Research Council, the European Commission, the OECD, and other organizations.

Preface

This book, co-edited by Carayannis, Kaloudis and Mariussen – the first of an Edward Elgar Series on Science, Technology, Innovation and Entrepreneurship (EE STIE) co-edited by Carayannis (GWU) and Kaloudis (NIFU STEP) – was the 'happy accident' of a transatlantic, trans-disciplinary collaboration on research about research on policies and practices on STIE.

A team of Europe- and US-based authors engaged in a 'creatively destructive' process of idea and concept co-opetition (a hybrid form of collaboration and competition), co-evolution and co-specialization to attempt to co-develop a more coherent and unified language of thought about, and thus shed more light on, the socioeconomic dynamics of knowledge production, diffusion and use from both a public and a private sector policies and practices perspective. These entrepreneurs of the mind have been architecting the constituent elements of a conceptual research modality to better deal with early 21st-century challenges and opportunities in the knowledge economy and society.

The implications are significant in terms of a number of current and even pressing matters that pertain to education, economic development, research and development policies and practices in the glocalizing socioeconomic and technological context of the early 21st century such as:

a. How does one achieve higher levels of efficacy in funding education in terms of both civil and economic development?
b. How does one inspire and achieve more efficacy in public–private partnerships in research and technological development so that more critical technologies are developed, resulting in more breakthrough innovations with limited resources?
c. How does one manage to leverage the glocalization process so that more people benefit more from it rather than becoming its victims?

We do not address such questions directly. Rather, we attempt to provide a conceptual framework and process to help pose those questions so that they can be better and more effectively addressed, namely a system-centric approach as indicated in Figure 1.1 in the book. We thus attempt to evolve the dialectic dynamic and exploratory frameworks to enable a more coherent and congruent understanding and facilitate a more effective dialogue across public and private sector domains and boundaries. This emerging and more fitting conceptualization framework recognizes that knowledge stocks and flows form, evolve, and dissolve through time and observe their particular rules and constraints different from those ruling tangible assets domains and traditional (localized) economies and societies.

We hope that this first book of the EE STIE series will serve as a useful building block towards a further exploration and a better understanding of the role of diversity and heterogeneity issues in knowledge systems.

Acknowledgements

Many thanks to the Research Council of Norway for funding the research done by NIFU STEP staff presented in this book. Thanks go also to John Taylor who helped the editors with proof-reading a large number of the chapters in the book.

To George E. Carayannis, Esq. and
Maria G. Karagiannis, MD

1. Introduction

Elias G. Carayannis, Aris Kaloudis and Åge Mariussen

In modern research and innovation policies there is a recurrent issue that clearly is purely understood and difficult to grasp, namely the issue of how to address the variety and heterogeneity of knowledge systems. Our understanding of causes and effects of this variety is still very poor. This is true at all levels of analysis, the micro (firm level), meso (regional, sectoral, institutional) level as well as the macro (national and global) level. This book attempts to bring the analysis a step further conceptually and analytically.

CONCEPT OUTLINE

System theories address complex phenomena often characterized by heterogeneity. Heterogeneity is *the quality of being diverse and not comparable in kind* (Webster's Dictionary). It is useful to remember the Greek etymological roots of the word to mean literally 'possessing different genes'. Darwin and his followers have extensively analysed how micro-level genetic heterogeneity, mediated by processes of selection, has created varieties of species co-existing, co-evolving and co-specializing in natural ecosystems, thus feeding back into new combinations and re-combinations of genes. However, as forcefully argued by Helge Godø in Chapter 2 the evolution of the knowledge economy and society cannot be understood simply through loose ad hoc metaphors to ecologies in nature.

Instead, this book develops basic building blocks of a new understanding of how heterogeneity, selection and diversity as properties of knowledge systems explain how entrepreneurship and innovation work.

The key message of the book is that heterogeneity and diversity should be seen as intrinsic and indispensable properties of knowledge systems. We address the concept of *heterogeneity* in a multi-disciplinary fashion, including perspectives from evolutionary economics and innovation system studies. We relate this attempt to existing theories in a broad range of fields. In doing so, we explore some of the ways in which rationalities, identities, preferences and intentions are related to national, sector and regional economic structures, firm demography, technological paradigms, institutions and policies.

In Chapter 5 (see Figures 1.1–1.3) Carayannis discusses how globalization serves as both a catalyst of accelerated development as well as an agent of chaotic disruption resulting in socioeconomic and political dislocations. The emerging gloCalizing (i.e. simultaneously globalizing *and* localizing) (Carayannis and von Zedwitz, 2005; Carayannis and Alexander, 2006) frontier of converging systems, networks and sectors of innovation confronts us. As a result, we need to re-conceptualize the dynamics of specialization (firm, sectoral, regional, national) in this new and emerging context of the gloCal knowledge economy and society.

As opposed to mainstream economics, evolutionary economics and innovation studies place the issue of firm heterogeneity at the centre of economic development. This is because *entrepreneurship* and *innovation* activities are understood as *heterogeneity-inducing micro-processes*, continuously reshaped due to market selection mechanisms.

Carayannis postulates that one approach to such a re-conceptualization is the 'Mode 3' system consisting of 'Innovation Networks' and 'Knowledge Clusters' for knowledge creation, diffusion and use (Carayannis and Campbell, 2005). This is a *multi-layered, multi-modal, multi-nodal and multi-lateral* system, encompassing mutually complementary and reinforcing innovation networks and knowledge clusters consisting of human and intellectual capital, shaped by social capital and underpinned by financial capital. In terms of enriching theory, this book attempts to promote the understanding of the role of heterogeneity and diversity in the inter-linkages of rationalities identities, preferences and intentions to macro-level phenomena such as institutions, national systems, regional city economies, the evolution of technological paradigms and the ways in which multi-level innovation systems work. Systems are systems because they reproduce themselves. At the same time, through heterogeneities of input and dynamic processes of co-opetition, co-evolution and co-specialization, systems generate outputs that differ from inputs. This has repercussions for the next round of inputs. These dynamic sequences of input – process – output – input – process, and so on, are called evolution.

A core issue in discussing evolution is the tension between structure and agency. Structures reproduce systems, whereas agents exploit and sometimes even transform them by looking for and exploiting heterogeneities. In that way, actors seek and find new combinations of input factors, and generate new processes. On the other hand, structural forces may send the

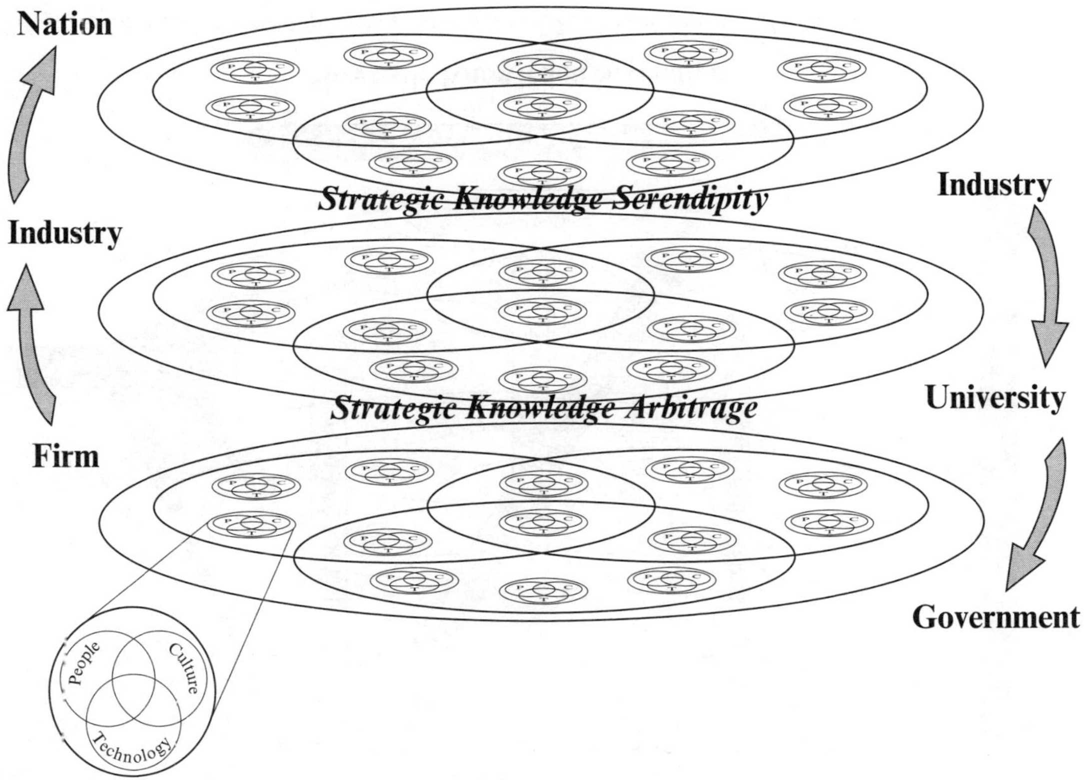

Source: Adapted from Carayannis and Campbell (2007, Figure 7.2).

Figure 1.1 Strategic knowledge, serendipity and arbitrage: a cross-sectoral view

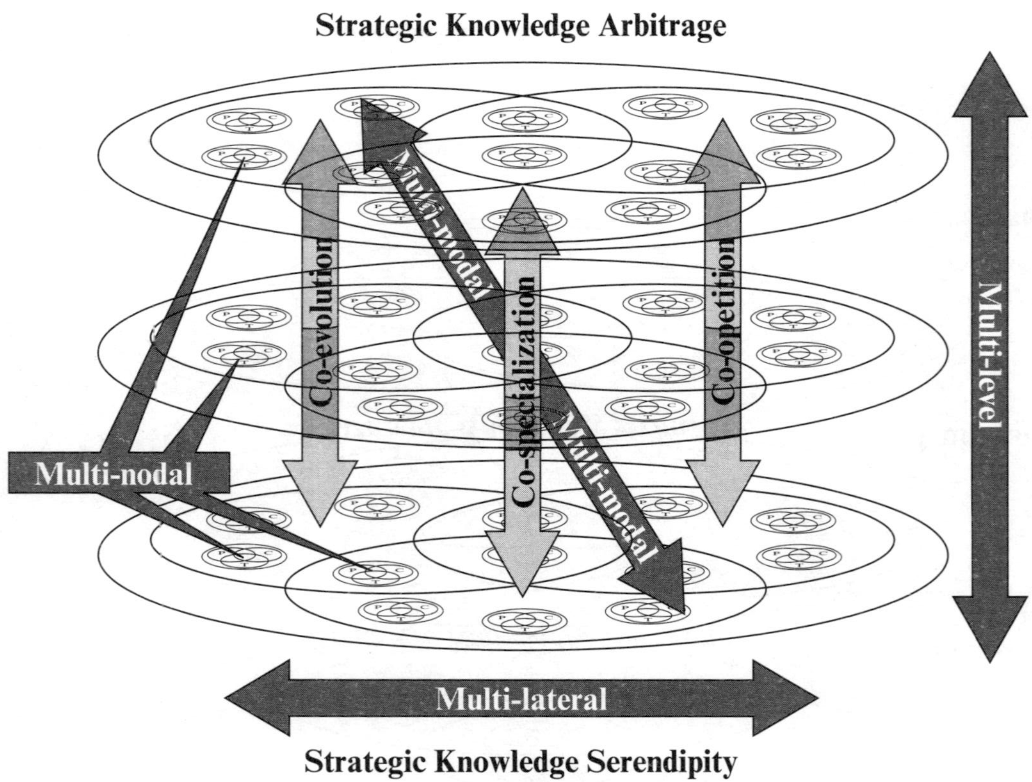

Figure 1.2 *Strategic knowledge, serendipity and arbitrage: multi-modal, multi-nodal, multi-lateral, multi-level co-opetition, co-specialization and co-evolution processes*

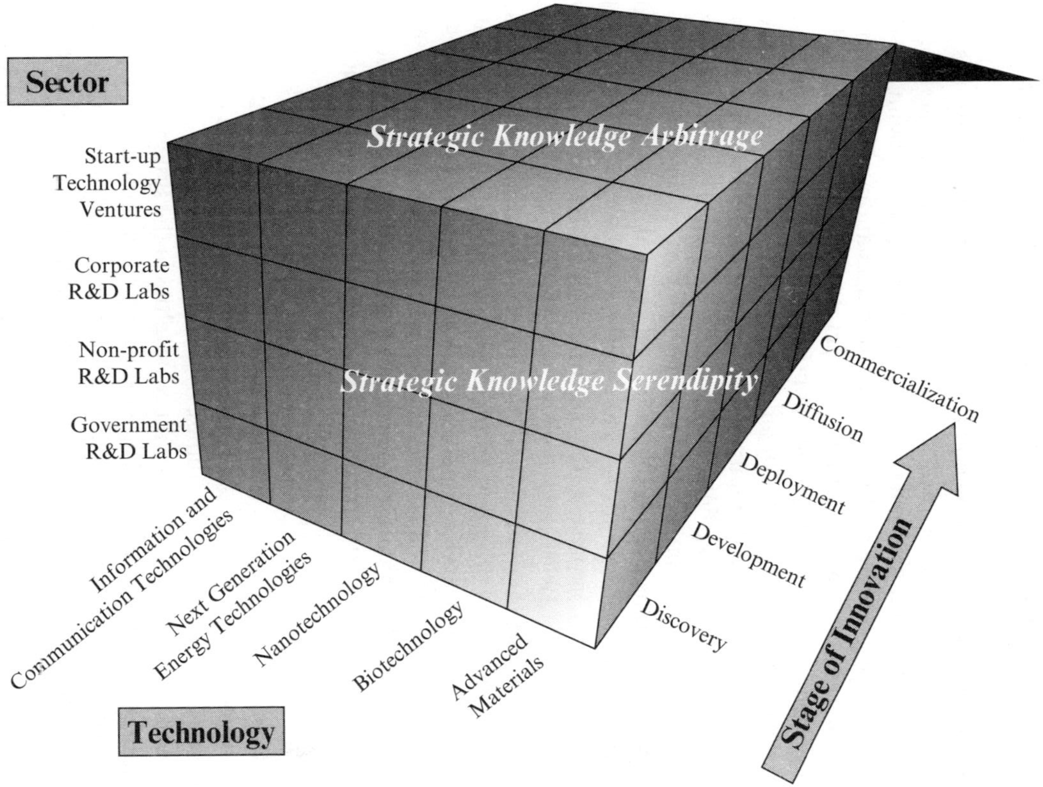

Figure 1.3 *Strategic knowledge, serendipity and arbitrage: a cross-sectoral and cross-technological view across the stages of technology innovation*

system into cycles where heterogeneities are gradually reduced, and evolution narrows down into a slim trajectory.

HETEROGENEITY AND INNOVATION SYSTEMS

Schumpeter is a natural starting point for discussing heterogeneity. His perception is that innovation is a process of combining various resources in specific ways. The combination of various resources, or combining 'materials and forces' as he phrased it, is a key to understanding his concept of development. According to Schumpeter, development is something that basically comes 'from within' the economic system, and is related to changes in the way production is organized; that is, resources are combined in new ways. Based on this, Schumpeter introduced the concept of *new combination* (1934/1996, pp. 65–66) and defined development by the introduction of a new combination.

A number of authors have discussed the principles and summarized the mechanisms of evolution. The seminal work of Nelson and Winter (1982) may represent the starting point of the evolutionary approach (Saviotti, 1996), and their approach was based on the three building blocks of organizational routines, search behaviour and the selection environment (Van den Bergh, 2004). A number of authors have followed up on this and summarized the principles in different ways (see for instance Aldrich, 1999; Carlsson and Stankiewicz, 1991; McKelvey, 1997; Peneder, 2001; Saviotti, 1996). Among them Edquist (1997) has stated that evolutionary theories often include the following elements:

1. The point of departure is the existence of reproduction of entities like genotypes in biology or a certain set-up of technologies and organizational forms in innovation studies.
2. There are mechanisms that introduce novelties in the system (i.e. mechanisms that create diversity). This includes significant random elements, but may also produce predictable novelties (e.g. purpose-oriented development work). In biology the novelties are mutations and in our context they are innovations.
3. There are mechanisms that select among the entities present in the system. This increases the relative importance of some and diminishes that of others. The selection process reduces diversity and the mechanisms operation may be the 'natural selection' of biology or the 'market selection' of competition as regards technical change. Together the selection mechanisms constitute a filtering system that functions in

> several stages and leads to a new set-up of, for example, technologies and organizational forms. There might also be feedback from the selection to the generation of new innovations.

The simple version of this is that a population, system or technology develops through mechanisms of variation and selection. Diversity is a necessary condition for change; that is, evolutionary change depends on the existence of diversity in economic actions (Peneder, 2001). Not only does this mean that an initial variety is required. The continuing creation of variety, for instance through new information, is also required, as the initial diversity will be 'consumed' during the process of selection (Peneder, 2001). This means that the variation and selection processes are working continuously and in some kind of interaction, it is no stage model or sequential model, but rather 'a sort of harmonica movement' (Van den Bergh, 2004).

Diversity is thus a necessary condition for evolution. The formation of new firms may be regarded as the manifestation of diversity, and it is the variety in the system that governs the pace and direction of change (Metcalfe, 2004). However, this does not mean that it will make sense to focus on diversity in isolation. Diversity per se will not create evolution, it is how the 'system' reacts to diversity that is important, and these reactions may be described through the type of selection processes that are at play. This means that evolution is determined by efforts to exploit diversity, and these efforts will depend on available competences and the system's learning processes (Cohendet and Llerena, 1997). *The continuous interplay between variety creating and selection mechanisms governs evolution.*

The discussion of the relation between *randomly generated diversity and structures* in innovation system theory was opened again by Nelson and Winter (1982). They were concerned with *differences in productivity between national economies*. In their attempt to explain contemporary *differences*, they emphasized the dualism between two phenomena to be located at two different points in time: 'variation', followed by 'selection'. The corollary was the '*co-evolution of technology and institutions*' (Nelson, 1991):

> Technology and the structure of industry co-evolve, and this process leads to growth in productivity, which is a statistical property of the system as a whole. (Nelson, 1991, p. 21)

His basic model of evolution took as a point of departure '*systematic selection where somewhat random variation plays a central role*' (Dosi et al.,

2000, p. 22). These were more or less fixed structures, which Nelson and Winter – based on Abernathy and Utterback – referred to as dominant *technological paradigms*. A dominant technological paradigm may be considered as a special case, a sectoral innovation system, which selects only what fits into the paradigm from the beginning. Hence, the technological paradigm starts to close. Abernathy and Utterback were criticized by Blauwhof, who, based on Hughes and Latour, pointed out that in the invention process, prior to the phase of closing in on mature products described by Abernathy and Utterback, the networks of the entrepreneurs and innovators were wide open.

By including the process of invention or innovation, Blauwhof argues, Hughes and Latour identified a communicative process (*interactive learning*) where different forms of knowledge were integrated through processes of 'translation', which enabled new knowledge combinations. In this early phase of inventions, the market product (the innovation) was an abstract idea, which was redefined and transformed by actors trying to find some way into the market. The 'invention system' was *open*, as different options were tested. Unlike the phase of invention, the mature product within an existing technological paradigm emerges with a fixed set of different forms of knowledge, linked within a specific structure. In this case the internal complexity of the innovation system may be larger than in the phase of invention, but the 'strengths of loose ties' are not as prominent as before. This is the parallel in the learning economy of forming a new species. In nature, part of a population that is specializing in a new direction may sooner or later discover that it has lost the capacity to mate with the population from which it once came. This loss of mating capacity is the birth of a new species.

In looking for species of specialized human knowledge, then, we must regard strategies of specialization in the context of the external threats of the market, and the capacity for agility, adaptation and turnaround. Given this rapid speed of destruction, even highly integrated clusters may – at least if they are forced to by the market – open up for new forms of knowledge. What enables this strategy is the *multi-dimensionality* of human knowledge systems. Humans may embed their various forms of specialized knowledge *in layers*. These layers may be interrelated through points of dense interactivity, where interactive learning is possible, such as organizations, regional clusters, or single humans. Here, complex processes of interactive translation and communication across different specializations are possible. For instance, two widely diverse knowledge systems may be mating inside a single human body, resulting in an entrepreneurial achievement.

Indeed, the drive towards increased specialization has resulted in the evolution of specialists who reap the benefits of crossing borders and initiating unusual or unheard of acts of mating – the *Schumpeterian entrepreneur*. This peculiar form of socializing of humans – and their entrepreneurialism – enhances the adaptability of human knowledge to changes. In nature, this option is lost – once the new species is formed, and the barriers against mating are established, there is no return. For systems that are contextualized by the global market, staying specialized should always – as pointed out by Blauwhof – be balanced by diversity – creating buffers, enabling rapid mating with new forms of knowledge if times are changing.

In Chapter 2 of this book, Helge Godø discusses innovations as *man-made* activities in opposition to ideas regarding innovations as 'random mutations'. Godø is concerned with the role of human will and purposeful behaviour. Purposeful behaviour, according to Godø, refers to design, which, he argues, is a core element in technological evolution and innovation. Design, in this way, moves the issue of evolution of paradigms into something different from both the evolution of the species, and the neo-classical logic of markets. It moves the issue into the realm of politics. The political implications of recognizing that innovations are made on purpose, for a purpose, according to Godø, may be beneficial for creating actions, and providing leadership and organization for innovation efforts, as suggested in the conceptual framework of innovation regimes. In the current political climate, the holders of power claim that the dynamics of market-oriented innovation will ensure that heterogeneity will flourish; that is, the market is a selection environment that 'by nature' will encourage people to be creative because the market represents a pressure towards innovation. However, the markets are aversive, even hostile, to radical innovations. More often than not, radical innovations need some type of strong political will or advocacy. Markets are incapable of initiating innovations in these areas – and do not possess the type of imagination or creativity that would foster heterogeneity. If society wants to be creative and imaginative, it has to muster the political will for this – innovations and heterogeneity are a matter for policy and associated agency.

More generally, innovation system theory emerged out of micro-level studies of technological systems, as well as middle- (meso-) and macro-level studies of innovation systems and innovation through interactive learning (Freeman, 1988; Lundvall, 1992; Nelson, 1993; Edquist, 1997), and through several EU and OECD publications (OECD, 1999), where the NIS perspective was promoted. A critical and comprehensive review of this literature is presented in a recent publication by Miettinen (Miettinen, 2002). Despite the fact that 'innovation system' has been on anyone's lips for the last ten years, 'system' is more often than not used as a heuristic device in the literature. To Lundvall, 'system' simply explained interactivity, seen in contrast to linear knowledge transfer. Schienstock and Hämäläinen (2001) define 'innovation system' with reference to the function of knowledge conversion (Schienstock and Hämäläinen, 2001), understood as new knowledge creation, diffusion, and commercial utilization, in short, the knowledge process (see also Carayannis and Campbell, 2005).

Finn Orstavik, in Chapter 6, indicates how innovation systems analysis hitherto has been developed mainly by institutional and evolutionary economists, but that such analysis has its more remote origins in classical social science. Orstavik argues that advances in innovation systems theory have been obstructed by theorists making inadequate assumptions about the nature of systems. Progress must be based on a better understanding of terms such as knowledge, communication and learning. By focusing on interactive learning, for example in the form of user–producer interactions, Bengt-Åke Lundvall has moved the analytical focus from overall institutional structures towards the actual core of innovation processes. But we claim that only a first move towards coherent and robust theorizing has been taken. A crucial resource for making further advances theoretically, we believe is found in Luhman's general theory about *social systems*. Building on the notion of *reflexivity* we develop a concept of innovation as a human, sociotechnical and organizational endeavour of 'reflexive transformation'. The fundamental difference between this approach and the national innovation system thinking model is that we move out of the whole-part systems discourse, and into the new paradigm of systems theory where systems–environment is the fundamental distinction. In this way, we eliminate resource-consuming discussions about which elements should be considered part of an innovation system, and which should not.

System–environment is also related to the degree of openness/closedness of a knowledge system. However, to *open or close* may be seen as optional strategies. The standard argument for closing in the boundaries of the knowledge system; that is, enhancing specialization is:

1. Reducing external transaction costs. Instead of investing in the transaction costs involved in contacting external sources of knowledge, the system may focus on its own specialization, thus . . .
2. Avoiding internal complexity. Closing in and specializing may be seen as a strategy to avoid the internal complexity that is necessary to be able to relate to and integrate external knowledge. It is better and more efficient to make it simple, and stay specialized within a narrow niche.

3. Specialization may be profitable. Specialization in the function of the system may prove to be a profitable strategy that is rewarded by the market, as the specialist may avoid price competition from other, less sophisticated competitors.

Strategies of specialization have their downsides as well.

1. Exposure to random events. A closed system is a system with no knowledge of its environment. It may be exposed to random and rapid destruction triggered by the market.
2. Loss of 'adaptive capacity'. A too narrow specialization may run the risk of turning the system into a unique species, which loses the mating option that may be necessary when the niche is made obsolete.

Furthermore, in focusing on system learning dynamics, we have to pay attention to the relation between '*Creative Accumulation*', with a low level of heterogeneity, and '*Creative Destruction*', with a high level of heterogeneity. One could consider '*Creative Accumulation*' (see Schumpeter's Mark I and II comments, 1943/1996) as a manifestation of co-specialization, an interim point between '*Creative Destruction*' (Schumpeter, 1943/1996) and what we call '*Destructive Creation*'. This move is possible through opening the issue of the specific *heterogeneities of human knowledge*. Heterogeneities of *human knowledge* – as opposed to forms of biological life – are feeding into a particular form of *knowledge 'mating' or knowledge 'osmosis'* (Carayannis, 2000–2005; Carayannis and von Zedwitz, 2005), which, as described by Niklas Luhmann (1984, 1994), is the act whereby humans share knowledge with each other – through shared *understanding* of what they are doing, enabling expectations and, hence, learning and adapting.

In saying so, however, we must at the same time remember that *knowledge* in this context is not an abstract substance. We are not interested in knowledge from the perspective of *stored symbols and texts*, whether it is found carved into a stone, in a library or in a hard-drive. Our interest is in *living knowledge*, used in practice for policy making and economic useful purposes, in contexts of shared understanding. Niklas Luhmann (1984, 1994) argues that shared understanding is a basic form of autonomous human self-organization, which may be seen as a form of life. This should not be regarded as allegorical. Quite the contrary, shared understanding – or social systems – is created interactively; the systems evolve through reproduction – and they die when they go out of practice.

Living knowledge is *embedded* in bodies, brains, spaces, institutions, organizations, communities of practice, as well as communities sharing formalized forms of knowledge. This is why, in analysing the heterogeneities of knowledge, we must relate to the deeper layers of organizational, regional and institutional arrangements within which different species of interbreeding knowledge are embedded. As argued by Lam (2000), these forms of institutional embedding of knowledge are crucial in *structuring* the relations between different forms of knowledge in the processes of interactive learning leading to innovation. We wish to better understand the tension between *heterogeneity* – which opens up the way for major changes generated by random events (radical innovations) – and *structure*, which tends to give evolution a specific direction. This question is accordingly similar to another problem, that of the relation between an open and a closed system. Whereas a closed system follows its own, internally defined path – until it eventually is struck by some unforeseen disaster – an open system may adjust to changes in the environment. This, again, is a part of an even wider debate.

METHODOLOGICAL CONSIDERATIONS

Knowledge systems are multi-level. At the core, we find mutually driving, complementary and reinforcing processes of *co-opetition* (a hybrid form of collaboration and competition), *co-specialization* and *co-evolution* (C3) (Carayannis et al., 2003; Carayannis and Alexander, 2004; Carayannis and Coleman, 2005; Freeman and Soete, 1997; Nowotny, 2006; Lundvall, 2006) (see Figure 1.4). In particular, we view heterogeneity as both a cause *and* an effect of the input, process and output (IPO) innovation stages (see Figure 1.4).

Dynamically, as Figure 1.4 illustrates, we depict a system for adding value via interconnected and interacting stages of socioeconomic being and becoming. Core phases are the concept of C3 as discussed above, which drives the creation, diffusion and use of knowledge. This system consists of multiple layers (macro, meso, micro) as well as modules (*input*, *process*, *output*). The presence of heterogeneity in those layers and modules acts as both cause and effect for driving the value-creating, diffusing and potentially destroying, processes of *co-opetition*, *co-specialization* and *co-evolution* (C3) as shown in Figure 1.4 (Carayannis and Campbell, 2005). *Input heterogeneity* refers to the variety and diversity of the key inputs to economic activity. Intrinsic in all these inputs is knowledge, which has been increasingly the key source of value adding of most human endeavours. *Process heterogeneity* reflects the variety and diversity intrinsic in the ways that the key inputs to economic activity are leveraged, allocated, re-

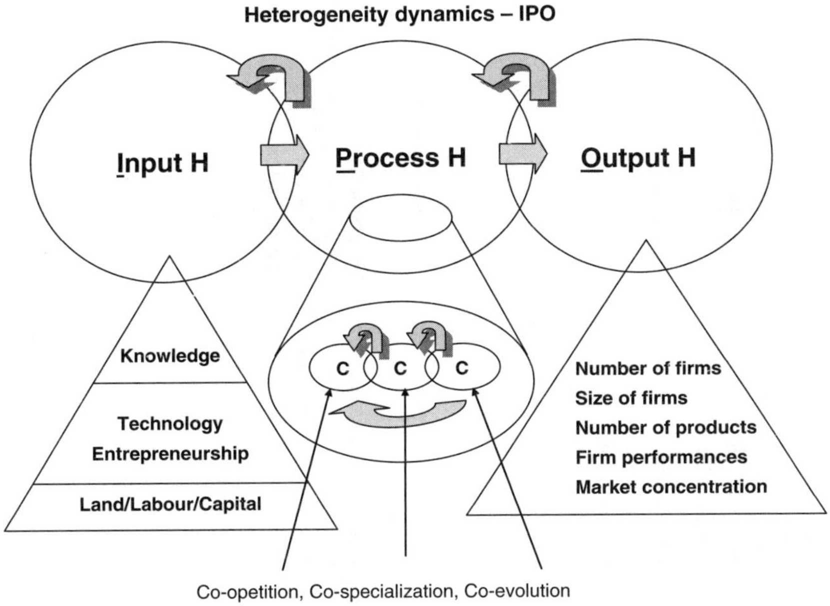

Figure 1.4 Heterogeneity dynamics – input/process/output

combined and recreated as part of the processes of technology innovation and entrepreneurship aiming at the maximization of value added. *Output heterogeneity* reflects the diverse ways and means that the value added of economic activity combining and leveraging the key inputs discussed earlier, is captured and exploited.

Table 1.1 provides an overview of the chapters and how they relate to the model presented in Figure 1.4.

Table 1.1 Conceptual topology

	Key system characteristics	Input heterogeneity diversity	Process co-evolution, co-opetition, co-specialization	Output heterogeneity diversity
Concepts and theory	Technological evolution, human intentionality, policies (Chapter 2)	Human intentionality policy decisions	Regulations co-evolving with innovation Co-opetition through innovation regimes	New path creation
	Diversity and heterogeneity in economic thought (Chapter 3)	Competitive rivalry	Market based innovation	
	Rationality and institutions (Chapter 4)	Institutional heterogeneities	Differentiation of rationalities	Diversities of capitalism
	Innovation systems constituted through communication (Chapter 6)	Diversities of innovation systems understood as development coalitions sharing understanding (co-opetition)	Interactive learning Co-evolution of machines, routines and rules	Diversification of production systems (new forms of co-specialization)
	The conceptual foundation of diversity and heterogeneity in innovation studies (Chapter 5)			
	Entrepreneurship and heterogeneity (Chapter 7)	Heterogeneity of opportunities and knowledge base within	Entrepreneurial (Mark I) and formalized (Mark II) regimes, operating	New firms, combinations of resources,

		entrepreneurial communities and technological regimes	through recognition, exploration and exploitation	technologies and products
	The role of institutions: Trademarks and diversity of products (Chapter 8)			Understanding the role of institutions in output heterogeneity
Analysis	Firm populations (Chapter 8)	Heterogeneity of firm population (size, ownership, age, etc.)	Diverse patterns of development	Survival – closure Transformation (takeover, spin-off, spinout, move)
	City economies (Chapter 9)	Specialization and differentiation in big city economies	Co-evolution and co-specialization of KIBS and their customers KIBS as knowledge translators	Innovation and diffusion
	Small national systems (Chapter 10)	Diversities of institutional complementarities	Co-evolution of institutions and industrial sectors	National specialization, small country squeeze
	EU15, USA and Japan The knowledge specialization of the EU15 (Chapter 11)	R&D investments	Patenting Scientific profiles and sectoral technological profiles (patents) of the EU15 compared with the USA and Japan	Economic specialization measured in value added, employment and exports. The EU's sectoral economic specialization (value added, employment, exports) compared with the USA and Japan

REFERENCES

Aldrich, H. (1999), *Organizations Evolving*, London, Thousand Oaks, CA and Delhi: Sage Publications.

Carayannis, E.G. (2000–2005), 'GWU lectures on technology, innovation and entrepreneurship', unpublished lectures 2000–2005, Graduate School of Business, George Washington University.

Carayannis, E.G. and J.M. Alexander (2004), 'Strategy, structure and performance issues of pre-competitive R&D consortia: insights and lessons learned', *IEEE Transactions on Engineering Management*, **52**(2), May, 226–32.

Carayannis, E.G. and J.M. Alexander (2006), *Global and Local Knowledge: Glocal Transatlantic Public–Private Partnerships for Research and Technology Development*, Basingstoke: Palgrave Macmillan.

Carayannis, E.G. and D.F.J. Campbell (eds) (2005), *Mode 3: Knowledge Creation, Diffusion and Use in Innovation Networks and Knowledge Clusters: A Comparative Systems Approach across the US, Europe and Asia*, Westport, CT: Quorum Books/Greenwood Press.

Carayannis, E.G. and J. Coleman (2005), 'Creative systems design methodologies: the case of complex technical systems', *International Journal of Technovation*, **25**(3), May, 831–40.

Carayannis, E.G. and M. von Zedwitz (2005), 'Architecting glocal (global–local), real–virtual incubator networks (G-RVINs) as catalysts and accelerators of entrepreneurship in transitioning and developing economies: lessons learned and best practices from current development and business incubation practices', *International Journal of Technovation*, **25**(2), February, 95–110.

Carayannis, E.G. and D. Campbell (2007), 'A "Mode 3" systems approach for knowledge creation, diffusion and use: towards a 21st-century fractal innovation ecosystem', in E.G. Carayannis and C. Ziemnowicz (eds), *Re-Discovering Schumpeter: Creative Destruction Evolving into 'Mode 3'*, Basingstoke: Palgrave Macmillan, Chapter 7.

Carayannis, E.G., D. Evans and Mike Hanson (2003), 'A cross-cultural learning strategy for entrepreneurship education: outline of key concepts and lessons learned from a comparative study of entrepreneurship students in France and the US', *International Journal of Technovation*, **23**(9), September, 757–71; 2003 Recipient of *Emerald Management Review*'s Citation of Excellence for Research Implications.

Carlsson, B. and R. Stankiewicz (1991), 'On the nature, function and composition of technological systems', *Journal of Evolutionary Economics*, **1**, 93–118.

Cohendet, P. and P. Llerena (1997), 'Learning, technical change, and public policy: how to create and exploit diversity', in C. Edquist (ed.), *Systems of Innovation: Technologies, Institutions and Organizations*, London and Washington: Pinter, pp. 223–41.

Dosi, G., R.R. Nelson and S.G. Winter (2000), 'Introduction', in G. Dosi, R.R. Nelson and S.G. Winter (eds), *The Nature and Dynamics of Organizational Capabilities*, Oxford: Oxford University Press, pp. 1–22.

Edquist, C. (ed.) (1997), *Systems of Innovation: Technologies, Institutions and Organizations*, London and Washington: Pinter.

Freeman, C. (1988), 'Introduction', in G. Dosi, C. Freeman and R. Nelson, *Technical Change and Economic Theory*, London and New York: Pinter, pp. 1–8.

Freeman, C. and L. Soete (1997), *The Economics of Industrial Innovation*, 3rd edn, London: Pinter.

Lam, A. (2000), 'Tacit knowledge, organizational learning and societal institutions: an integrated framework', *Organization Studies*, **21**(3), 487–513.
Luhman, N. (1984), *Social Systems*, Stanford, CA: Stanford University Press.
Luhman, N. (1994), *Die Wirtschaft der Gesellschaft*, Frankfurt am Main: Suhrkamp Tassenbuch Wissenschaft.
Lundvall, B.-Å. (ed.) (1992), *National Systems of Innovation: Towards a Theory of Innovation and Interactive Learning*, London: Pinter.
Lundvall, B.-Å. (2006), 'One knowledge base or many knowledge pools?', DRUID Working Papers 06-06, Copenhagen Business School, Department of Industrial Economics and Strategy/Aalborg University, Department of Business Studies.
McKelvey, M. (1997), 'Using evolutionary economics to define systems of innovation', in C. Edquist (ed.), *Systems of Innovation: Technologies, Institutions and Organizations*, London and Washington: Pinter, pp. 200–222.
Metcalfe, J.S. (2004), 'The entrepreneur and the style of modern economics', in G. Corbetta, M. Huse and D. Ravasi (eds), *Crossroads of Entrepreneurship*, Boston/New York/Dordrecht: Kluwer Academic Publishers, pp. 33–51.
Miettinen, M.R. (2002), *National Innovation System: Scientific Concept or Political Rhetoric*, Helsinki: Edita.
Nelson, R.R. (1991), 'Diffusion of development: post-World War II convergence among advanced industrial nations', *American Economic Review*, **81**(2), May, 271–5.
Nelson, R.R. (ed.) (1993), *National Systems of Innovation: A Comparative Study*, Oxford and New York: Oxford University Press.
Nelson, R.R. and S.G. Winter (1974), 'Neoclassical vs evolutionary theories of economic growth: critique and prospectus', *Economic Journal*, **84**, 886–905.
Nelson, R.R. and S.G. Winter (1982), *An Evolutionary Theory of Economic Change*, Cambridge, MA and London: The Belknap Press.
Nowotny, H. (ed.) (2006), *Cultures of Technology and the Quest for Innovation*, New York and London: Berghahn Books.
Peneder, M. (2001), 'Dynamics of initial cluster formation: the case of multimedia and cultural content', in OECD (ed.), *Innovative Clusters. Drivers of National Innovation Systems*, Paris: OECD, pp. 303–13.
Saviotti, P.P. (1996), *Technological Evolution, Variety, and the Economy*, Cheltenham, UK and Brookfield, VT, USA: Edward Elgar.
Schienstock, J. and T. Hämmäläinen (2001), *Transformation of the Finnish Innovation System: A Network Approach*, SITRA report series 7, Helsinki: SITRA.
Schumpeter, J.A. (1934/1996), *The Theory of Economic Development*, London: Transaction Books.
Schumpeter, J.A. (1943/1996), *Capitalism, Socialism and Democracy*, London and New York: Routledge.
Van den Bergh, J.C.J.M. (2004), 'Firm behaviour and organisation from an evolutionary perspective', in H.L.F. de Groot, P. Nijkamp and R.R. Stough (eds), *Entrepreneurship and Regional Economic Development: A Spatial Perspective*, Cheltenham, UK and Northampton, MA, USA: Edward Elgar, pp. 15–45.

2. Technological evolution, innovation and human agency

Helge Godø

INTRODUCTION: TECHNOLOGICAL EVOLUTION AS 'INTELLIGENT DESIGN'

Technological evolution – and the role of innovation and diffusion as an element in this process – cannot be understood, much less explained, without taking into account the role of human will and purposeful behaviour. Herbert Simon's (1969) concept of the 'science of the artificial' and his emphasis on design as distinct from the natural, aptly sets the focus on this fundamental aspect. Understanding and explaining human will and purposeful behaviour is essential for making evolutionary approaches relevant to the study of technological innovation and development. Hence, in explaining technological evolution – and, more broadly, the development of society and social change – we need to take into account how political agency and advocacy, human desires and passions, rationality, visions and goals become embedded in the complex process of creating technical novelties, and how these are disseminated, negotiated and shaped in the process of diffusion, and subsequently contribute to social change. This point has been emphasized by numerous theorists, such as John Ziman (2000, p. 6) who writes ' "Design" is central to modern technology. How can that be reconciled with "evolution", which both Darwin and Lamarck explained as a process through which complex adaptive systems emerge *in the absence* of design?' (Ziman's emphasis). Elster (1983) made a point that aligns closely with this: he suggests that in evolutionary approaches to social development (including technological development), social science essentially employs an intentional mode of explanation; that is, the idea that actions are intentional and related to future desirable states. Hence the term 'intelligent design' should be redefined as the dynamic that causes humans to create technology. This, more than anything else, is fundamental in the dynamic of technological evolution.

The basic assumption in the statements above is that technological innovations, more often than not, are created *on purpose, for a purpose*. The phenomenon of serendipity is often misinterpreted as evidence that chance – just like the process of haphazard selection and mutation in nature – somehow guides the evolution of technology. Although this statement may seem to be a truism, the full implication of this is not clear in evolutionary approaches to technological development – in particular, research attempting to develop the approach of evolutionary economics. Instead, this type of scholarship seems to be directed at a critique of neo-classical economics, yet simultaneously apologizing for why the evolutionary economic approach has not yet developed models and mathematical explanations equivalent to those found in orthodox economic theory. Also, what may be termed as a categorical confusion seems to confound explanations of dynamics in current evolutionary economic approaches. By using a highly stylized notion of the firm as the basic unit in an evolutionary process, the main focus on the innovation process is set on these and their ability to generate profits and other economic indicators of performance and viability, as tokens of processes that may be explained by bio-evolutionary rules and logic.

The intellectual seductiveness of an evolutionary approach rests on its potential for a rational, scientific explanation of development in general; that is, a comprehensive, all-embracing theory of why and how the world we live in changes and develops, somewhat akin to a 'theory of everything'. Explaining development of human society has always been a challenge. Although one may observe social change, explaining why this happens, and why other aspects of society change slowly or seem to remain stable, is non-trivial. Perhaps the main challenge in using evolutionary approaches for explaining technological development and socioeconomic and cultural development in general is *not* to find smart ways of applying bio-evolutionary concepts and theories, but is associated with the fundamental recognition of evolution itself. This implies that as our world evolves and develops, new dynamics and forces emerge that steer and constrain evolution. These will interact with existing dynamics and forces and, by this, change some of the 'rules of the game' in evolution, and qualitatively transform these into a 'new game' with new rules of evolution. Although human beings in a biological and genetic perspective are almost identical to most other species, and we share an ancestry in a universe that has been comparatively stable for billions of years, during the last 20 000 years humans have created technology, institutions, knowledge, social and cultural systems that must be explained as a result of other factors than those which explain bio-evolutionary processes. Also, the emergence of these new evolutionary forces will change or transform some of the antecedent evolutionary forces, and by this dialectic, change the overall dynamic of evolution. The factor of human will and purposeful behaviour is fundamental in this and in explaining technological evolution.

In order to justify this argument, the chapter first presents a discussion and critique of the use of bio-evolutionary concepts for explaining technological evolution; that is, why transposing bio-evolutionary theory to socioeconomic and technological development is problematic. Following this, an alternative approach – a translation process approach – is presented. Using this, I attempt to explain how this may be fertile ground for explaining technological evolution, and the essential role of policy and political agency in technological development. This point is important for making evolutionary approaches relevant for policy making: policy matters.

TRANSPOSING BIO-EVOLUTIONARY THEORY TO SOCIOECONOMIC AND TECHNOLOGICAL DEVELOPMENT

Various disciplines in social science have attempted to employ evolutionary approaches since Darwin. In post-World War II research, mainly social and cultural anthropological researchers, together with some researchers in history and archaeology, were pioneers in developing and exploring evolutionary approaches (Harris, 1969). In their research, they were interested in explaining how societies evolved from decentralized, dispersed social units such as tribes and bands, into chiefdoms and nation states, and, conversely, why some societies apparently chose to remain as hunters and tribesmen. Furthermore, their focus has been on why and how large civilizations (e.g. Ancient Egypt, Maya, etc.) have emerged – and why some suddenly disappeared – and the role of technological development, such as how these societies were able to 'harness energy' and support population growth by agriculture and irrigation in these developmental trajectories (Sahlins, 1968, 1972). Some researchers were more courageous and proposed theories of developmental 'stages', such as Rostow (1971), or Karl Wittfogel in his theory of hydraulic society (1957). Although Rostow and Wittfogel were vehemently anti-communist, many of these theoretical approaches were inspired by Marxist ideas and approaches in which materialism, development of technology, industry and infrastructure played an important role. In another corner of this landscape, the works of Joseph Schumpeter (1934/1974) and his concept of the entrepreneur also had some influence, as evident in the works of Fredrik Barth (1972). However, the 'breakthrough' – or more aptly the modern revival of an evolutionary approach – was apparently sparked off by the 'oil crisis'[1] in 1973 and the subsequent 'stagflation' that became pervasive in the world economy; that is, simultaneous high rates of inflation and unemployment. This period also coincided with a number of other events and changes of which the following may be relevant:

- The commencement of a long period of 'stagflation' in most OECD countries (except Japan).
- A crisis of legitimacy of the 'steady state' notion of the economic system and the possibility of governing economic systems. This crisis led to a rehabilitation of market-oriented ideologies and economic policy.
- The emergence of environmental movements and advocacy as a strong political factor. The notion of 'limits to growth'[2] gained followers, while the technocratic approaches to welfare and society were increasingly questioned.
- A crisis of legitimacy for large-scale technological systems, specifically nuclear power, as public opinion became increasingly sceptical and concerned about the dangers of nuclear contamination.
- An increasing acknowledgement among economists that knowledge, education, technological development and R&D are fundamental for economic growth.

One of the proponents of this revival in evolutionary approaches, the Belgian Marxist and economist Ernst Mandel (1981), observed that the interest in 'long waves' in the economy and their relationship to technological development and innovations seems to be inversely correlated with business cycles; that is, that these attracted highest attention during depressive periods in the economy. However, in the 1990s, the evolutionary approach to technological development seems to have gained momentum – a new, somewhat heterodox and heterogeneous research tradition emerged, trying to gain recognition and strength as a legitimate approach in various social and technological disciplines. This research has generated a number of significant theoretical works and gained increasing academic and political recognition. Simultaneously, a number of books attempting to present a comprehensive understanding of technological evolution were published. A sophisticated book written by Paolo Saviotti (1996), is representative of this, hence some aspects and key concepts presented in the book will now be discussed.

Saviotti (1996) suggests that a 'structural and goal similarity' exists between evolutionary approaches to biology and evolutionary economics. He elaborates by pointing to a correspondence between entities observed, for example animal species in biology that correspond to institutions or technologies in evolutionary economics, and the knowledge goals of both approaches, because they both attempt to understand and explain 'long-run behaviour in terms of qualitative change' (1996, p. 34). This, according to Saviotti, has also provided them with a common theoretical heritage based on systems theory and non-equilibrium thermodynamics. Using this as a basic assumption, Saviotti suggests that a set of bio-evolutionary key concepts is fertile ground for evolutionary analysis and an explanation of socioeconomic development. These will be reviewed and discussed below.

The concept of *variation* or, more specifically, the *generation of variety*, is considered fundamental for the introduction of novelties. Hence, according to Saviotti (1996, p. 42), innovation may be considered equivalent to mutations. Specifically, technological innovations that may be classified as 'radical' are outcomes of mutations. However, Saviotti simultaneously points out that innovations are generated by what he calls 'search activities'; that is, purposeful, intentional actions by humans in which the anticipation of obtaining an innovation is important as a dynamic factor. In a socioeconomic context, the generation of technological variety is ruled by dynamics that are qualitatively different from a biological generation of variety.

Selection, which acts or responds to variation-generating mechanisms, invariably implies a relationship with an environment; that is, the arena in which selection takes place. According to Saviotti, selection as a mechanism is a similar feature of both biological and economic systems, in the latter because 'firms are selected on the basis of their adaptation to the environment in which they operate' (p. 43). This understanding of innovation processes (which also ignores the role of diffusion of innovations as a mechanism of change in the economy) deviates from insights provided by research on how selection of technology takes place; that is, how diffusion occurs.

The classical, almost paradigmatic, approach to explaining technology diffusion processes has been elaborated by Everett M. Rogers (1995) in his seminal book, *Diffusion of Innovations*. Rogers distinguishes between at least three different types of diffusion process, depending on how decisions are taken in terms of adoption:

1. *Optional innovation decisions*, which usually involve decisions taken by individuals or small social units (e.g. families), typically in innovation diffusion processes involving autonomous objects or novelties that are independent of others, such as the decision to start using a PC instead of a typewriter, contraceptive pills, contact lenses, hang-gliding, and so on. A large portion of diffusion studies focuses on these types of decisions because they are important for understanding consumer choice.
2. *Collective innovation decisions* – decisions that implicate all parties, typically political decisions taken to introduce or adopt a novelty subsequent to a consensus, majority vote or referendum, such as when a community decides to introduce fluoridation.
3. *Authority innovation decisions* – in which a person, usually as a representative of some kind of private or public authority, may decide that

a community or organization should be encouraged or compelled to adopt a novelty, typically justified by reference to a new law, a new policy or regulation, as evident when crash helmets were made compulsory for motor cyclists, as a 'top-down' decision.

The first type of diffusion decision (optional innovation decisions) is typical of consumer behaviour and choice. According to Rogers (1995), a successful innovation diffusion process may be statistically depicted by an S-shaped curve. The speed and outreach of the diffusion process is influenced by the following factors:

- *Relative advantage* of the novelty to the user (including economic factors such as costs).
- *Compatibility* of the novelty to existing solutions; that is, its degree of interoperability with other existing technologies.
- *Degree of complexity* inherent in the novelty, such as how much training and skill adaptation the novelty requires in order to function according to its potential, which also encompasses how the user interface is designed.
- *Trialability*, the degree to which a novelty may be experimented with (e.g. hands-on experience) is important.
- *Visibility* of the novelty – that is, the immediate, obvious and intuitive appeal, such as is evident in the success of 'post-it' note pads.

What makes Saviotti's notion of selection problematic is his focus on the firm as a basic unit in the variety generation and selection process. Of course, firms are important in most modern economic systems; however, they are only one of many actors and institutions involved in technological innovations and development.

Reproduction and *inheritance* are fundamental factors according to biological theory, and 'success' in this may be measured in terms of reproductive capability and population growth. As a result of socioeconomic and technological improvements, human societies have been very successful, as measured by the population growth during the last century – perhaps too successful. However, reproduction in almost all societies is a matter of strict socioeconomic, political and cultural regulation. This covers a wide range of aspects, from birth control to ideas as to how reproduction should be undertaken – and, sometimes, strict sanctions for any deviation. This is very different from reproduction in nature. Applying the notion of reproduction to a population of firms or technological artefacts may be elegant in a poetic sense, but does not really explain the evolution of these phenomena. One may interpret learning, knowledge transfer and institutional longevity as the societal equivalent of reproduction and inheritance, but in a very metaphorical sense – and far removed from the sexual and bio-reproductive dynamics in the biological approach.

As far as the concept of *fitness and adaptation* is concerned, quoting Hodgson (1993), Saviotti defines fitness and adaptation as the propensity of a unit to be successful in a given environment. This notion was originally launched by Darwin, but later inspired Herbert Spencer to coin the famous term 'survival of the fittest'. The seductiveness of the idea of 'fitness' as a measurement of economic success of firms and individuals is, of course, rampant, as evident in various 'socio-Darwinist' ideals. But its fertility for explaining technological development is not obvious, as evident in numerous historical cases which show that political power, more than the 'invisible hand' of markets, determines fitness and adaptation.

Another key concept is *elementary interaction*. Probably as part of his critique of neo-classical economics, Saviotti suggests that: 'In economics, competition is virtually the only type of interaction studied' (1996, p. 47) – in contrast to biology in which, according to an authority quoted by Saviotti, in addition to competition, there are:

- *commensalisms*: a type of cooperative behaviour that makes other species grow simultaneously with own growth, somewhat related to the idea of symbiosis and synergy, and
- *predations*: species are connected in a food chain and a predator's behaviour inhibits the growth of the population of the victims' species; hence the idea of an ecological balance.

Again, transposing these concepts to an analysis of the development of economic systems and technology is, of course, feasible in a metaphorical, poetic sense, although in doing this one may introduce euphemisms rather than accuracy and precision. Hence, the observation that, over time, the number of firms in a sector may decrease as the 'winners' grow in size and increase their number of employees and production output (Abernathy and Clark, 1985), and as a *dominant design* (Utterbach and Suarez, 1993) becomes established. In such processes, mergers and acquisitions may be interpreted as 'predation' – as evident in cases of hostile, leveraged buy-outs. But the analogy with the biological world is faulty because few, if any, firms derive their income from devouring other firms such as when big fish live by eating smaller fish. That is to say firms cooperate and relate to each other in various forms of division of labour, or may compete, forcing others out of business, but not by predation as in a food-chain in nature. Even though predation may be interpreted as a salient feature of aggressive nation states, characterizing the phenomena and strategies of imperialism, colonialism, war, 'pre-emptive strikes' (e.g. the recent wars in Afghanistan, Iraq, etc.), the analogy to nature and its food-chains is, at best, poetic. Furthermore, transposing the biological idea of commensalisms and competition to the socioeconomic scene, one has to explain the role of institutions that are very different from natural environment. Markets – the institution for competition and the division of labour – are usually heavily regulated, politically and through legal measures. Without laws, rules, supervision by authorities, political and military power, markets will cease to exist. All these 'artificial' systems and institutions are absent in nature.

As pointed out, in looking at key concepts and theories derived from bio-evolutionary theories, transposing these to explanations of innovations does not really contribute to a deeper understanding. In fact, although this may be labelled as a type of reductionism, it is, in fact, more serious because important factors in socioeconomic and technological development are ignored in order to make explanations that fit into a bio-evolutionary schema.

As pointed out by Saviotti (1996, p. 50) and many other theorists (for example, Hodgson, 1993; Ziman, 2000) who have attempted to use evolutionary approaches in explaining techno-economic development, what may be termed 'the purposeful character of human action'; that is, intentionality, human will, design, and so on, make evolutionary approaches to innovations problematic. Once factors related to intentionality are introduced, key concepts in bio-evolutionary analyses take a different meaning: 'Selection' is no longer the result of haphazard events, but more or less rational choice; that is, human will. Of course, human choice is constrained by a number of factors such as institutions, norms, mental conceptions and knowledge (e.g. Herbert Simon's 1969 notion of 'bounded rationality' and Elster's 1983 elaboration of this), other people's choice, opportunities, and so on. Choice manifests itself differently depending on the unit and size involved (individual, group, society, nation, etc.), and interaction with other units; that is, the complexity of these systems. Generation of variety, another key concept in bio-evolutionary theory, also becomes a different phenomenon once the factor of human purposeful action is taken into consideration: humans have the capability to create and increase variety – by creating innovations and design. This is different from bio-evolutionary variety-creation dynamics such as mutations. Although some advocates of the gradualist approach to techno logical innovations (Basalla, 1988) suggest that innovations are always based on antecedents and precedents, others support the notion of discontinuity (Freeman and Perez, 1988; Godø, 2004; Mokyr, 1990). In the latter, the concept of 'radical innovation' is a logical extension; that is, that humans have the capability of designing and inventing technologies or other solutions which have not existed previously. Hence, these become 'new to the world'. This is different from the biological process involved in mutation; even adherents of the gradualist approach to technological evolution, with their emphasis on the importance of antecedents and precedents as a source of innovation, acknowledge that human knowledge, design and imagination are fundamental in these creative processes. The same applies to reproduction and inheritance, which in modern society – as a result of human will, knowledge accumulation, culture and institutions – is radically different from the genetics and dynamics related to bio-evolutionary approaches, even if genetics plays a role also in the reproduction of human societies.

LOST AND FOUND IN TRANSLATION?

In his critique of mainstream diffusion models such as Rogers, Bruno Latour (1987) has developed a conceptual framework of 'translation processes' for characterizing and explaining the stages involved in innovation processes. Although the logic of this is somewhat esoteric in explaining the diffusion of innovations processes, Latour's approach may have some interest for evolutionary approaches. Using Elster's tripartite distinction of explanations in physics (causal explanations), biology (functional explanations) and social sciences (intentional mode of explanations), by adapting Latour's idea of a translation process one may distinguish transformation and growth in evolutionary dynamic factors that make each domain different from the other. Figure 2.1 is an attempt to illustrate this idea. Although the development of the universe (15 billion years from the 'Big Bang' to present time) may be explained as an evolutionary process as seen in the 'Standard model', this evolutionary perspective has to be 'translated' into radically different concepts in order to explain bio-evolutionary development. Likewise, in explaining social development, including the development of technology and knowledge, bio-evolutionary concepts have to be 'translated' into something radically different for explaining the development of society and human culture. What all explanations have in common is a diachronic perspective, which means that over time changes are introduced, emerge and evolve – and that these dynamics feed into new changes and developmental processes. In spite of phenomena related to dynamics such as chaos and bifurcation, most of these processes are stable and predictably cyclic for long periods of time, such as biological life cycles (birth → maturity → reproduction → decay and death), or the Earth's rotation around the sun, or, for that matter, cultural institutions that follow these rhythms. However, when translating into an evolutionary approach for social development, the factor of 'purposeful character of human actions' enters as uniquely distinct[3] from the other domains. Politics and

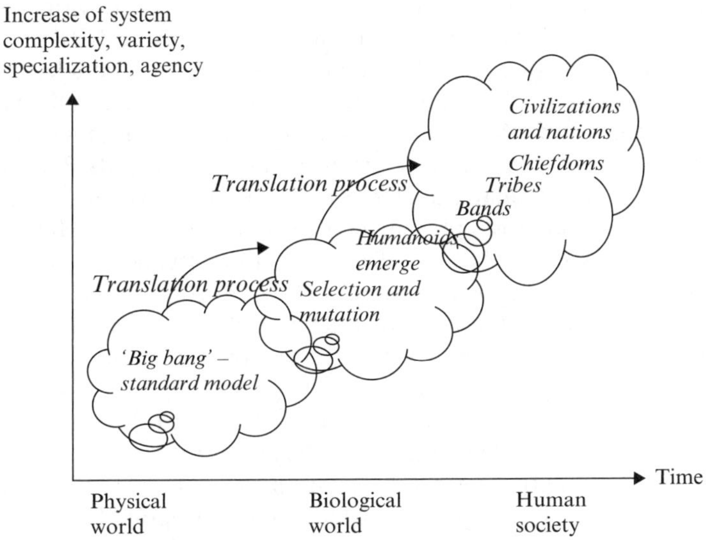

Figure 2.1 Tentative model of translation processes in evolution

strategy in the broad sense of these terms, then become relevant in terms of innovation in so far as this most aptly designates what may be termed the 'purposeful character of human action'.

INNOVATION DYNAMICS: RADICAL → INCREMENTAL → RADICAL

In attempting to explain why numerous radical innovations emerged in the telecom sector during the 1980s, we may suggest that these were created *on purpose, for a purpose*. Furthermore, that these innovations were successful because of strong *innovation regimes* in the telecom sector at that time (Godø, 2000). The mobile communication system, GSM[4] (initially developed in Europe) and the Internet (basically developed in the USA) are cases that illustrate both radical innovation and the role of innovation regimes. Following Krasner (1985), an innovation regime may be defined as principles, norms and ideology, rules and decision-making procedures forming actors' expectations and actions in terms of the future development of a technology. In the ICT sector, the innovation regimes are constituted in various cooperative organizations and institutions, such as collaborative R&D projects, consortia or networks; that is, technology-oriented social networks with a substantial participation of researchers and management from the R&D institutes in the industry. In the GSM development, the European standardization organization, CEPT, (now ETSI) played an important role, together with the R&D communities of the telecommunication operators and European equipment manufacturing industry. With the development of the Internet, which was contemporary with the GSM development, the US Department of Defense with its R&D agency DARPA[5] played a similar leadership role, uniting universities and R&D institutes in a common, prolonged development effort (Mowery and Simcoe, 2002). A salient feature of an innovation regime is its orientation towards future technological solutions; that is, creating consensus on future technologies.[6] Because standards are often involved, this may also be termed *anticipatory standardization*. Needless to say, the innovation regimes outlined above are *sector specific* and in many cases *highly international*, so are difficult to analyse within a national framework. For this reason, national policy measures may have an impact to the extent that these are conducive to innovation activities undertaken within the innovation regime. The notion of an innovation regime is related to the conceptual framework of a *technological regime*. However, the latter concept fails to explain change; that is, how technology evolves and establishes a new technological regime and a new economic sector as evident in the emergence of ICT as a new sector during the 1990s. Rip and Kemp (1998, p. 43) define a technological regime as 'the rule-set or grammar embedded in a complex of engineering practices, production process technologies, product characteristics, skills and procedures, ways of handling relevant artefacts and persons, ways of defining problems – all of them embedded in institutions and infrastructures'. Within this type of notion, some theorists have developed the concept of *dominant design* (Utterbach and Suarez, 1993) to designate an observed technological stability and predictability in the development patterns, what others call *technological trajectory* (Dosi, 1988) or *technological signposts* (Sahal, 1985). Sahal suggests that whereas incremental innovations emerge as result of *necessity*, by contrast radical innovations emerge because of *chance*. This understanding is endemic in many evolutionary approaches to technological innovations, as evident in Mokyr (1990) and Dosi (1988), the latter also attributing radical innovations to scientific breakthroughs (i.e. tacit acknowledgement of a linear innovation model).

Following this, complex dynamics involving technological regimes, innovation regimes and users/markets in a diffusion of innovations process contribute to the development of technology. Needless to say, numerous actors of different categories are involved in such processes, such as entrepreneurs, who work in a number of different roles, and politicians, bureaucrats, and so on. Furthermore, researchers and inventors, and users and other actors who constitute markets, are also important actors in these processes.

In their work and decisions, these individuals are guided by aesthetic and emotional values, in addition to, or in conjunction with what is usually recognized as 'rationality'. In expounding this, Loasby (2002) draws on the insights from Adam Smith (1723–90) and Frank Knight (1885–1972), the latter for his famous distinction between risk and uncertainty. Perhaps most elusive in this is the role of aesthetics, which may be understood as a type of emotion that motivates choice and creative activities, and which is fundamental in innovation processes. Adam Smith was aware of these factors and, according to Loasby's interpretation: 'a combination of imagination and ex-ante selection guides the invention of "connecting principles" which sort phenomena into categories and link these categories by an explanation which is sufficient to "soothe the imagination"' (2002, p. 1231). This point is elaborated by Joel Mokyr (2002) in his distinction between useful knowledge, Ω, and what he calls a 'set of feasible techniques' – a dichotomy that also bears on Knight's distinction between risk and uncertainty. A coupling of the latter with imagination, aesthetics and knowledge may be seen as a strong motivator for creating innovations (Basalla, 1988). Although we, as human beings, in our analysis of nature and biology, may marvel at the beauty and ingenuity (both aesthetics) of what we observe, and even hypothesize that aesthetics somehow must be a dynamic in bio-evolutionary processes (see Carroll, 2005), as a factor aesthetics is uniquely human and closely linked to the phenomenon of 'purposeful character of human actions'; that is, political action and agency in the broad sense of these terms.

KNOWLEDGE, INNOVATIONS AND POLITICAL AGENCY

Knowledge and language are basic to the development of human society, and for technological development and the creation of innovations. Of course, all biological organisms possess knowledge and are able to communicate; that is, have some type of language. Usually this is 'hard-wired' and generated by the genetic makeup of these organisms. However, knowledge and language are qualitatively different and more advanced in human societies in the sense that abstraction, symbols and codification are radically more developed compared to other organisms. More generally, humans depart from monkeys because we are able to *anticipate and conceptualize* new undetermined futures that are different from those in which we live, *and to act on these anticipations*. The basic reason why we can do this is our ability to create, store and disseminate knowledge. In addition, the use of language gives us abstract concepts and capabilities that we are able to manipulate and use to create new designs. This is essential for making 'will' and 'purposeful behaviour' possible in the first place.

Saviotti (1996) suggests that the phenomenon 'purposeful character of human actions' is an unresolved issue in the evolutionary approach to technological development, apparently because this does not exist in the bio-evolutionary conceptual framework. In the perspective of translation, it is apparent that in each translation new factors come into play, and because of these the dynamics of evolution change, causing qualitative changes to factors that were important in previous translations. As the phenomenon 'purposeful character of human actions', or political agency in a broad sense, comes into play, this will influence and change all other dynamics that played a role in the former translations. Hence, the dynamics of generation of variety becomes an issue of human action, hence a matter of technological development (culture, institutions, economic systems, etc.). Needless to say, recognition of this makes analysis of technological evolution much more complex. Still, increase in complexity is a general, almost imperative logic in evolution. The suggestion that innovations are created on purpose and for a purpose does not violate the basic rules of evolution in human social systems. In fact the opposite, assuming that the bio-evolutionary dynamics of chance (mutation) and randomness as an explanation of 'purposeful character of human actions' is not only a logical contradiction, but violates empirical observations of how technological innovations are created.

A quick glance at history provides support for this suggestion. The case of China during and after the Ming Dynasty illustrates in a negative sense the significance of policy for innovation activities in a society. Prior to 1500, China was the most technologically and culturally innovative and advanced society in the world. This resulted in a number of important technological innovations, the most basic and well-known of these were gunpowder, paper, navigation and printing – and a vast reservoir of knowledge associated with all the inventions and subsequent innovations originating in China (Fairbank et al., 1973; Mokyr, 1990; Needham, 1964). Somehow, an 'anti-innovation switch' was thrown during the early Ming Dynasty around the year 1500, after which China became stagnant from a developmental perspective. This shift has been the object of a number of studies because no formal decision or explicit policy statement was made for this change. Although this shift remains enigmatic, most of the studies seem to point to the political power structure and its ideology, which favoured some idea of social harmony and stability, hence preserving the status quo, and as a result a culture of complacency became inherent. The holders of power in China were able to maintain this policy for four centuries because the Ching Dynasty that succeeded the Ming Dynasty, and which lasted until 1911, upheld this policy. Some would suggest that this policy was continued by the Communists during Mao Tse-tung's regime until he died in 1976. Counterfactually, one may suggest that had China been successful in maintaining its society as a completely closed system, this policy of anti-innovation would have been viable for a long time. However, pressure by outside imperialists and colonialists (Europeans, and later Japanese) who were in possession of superior military technology, made this policy unsustainable. Now, in 2006, as the result of radical policy changes inaugurated by Deng Xiaoping in 1978 after Mao's death, China has again become very innovative – so innovative that many nations consider China as a threat. A parallell to China's development may be observed in Japan during the reign of the Tokugawa shogunate (1603–1868) which in 1721 decreed a 'Ban on Novelty', a law that was intended to prohibit innovations. Like China, Japan was subject to external military pressure, specifically by the USA in the middle of the 1850s. However, in Japan, powerful political and military groups – mainly samurais – were successful in launching a radical modernization process from 1868, a process that made Japan technologically strong in a short period of time. The success of Japan inspired many revolutionary movements in colonized countries throughout Asia, including Indonesia, China, Vietnam, and others. Ironically, these countries soon became victims of Japan's initial brutal success during World War II. However, European supremacy, and that of the USA in Asia, gradually diminished after the war. These historical cases briefly illustrate how politics and technological development – and technological *non*-development – are closely linked.

The implications of 'purposeful character of human actions' (i.e. intentionality, design, ex ante thinking as a factor in technological innovations and evolution) is that political factors and political agency are important and represent a non-trivial challenge for research. The implication of this in terms of policy, bearing in mind that even denying policy intervention by relegating innovation dynamics to market competition, is also a type of policy that depends on the functioning of political institutions just as much as other types of policy. Paradoxically, recognition of this may imply that some insights from bio-evolutionary dynamics, such as the importance of variety generation and heterogeneity, may be translated into policy actions that are conducive to specific goals and visions that societies hold.

The political implications of recognizing that innovations are made on purpose, for a purpose, may be beneficial for creating actions, and providing leadership and organization for innovation efforts, as suggested in the conceptual framework of innovation regimes. In the current political climate, the holders of power claim that the dynamics of market-oriented innovation will ensure that heterogeneity will flourish; that is, the market is a selection environment that 'by nature' will encourage people to be creative because the market represents a pressure towards innovation. As numerous studies have shown, this is correct in so far as incremental innovations (i.e. the type of innovation activities that involve comparatively little uncertainty) are concerned (Mowery and Rosenberg, 1982). Over time, numerous incremental innovations may accumulate to contribute to substantial technological progress. However, the markets are aversive, even hostile, to radical innovations, somewhat analogous to the political situation of a closed system as in China during the Ming and Ching Dynasties. More often than not, radical innovations need some type of strong political will or advocacy. As is evident in many radical innovations during the past century, the sources of these have usually come from outside the markets, such as the military, medical community, or monopolist organizations such as the telecommunications sector prior to the deregulations of the 1990s. Contrary to the short-sighted and inarticulate markets (markets are usually only able to say 'yes' or 'no' or 'more'), these organizations outside the markets have the capability of creating innovation regimes, if and when they want to. Their ability to succeed is, of course, a different matter. This is particularly the case when regimes are more political than technological, such as in political movements aimed at specific goals, for example gender equality, environmental protection, and so on. Markets are incapable of initiating innovations in these areas, and do not possess the type of imagination or creativity that would foster heterogeneity, which is another way of stating that if society wants to be creative and imaginative, it has to muster the political will for this – innovation and heterogeneity are a matter for policy and associated agency.

NOTES

1. The reason for this was that the OPEC increased the price of oil per barrel from US$3 to US$12 and imposed production limitations on its member nations.
2. The book, *The Limits to Growth* (Meadows et al., 1972/1989), commissioned by 'The Club of Rome', was initially published in 1972. Twelve million copies of the book have been sold, and it has been translated into 37 languages. Although considered controversial by many, the book became highly influential.
3. Zoologists and biologists, in particular the socio-biological movement, point to behavioural and organizational similarities between human and animals and other biological organisms. A recent example (van Schaik, 2006) attributes the use of simple tools (wooden sticks) among orang-utans in a Sumatran swamp to their 'culture' and this culture's ability to transmit technological knowledge and skills to the members of this society. More generally, the socio-biological approach has created controversy because advocates of this approach suggest that their explanations from biological life are valid and meaningful also for human society.
4. The acronym GSM is now based on 'Global Systems of Mobile Communications'. However, the acronym was originally coined from 'Groupe Special de Mobilité', the

official French name of a working group in the European standardization organization CEPT in the early 1980s. CEPT is the acronym for Conférence Européenne des Administrations des Postes et des Télécommunications.
5. DARPA is the acronym for Defense Advanced Research Projects Agency of the US Department of Defense.
6. In the development of the Internet, the system of RFC (requests for comments) played an essential role, in particular in the early, formative stages of its development. See Mowery and Simcoe (2002, p. 1374).

REFERENCES

Abernathy, W.J. and K.B. Clark (1985), 'Innovation: mapping the winds of creative destruction', *Research Policy*, **14**, 3–22.
Barth, F. (ed.) (1972), *The Role of the Entrepreneur in Social Change in Northern Norway*, Oslo: Universitetsforlaget.
Basalla, G. (1988), *The Evolution of Technology*, Cambridge: Cambridge University Press.
Carroll, S.B. (2005), *Endless Forms Most Beautiful – The New Science of Evo Devo and the Making of the Animal Kingdom*, New York: W.W. Norton and Co.
Dosi, G. (1988), 'Sources, procedures and microeconomic effects of innovation', *Journal of Economic Literature*, **xxvi**, 1120–71.
Elster, J. (1983), *Explaining Technical Change*, Cambridge: Cambridge University Press.
Fairbank, J.K., E.O. Reischauer and A.M. Craig (1973), *East Asia: Tradition and Transformation*, London: Allen and Unwin.
Freeman, C. and C. Perez (1988), 'Structural crisis of adjustment, business cycles and investment behaviour', in G. Dosi et al. (eds), *Technical Change and Economic Theory*, London: Pinter Press.
Godø, H. (2000), 'Innovation regimes, R&D and radical innovations in telecommunications', *Research Policy*, **29**, 1003–46.
Godø, H. (2004), *Virtual Keys in Cyberspace – Actors and Networks Creating New Technology*, Oslo: Abstrakt, Liber and Copenhagen Business School Press.
Harris, M. (1969), *The Rise of Anthropological Theory: A History of Theories of Culture*, London: Routledge & Kegan Paul.
Hodgson, G.M. (1993), *Economics and Evolution – Bringing Life Back to Economics*, Cambridge: Polity Press.
Krasner, S.D. (1985), *International Regimes*, Ithaca and London: Cornell University Press.
Latour, B. (1987), *Science in Action – How to Follow Scientists and Engineers Through Society*, Boston, MA: Harvard University Press.
Loasby, B.J. (2002), 'The evolution of knowledge: beyond the biological model', *Research Policy*, **31**, 1227–39.
Mandel, E. (1981), 'Explaining long waves of capitalist development', *Futures*, **13**(4), 332–8.
Meadows, D.S., D.L. Meadows, J. Randers and W.W. Behrens III (1972/1989), *The Limits to Growth: A Report for the Club of Rome Project on the Predicament of Mankind*, New York: Universe Books.
Mokyr, J. (1990), *The Lever of Riches – Technological Creativity and Economic Progress*, Oxford: Oxford University Press.

Mokyr, J. (2002), *The Gifts of Athena: Historical Origins of the Knowledge Economy*, Princeton, NJ: Princeton University Press.
Mowery, D.C. and N. Rosenberg (1982), 'The influence of market demand upon innovation: a critical review of some recent empirical studies', in N. Rosenberg (ed.), *Inside the Black Box: Technology and Economics*, Cambridge: Cambridge University Press.
Mowery, D.C. and T. Simcoe (2002), 'Is the Internet a US invention? An economic and technological history of computer networking', *Research Policy*, **31**, 1369–87.
Needham, J. (1964), 'Science and China's influence on the world', in R. Dawson (ed.), *The Legacy of China*, Oxford: Oxford University Press.
Rip, A. and R. Kemp (1998), 'Technological change', in S. Rayner and E.L. Malone (eds), *Resources and Technology* (Vol. 2), Columbus, OH: Batelle Press.
Rogers, E.M. (1995), *Diffusion of Innovations* (4th edn), New York: Free Press.
Rostow, W.W. (1971), *The Stages of Economic Growth: A Non-Communist Manifesto* (2nd edn), Cambridge: Cambridge University Press.
Sahal, D. (1985), 'Technological guideposts and innovation avenues', *Research Policy*, **14**, 61–82.
Sahlins, M. (1968), *Tribesmen*, Englewood Cliffs, NJ: Prentice Hall.
Sahlins, M. (1972), *Stone Age Economics*, Chicago: Aldine Atherton.
Saviotti, P.P. (1996), *Technical Evolution, Variety and the Economy*, Cheltenham, UK and Brookfield, USA: Edward Elgar.
Schumpeter, J.A. (1934 [1974]), *The Theory of Economic Development*, Oxford: Oxford University Press.
Simon, H. (1969), *The Sciences of the Artificial*, Cambridge, MA: MIT Press.
Utterbach, J.M. and F.F. Suarez (1993), 'Innovation, competition and industrial structure', *Research Policy*, **22**, 1–21.
van Schaik, C. (2006), 'Why are some animals so smart?', *Scientific American*, April.
Wittfogel, K.A. (1957), *Oriental Despotism: A Comparative Study of Total Power*, New Haven, CT: Yale University Press.
Ziman, J. (2000), 'Evolutionary models for technological change', in J. Ziman (ed.), *Technological Innovation as an Evolutionary Process*, Cambridge: Cambridge University Press.

3. Heterogeneity in economic thought: foundations and modern methods

Mark Knell

INTRODUCTION

Behavioural heterogeneity and technological diversity have been central issues within economic thought since at least the time of Adam Smith. Heterogeneity among competitors, which is one of the most important features of the modern market economy, has also been one of the most controversial problems driving the development of the theory of competitive equilibrium.[1] Modern neo-classical economists, including authors of virtually all textbooks, define competition as an end state that is synonymous with market structure. The assumption of perfect competition is necessary for the theory of competitive equilibrium, which requires that all goods and services are perfectly substitutable and that all economic actors have free and complete information. If this knowledge is not complete or product differentiation exists, then the market structure would be imperfectly competitive. By contrast, the classical economists, many of the early neo-classical economists, including the Austrian school, and evolutionary economists view competition as a process where competitive rivalry necessitates the need to be different and hence heterogeneous.

Numerous episodes in the history of economic thought emphasize the importance of behavioural heterogeneity and technological diversity among firms and agents. In most instances, they were associated with the economists' view of competition and the role of the firm within the economic theory. Our story begins with Adam Smith and the beginning of his magnum opus, *An Inquiry into the Nature and Causes of the Wealth of Nations*, first published in 1776. In the first three chapters of the first book, Smith explains how an ever-increasing division of labour could lead to expanding diversity and variety of goods and services. In subsequent chapters, he outlined a theory of value and distribution, which considered competition as a rivalrous process leading to a single uniform rate of profit that was independent of the heterogeneity and diversity in the economy. However, the theory, and subsequent developments in the nineteenth century, did not contain an adequate measure of value that was invariable to the heterogeneous inputs and outputs. The marginalist (or neo-classical) theory of value and distribution provided an alternative to the classical labour theory of value in the second half of the century, but it eventually made behavioural heterogeneity and technological diversity a problem for economic theory that could not be easily dealt with. Development of the idea of perfect competition and Alfred Marshall's concept of the representative firm are two related examples where the issue of heterogeneity across firms became a problem. This problem appeared in a lively debate in the late 1920s that questioned Marshall's theoretical edifice when increasing returns are present and in the empirical literature where heterogeneity among firms is a regularly observed phenomenon. Developments within different groups within the Austrian school of thought, namely by Fredrich von Hayek (1948; 1978) and Joseph Schumpeter (1912; 1942), promoted the idea that competition is a process of discovery, and success in providing classical economic theory with an invariable measure of value by Piero Sraffa (1960) helped revive the classical view of competition.

The main objective of this chapter is to describe some of these episodes in the history of economic thought and ask whether the classical and neo-classical theories can cope with business heterogeneity and product diversity. A central theme within this discussion is the business enterprise and the problem of increasing returns in competitive equilibrium. The debate on the concept of the representative firm eventually led to further discussions in the empirical literature on the firm, and most recently in the econometric methods used to measure firm activity. If there is any conclusion to this story, it is that behavioural heterogeneity and technological diversity are important issues of economic theory that remain part of the current debate. The following section provides the background for subsequent episodes in the history of economic thought, by focusing on one of the overarching issues underlying behavioural heterogeneity and technological diversity.

THE DIVISION OF LABOUR AND HETEROGENEITY

The classical economists were concerned with the growing heterogeneity and diversity of knowledge. Adam Smith (1776) knew that an ever more sophisticated division of labour was the main source of productivity growth, and that it also implied an increasing heterogeneity and fragmentation of knowledge across the many different tasks. His original example was the variety of tasks in a pin factory, however he also applied the idea to describe the difference between a street porter and a philosopher.[2] Smith recognized that an increasing division of labour could increase the dexterity of workers, save time lost in switching between different tasks, and lead to the invention of machines and types of organization that facilitate work. Driven by the extent of the market, specialization divided productive operations into their constituent elements, which both saved time in changing between different tasks and facilitated the introduction of equipment and machines. Loasby (1999) describes this process in modern terms as one that encourages the development of differentiated knowledge, and therefore a set of distinctive and heterogeneous capabilities.[3] Both the size and the growth of the market determine the division of labour and heterogeneity in the economic system. International trade becomes important in this context because it not only increases the size and growth of the potential market but, as a vent for surplus, it also gives rise to specialization across countries as businesses subdivide tasks into well-defined activities and products.[4]

Causality between productivity growth and the division of labour ran in both directions in Smith's theory. Smith was careful to bring out its two-way link with the division of labour in his discussion of the accumulation of capital. When the use of machinery facilitated and abridged labour, he pointed out that the accumulation of stock comes, in the nature of things, before the division of labour. As the division of labour becomes more sophisticated and heterogeneous, the incentive to innovate becomes greater. Further, subdivision of tasks requires more capital to keep busy all the different kinds of workmen. The more the productive hands employed, as a result of the higher division of labour, the more the capital required for wage advances as well as to provide them with proper tools and equipment. In this context, capital accumulation makes the heterogeneity of labour and capital inputs a progressive cumulative process, whereby the accumulation drives the division of labour while at the same time defining its limits, as Young (1928) pointed out.

The idea of the division of labour rarely appeared in the economic discourse after Adam Smith. Marx understood the technical necessity of the division of labour, but he described the process as a form of alienation where workers become increasingly specialized and the work repetitious. In economic theory the idea became absorbed into the idea of increasing returns, which is shorthand for the potential returns to further specialization when there is growth in a particular sector, a specific kind of capital, or the economy as a whole. Alfred Marshall (1890) developed this idea further in his *Principles of Economics*, but also became the target of much criticism in the late 1920s by Allyn Young and Piero Sraffa. By shifting Smith's analysis away from activities and towards the firm, Young (1928) explained how firms within an industry can become more heterogeneous because of the ever more sophisticated division of labour. Several papers by Nicolas Kaldor (Targetti and Thirlwall, 1989) from the 1950s to the 1970s and since the late 1980s Paul Romer (1987, 1990), considered Smith's and Young's idea of the division of labour to generate endogenous growth (Lavezzi, 2003), and W. Brian Arthur (1994) explored the issues of multiple equilibrium, path dependence, lock-in and possible inefficiency when increasing returns are present.

COMPETITION AMONG HETEROGENEOUS ECONOMIC AGENTS IN THE CLASSICAL ECONOMISTS

Competition plays an important role both in providing economic agents (individuals and firms) the incentive to recognize, develop and use the boundless opportunities provided by the economies of scale and specialization as well as in determining equilibrium prices (Richardson, 1975). Smith saw competition as a rivalry between independent and heterogeneous economic agents, all pursing their own self-interest, but resulting in the common good of society as a whole as if they were guided by the invisible hand. He, together with the other classical economists, also saw the important role that this dynamism plays within the circular flow and in the generation of a physical real surplus that promotes consumption, growth and international trade. The development of the classical theory of value and distribution is an important episode in the history of economic thought that shows how theory can avoid the kind of rigid behavioural assumptions that conceal the behavioural heterogeneity in the economy (see Löwe, 1965).

On the surface, the classical equilibrium and evolutionary views of competition appear incongruent because the classical theory of value and distribution considers the technology and organization of production as given. Garegnani (1984) identified three forms of data that the classical economists (including Adam Smith, David Ricardo and Piero Sraffa) and Karl Marx typically start from to derive the prices of production: (1) the technical conditions of production; (2) the size and composition of total output; and (3) the wage rate(s) or rate of profit. In this theory, wages and profits represent a fundamental asymmetry or equilibrium condition, which means that these data are sufficient to determine either the rate of profit or wage rate and the relative prices underlying the cost-minimizing system of production, given the levels of output. In short, the classical economists assume that economic agents follow a cost-minimizing behaviour, but that this behaviour can include a multitude of different strategies.

The prices of production in classical theory reflect the conditions of reproduction of the economy as a whole; that is, the cost of production plus profits at the 'ordinary' or 'average' rate of return. These economists saw these prices not only as purely theoretical, but also as 'centres of gravitation' around which market prices fluctuated in a way analogous to Newton's gravitational forces. This suggested that the rate of profit could appear very different across firms and industries in the economy at any moment in time. Thus for the classical economists the uniform rate of profit describes the *outcome* of the competitive behaviour of heterogeneous actors in the market, whereas profit-seeking entrepreneurs, or undertakers as Smith called them, minimize the cost of production *because* of the competitive process.

In the classical theory of value and distribution, equilibrium prices are 'centres of gravitation' or 'attractors' of market prices. As attractors, they imply that self-interested individuals search for the most profitable opportunities and minimize the costs of production. The outcome of this process of competition and selection is a cost-minimizing system with a uniform rate of profit and uniform rates of remuneration for each particular kind of input in the process of production, such as the different kinds of labour and materials used in production. This process of search and selection explains movements of capital and labour *across industries* in the absence of significant barriers to entry and exit. The classical economists saw these 'natural' or 'normal' prices as the *result* of the actions of profit-seeking producers concerned with minimizing production costs under a system of perfect liberty.[5] In other words, 'normal' prices or prices of production describe the forces governing the competitive economic system, whereas market prices reflect influences that are accidental or temporary.

The object of the theory is to describe how higher than average profits in one industry would act as an attractor of capital and labour from other industries. Adam Smith described a process where individuals reallocate their land, labour and capital to uses that are more profitable. In *The Principles of Political Economy and Taxation*, David Ricardo (1817, p. 91) described this process as being driven by 'the desire, which every capitalist has, of diverting his funds from a less to a more profitable employment'. This was because 'it is this competition which so adjusts the exchangeable value of commodities, that after paying . . . [wages and the other costs of production], the remaining value or overplus will in each trade be in proportion to the value of the capital employed'. In other words, finance capital moves to those sectors with a higher than average rate of profits, which over time allows the physical capital to become mobile through the process of investment and depreciation. These early economists were mainly interested in the intersectoral flows of finance capital, where financiers gain profits by minimizing costs. Nevertheless, the classical economists had little to say about the strategic behaviour of the firm, and the marginalist economists ran into difficulties when they tried to include it. The method, however, does explain why entrepreneurs and business firms enter and exit a particular market.

When there is a division of labour, or increasing returns, profit-seeking entrepreneurs can minimize the cost of production through the accumulation of differentiated knowledge. Since firms that accumulate this kind of knowledge often see rising profit rates, entrepreneurs will move their finance capital to these sectors. Thus the uniform rate of profit as a tendency explains why technical change and technological learning takes place, which also implies that profit-seeking entrepreneurs minimize the cost of production by generating new and differentiated knowledge. Smith recognized this type of competition in book 5 of the *Wealth of Nations* (1776, p. 706) when he stated that 'the competition of producers who, in order to undersell one another, have recourse to new divisions of labour, and new improvements of art, which might never otherwise have been thought of'.

Karl Marx (1867) also recognized that competition among producers generated the surplus by increasing labour productivity and adopting new technology. In attaching importance to competition taking place in both production and exchange, he makes an important distinction between competition within and between industries. Within each industry, price competition is most important for increasing the firm's market share. Like in the other classical economists' theories, interindustry competition results in investment flows, bringing about a tendency for the rate of profit to equalize across industries, but he asserts that intra-industry competition aims at enlarging the market share and improving the profitability of firms. In Chapter 13 of the first volume of *Das Kapital*, Marx (1867) argues that capitalists search for ways to increase profitability and market share by introducing new products and processes. By lowering (and minimizing) the cost of production, individual entrepreneurs have higher than average profits until the new technique becomes widespread. Although the new technology allows the innovating firm to have above average profits temporarily, it plays a crucial role in the competitive process by contributing to the continuous change of the productive structure. The determination of production prices in each industry, and the tendency towards a uniform rate of profit, lead to continuous investments in the innovating industry as well as the diffusion of technology across industries.

The objective of an entrepreneur or business firm to increase the surplus value or profit contained in the commodity provides the rationale for technological accumulation, but it ultimately depends on the interaction between knowledge and the learning process acquired from production experiences. In volume three of *Das Capital*, Marx (1894) made an important distinction between perfecting and inventing new machines (or radical and incremental process innovations), engendered by the division of labour, and the introduction of new machinery that depends on other factors (Giammanco, 2002). One of these factors is the dependence of science on labour, which implies that some time must elapse before the firm integrates an invention into the production process (Rosenberg, 1974). Another factor is the interdependence of industries, which implies that innovations in one industry can be the cause of innovation in others.

Most importantly, behavioural heterogeneity and technological diversity are important drivers of economic growth and knowledge accumulation in the theories of both the classical economists and Marx, and this heterogeneity was understood to be independent of the equilibrium conditions. However, the early classical economists and Marx ran into a significant theoretical problem because they considered labour, itself heterogeneous, as a measure of value. The heterogeneity across firms and products made it difficult to make interspatial and intertemporal comparisons without an invariable measure of value, which eluded these economists until Sraffa (1960) proposed the standard commodity. With the standard commodity, Sraffa was able to demonstrate the equilibrium conditions necessary for obtaining the prices of production, irrespective of the heterogeneity and variety in the economy, and to provide the foundations for a theory of capital accumulation, technical change and economic growth in an economy with widespread heterogeneity and diversity. Indeed, Chiaromonte and Dosi (1993) showed that increasing diversity among economic agents results in simpler economic dynamics.

COMPETITION AMONG HOMOGENEOUS ACTORS IN THE NEO-CLASSICAL ECONOMISTS' THEORIES

Although the method used by the classical and the early marginalist (neoclassical) economists was essentially the same, the content of the theory was very different (Kurz and Salvadori, 1995). The early marginalist economists described a similar competitive process, but defined prices in terms of the scarcity of productive resources with respect to the demand for goods. Excess profits; that is, profits above the interest rate, vanished when there was free competition. This difference had an important influence on how the meaning of competition changed in the twentieth century. In contrast to the classical approach, the marginalist approach determined prices by the scarcity of productive resources with respect to the demand for goods. They started from a very different set of data (Garegnani, 1984): (1) the initial endowments; (2) the preferences of consumers; and (3) the technical alternatives from which cost-minimizing producers can choose. Thus, given the technology endowments, the prices of all factors of production are determined simultaneously and symmetrically through the intersection of supply and demand. Initially the theory appeared to be able to cope with the behavioural heterogeneity and product diversity because it maintained the classical method, but over time it became increasingly clear that the marginalist approach could not deal with the issue adequately. The development of the marginalist approach and the meaning of competition was an important episode in the history of economic thought that set the stage for the current discussion on firm heterogeneity.

In his book *Éléments d'Économie Pure*, Leon Walras (1874), a pioneer of the marginalist approach to value and distribution, changed the content of the theory but maintained a view of competition that was similar to the classical economists in stressing that value in exchange, when left to itself, arises spontaneously in the market as the result of competition. In doing so, he used the analogy of an auction to symbolize the market, and the competitive bidding process that goes on in such a market. Nevertheless, he also introduced the concept of perfect competition, but only for logical and mathematical convenience, considering it analogous to 'perfectly frictionless' machines. He claims to have made this assumption only for mathematical convenience because he later introduces the idea of tâtonnement to describe the process of discovering the equilibrium price. Nevertheless, the mathematical requirements of the theory pushed subsequent contributions towards merging the concepts of competition and the market, which then pushed the theory towards a new method.

This change in the content of the theory led to a change in the meaning of competition from a gravitational force to a state of perfection. For the classical and early marginalist economists, gravitational force brought about stability and order in the economy much as it does in physics. By applying this analogy to marginalist theory, it is possible to interpret the equality between prices and marginal costs as an attractor for the efficient use of scarce resources. Nevertheless, as Paul J. McNulty (1968) pointed out, the gradual refinement of the marginalist approach shifted the meaning of competition to one where competition became an ideal that was impossible to realize. The change in the meaning of competition to an idealized state also allowed the marginalist economists to equate competition with the market, which meant, as George Stigler (1957) suggested, the market lost all of its institutional characteristics. Although the early marginalist economists agreed that competition would be greater if more individuals were involved, they never considered the gravitational forces that attract prices and profits to their normal level to be the same as a market, but reasoned instead that they operated *within* a market.

Neither Walras nor Alfred Marshall, one of the most influential economists in the development of the marginalist approach, went so far as to define competition as a market structure. In anticipation of the marginalist approach, A.A. Cournot (1838) set up a mathematical model with two rival producers of a homogeneous product. Known as the Cournot 'duopoly model', he mathematically derived a deterministic solution to the quantities chosen by the rival producers in accordance with each other's anticipated reactions. In Chapter 8 of his book *Recherches sur les principes mathématiques de la théorie des richesses*, he introduces the case of *unlimited competition*; that is, where the number of producers is so large that the entry and exit of any producer will have a significant influence on the total quantity produced. Although Cournot did not explicitly equate competition with the market structure, he did show price is equal to marginal cost when there is unlimited competition. It was Stanley Jevons (1871) and F.Y. Edgeworth (1881) who provided the basis for equating competition with the market structure towards the end of the nineteenth century, and J.B. Clark (1900) and Frank Knight (1921) who refined the concept in the early twentieth century.

Jevons, another pioneer of marginalist theory, showed how the principle of diminishing marginal utility governed individual choice, but was not able to demonstrate how this idea could explain market equilibrium. By focusing on a pure exchange economy, Jevons identified competition as entirely an exchange phenomenon. This new focus, combined with the indeterminacy between market equilibrium and the 'degree of competition', led Edgeworth to attempt a systematic and rigorous definition of perfect competition. Towards the beginning of his book, *Mathematical Physics: An Essay on the Application of Mathematics to the Moral Sciences*, Edgeworth (1881) outlined three conditions for perfect competition: (1) an infinite number of individuals on both sides of the market; (2) no limitations on individual self-seeking behaviour; and (3) complete divisibility of commodities traded. Frank Knight (1921) later refined the conditions necessary for perfect competition to require that every individual possesses complete knowledge of the market conditions, there is perfect mobility with no costs involved, and there is an infinite number of independent traders such that no individual can influence the quantity of goods brought to the market. These assumptions imply that all goods and services in the particular market must be homogenous. The contributions by Edgeworth were perhaps the first attempt to merge the concepts of competition with the market as Stigler (1957) claimed. Nevertheless, the subsequent developments of the concept of perfect competition anticipated an important change in marginalist theory from the long-period method used by Walras and Marshall to the intertemporal method pioneered by Erik Lindahl and Fredrich von Hayek and the temporary method pioneered by John Hicks and Gerald Debreu. Debreu (1959) provided the most complete break with the long-period method by using the concept of perfect competition to compress the futures markets into the present and ensure that all markets clear simultaneously.

While the focus of the marginalist approach in the twentieth century was on perfect competition, Carl Menger (1871), and succeeding economists trained in the Austrian tradition, picked up some of the dynamic aspects of competition that concerned Smith and the classical economists. Also a pioneer of the marginalist approach to value and distribution, he not only provided the basic concepts of marginalist theory independent of the others, but also was different from them in emphasizing the concept of subjective value, the spontaneous evolution of institutions, and disequilibrium. By focusing on the subjective nature of knowledge, Fredrich von Hayek and Joseph Schumpeter described competition as a process of discovery, where entrepreneurs would gain higher than average profits by producing new goods, or producing existing goods in new ways. Hayek (1945, p. 94) viewed competition not as a state that assumes 'the data for the different individuals are fully adjusted to each other' but as a process that 'necessarily involves continuous changes in the data for the different individuals'.[6] At the same time, he asserted that equilibrium is an outcome of the formation of a spontaneous order through individual actions; that is, the economy is both self-organizing and self-reproducing. In this context competition is a 'process of discovery' that is endogenous to the economic system, an idea reminiscent of Adam Smith's division of labour and Marx's linkage between competition and technological change (Eatwell and Milgate, 1994, p. 84).

MARSHALL AND THE REPRESENTATIVE FIRM

Alfred Marshall (1890) also adopted the same method as the classical economists, but determined both the market and normal prices through the relationship between demand and supply.[7] One important difference is that he distinguishes between short-period normal prices that refer to less than a year and long-period normal prices that refer to several years. The long-period normal prices include capital goods, which corresponds to the prices described by Walras. While Marshall maintains the same method as the classical economists and Walras, he also introduces a short-period equilibrium of an industry in which the capital stocks and productive capacities of firms are considered as given, and a long-period equilibrium in which firms can adjust the size of their capital stocks and productive capacities. This difference led to another episode in the history of economic thought that had important consequences for the way economists perceived behavioural heterogeneity.

To make the notion of short-period normal prices more applicable to concrete issues and focus the analysis on an industry, Marshall (1890) introduced the idea of partial equilibrium analysis and the representative firm. Partial equilibrium analysis focused on the determination of prices and quantities in a specific market, independent of the effect that this market might have on supply and demand, and hence prices in other markets. To deal with economies external to firms, Marshall (1961, book IV) introduced the representative firm in the second edition of the *Principles of Economics*. He defined the representative firm as 'an average firm' with 'access to the economies, external and internal economies, which belong to the aggregate volume of production'. This idea of an average firm was in fact Marshall's way of recognizing the vast behavioural heterogeneity across actual firms in the economy, but he also considered it essential to his discussion of normal value in relation to supply and demand. Marshall also claimed that he used the concept to simplify problems that were relevant to the long-period equilibrium, particularly with issues related to the growth of firms (internal economies) and the growth of knowledge (external economies), but it meant that the long-run costs of a firm should be equal to the industry. It also meant that an industry would be composed of *identical* firms with the same behaviour, same size and earning the normal rate of profits (Opocher and Steedman, 2006).

Marshall's theory, however, contained an inconsistency that made variable costs, including increasing and decreasing returns to scale, incompatible with free or perfect competition. The problem with the theory was that whatever happened in one market or industry had no effect on the prices of goods in other markets. One could accept this argument if the division of labour could be contained within the industry being investigated, but Piero Sraffa (1925) argued that this was contrary to the facts. This issue was the main topic of a debate that took place during the late 1920s in the *Economic Journal* (see Robertson et al., 1930). Starting from the point of view that diminishing returns (rising costs) and increasing returns (falling costs) originate from different causes, Sraffa (1926) showed that when variable costs are external to an industry, they will change the costs of firms in the other industries affected.[8] This incompatibility of economies external to the firm with economies internal to the industry violates the assumptions that underlie Marshall's partial equilibrium analysis because it is not possible to establish the supply curve for the firm. This meant that Marshall's view of partial equilibrium and the representative firm was untenable. One way to avoid this problem is to assume that all variable costs are internal to the firm, but this would mean that the firm would be a monopolist, in a non-competitive environment.

Allyn Young (1928) provided additional support to Sraffa's argument by pointing out that industrial stratification suggests a division of labour among firms and industries. Adam Smith's idea of increasing returns was 'progressive and propagates itself in a cumulative way', which implies that variable costs are always external to the industry. This essentially meant that behavioural heterogeneity was a response to changes in the market external to the firm. Despite the best efforts by D.H. Robertson and G.F. Shove to defend Marshall's partial equilibrium analysis and the representative firm, Sraffa demonstrated once more that increasing returns are inconsistent with internal economies, and that as a consequence the representative firm would change position within the industry as new firms enter and others leave (Robertson et al., 1930). This issue proved to be important for empirical studies of the growth of firms; however, Sraffa was mainly concerned with solutions to theoretical problems, including further development of the classical approach and economic growth and dynamics.[9] Although Sraffa (1960) chose to return to the classical approach and develop a more systematic approach to the theory, he also suggested that the introduction of imperfect competition into Marshall's partial equilibrium approach could be an alternative.

When looking for a solution to Marshall's problem, both Robinson (1933) and Chamberlin (1933) arrived at the same solution with regard to the equilibrium of each individual firm and market equilibrium, Chamberlin (1933), using the term 'monopolistic competition' provides a more general view of competition by including all market situations that lie between perfect competition and monopoly. The theory considers all individual firms as independent and there is free entry and exit to and from the market as in free competition, but postulates that although the goods produced by the various firms are differentiated, there are many close substitutes. He was also the first economist to take account of advertising as a selling cost. Chamberlin (1933, p. 56) defines differentiation as the 'characteristics of the product itself, such as exclusive patented features; trade marks; trade names; peculiarities of the package or container, if any; or singularity in quantity, design, colour or style [as well as] the conditions surrounding its sale'.

By contrast, Robinson (1933) introduced Cournot's concept of marginal revenue into the theory of the firm and assumed that it is downward sloping. Although less general, Robinson gets around the problem of aggregating heterogeneous commodities at the industry level and she showed that it was not possible to pay labour their marginal product if perfect competition does not exist. Recognizing much of the criticism in the *Economic Journal* discussion, Robinson also moved away from Marshall's representative firm, instead adopting A.C. Pigou's (1928) idea of the equilibrium firm. This idea allowed her to assume that all firms within an industry were identical as in Marshall's theory, but because they were in equilibrium all firms were 'representative'. Nevertheless, Robinson identified several different forms of competition that did not rely on price, including product innovation, product differentiation, packing and design, availability of credit, advertising, and marketing, all of which would generate behavioural heterogeneity in the economy. Eventually Robinson (1980) embraced the idea that knowledge is always 'imperfect' and that 'the full information required to make a correct choice can never be available'.

Despite using a concept of equilibrium that equated firms to the industry, Robinson and Chamberlin both explain why there is so much heterogeneity among firms, but to the neo-classical (marginalist) economists these forms of competition were seen as imperfections that needed to be removed before prices could be in equilibrium (Eatwell and Milgate, 1994).[10] One reason for this is that wages no longer measure the marginal product of labour when there is imperfect competition, which is an essential principle in the marginalist approach to value and distribution. Partly for this reason, the title of Robinson's book *The Economics of Imperfect Competition* (1933) carries with it a certain amount of sarcasm and it illustrated why Sraffa thought that 'Marshall's theory should be discarded'. Nonetheless, without a theory of the firm it becomes difficult, if not impossible, to relate competition to the issues of behavioural heterogeneity and technological diversity.

BEHAVIOURAL HETEROGENEITY AND MODERN APPLIED ECONOMIC THEORY

Marshall's representative firm contains the idea that a certain size firm with average access to internal and external economies is normal for the industry. This implies that differences in the size of firms should diminish over time as firms adjust their size to fit the industry's economy of scale. The idea of imperfect competition, however, suggests that firms would not adjust their size in this way. One of the first empirical criticisms of Marshall came from Josef Steindl (1945) who observed 'enormous differences in the size of enterprises' and that 'small firms will mostly move in a small circle only'. His criticism was not aimed at the lack of entrepreneurial ability as suggested by Marshall, but at the presence of risk, limits to borrowing, the high mortality rate and monopolistic domination. In essence, Steindl (1945, 1952) used empirical evidence to show a kind of persistence of heterogeneity among firms, which confirms the idea that the representative firm is not relevant to the theory of competition (Hölzl, 2003). The ongoing discussion on the growth of firms and market structure provides another episode in the history of economic thought that is important for understanding some of the contemporary discussions about firm heterogeneity.

While in Oxford during the early 1940s along with Steindl, Michael Kalecki (1945) critically described an important study in French on the size distribution of French firms by Robert Gibrat (1931), which provided the starting point for the ongoing discussion on the growth of firms and market structure. Using a skewed lognormal distribution, Gibrat found that all firms within an industry had the same chance of growing, regardless of their size.[11] This stylized fact generated a large and growing literature on the measurement of firm growth and heterogeneity and on the shape of the long-run cost curve (see Sutton, 1997; Lotti et al., 2003; Coad, 2007). The first wave of studies in the late 1950s and 1960s focused mainly on large manufacturing firms in the UK (Hart and Prais, 1956) and the USA (Simon and Bonini, 1958) and found little evidence that a relationship existed between the log size of firms and their average growth rates, which supported Gibrat's prediction. Using a larger sample, a study by Steven Hymer and Peter Pashigian (1962), however, found a strong negative dependence of the variance of these growth rates on size when smaller firms were included in the analysis, which suggests that there is much more heterogeneity in firm distributions across different sectors. Alternative distributions were also used in the analysis; Herbert Simon and Charles Bonini (1958) preferred the Yule family of distributions because it could incorporate firm entry better, Yuji Ijiri and Herbert Simon (1977) and Josef Steindl (1965) applied the Pareto distribution to US and Austrian firms respectively, Robert Axtell (2001) and Giovanni Dosi et al. (2007) applied a Zipf distribution to US and European firms respectively, and Giulo Bottazzi and Angelo Secchi (2002) applied the Laplace or symmetric exponential distribution to Italian firms.

The majority of subsequent studies found that small firms tended to grow faster than larger firms, suggesting that 'Gibrat's law' was too strong, and that the issue of firm survival be included in the analysis.[12] By the 1980s the analysis shifted focus towards the functional relationship between size and firm growth, and the problem of heteroscedasticity (different variance), suggesting that behavioural heterogeneity and technological diversity are essential. The more recent models focused either on persistent production efficiencies or innovation to explain firm heterogeneity as pointed out by Tor Jakob Klette and Arvid Raknerud (2002) and Giovanni

Dosi (2005). John Sutton (1998) demonstrates how differences in the innovative behaviour of large versus small firms can generate persistent differences in firm size and a concentrated market structure. Drawing on existing studies, Dosi (2005) provides considerable evidence that interfirm heterogeneity is extensive and persistent over time. The presence of persistent differential profit rates does not destroy the classical theory of value and distribution, but underscores the importance of technical change and technological learning. Recent empirical evidence on innovation and firm performance also indicates a similar skewed pattern of behaviour by firms and that this heterogeneity is persistent over time (Peters, 2005).[13]

This episode in the history of economic thought shows that Marshall's representative firm is not only theoretically indefensible, but also empirically unsound. James Heckman (2000) also emphasized this point in his Nobel address when he described the empirical importance of heterogeneity for macroeconomic theory and the interpretation of macro evidence.[14]

CONCLUDING REMARKS

Although the classical economists had little to say about the strategic behaviour of the firm, the neo-classical economists ran into difficulties when they tried to include it. As Kurz (2006) emphasized, the neo-classical economists took 'great pains' to bypass the issue of growing behavioural heterogeneity and technological diversity by making some 'quite remarkable assumptions'. The twin ideas of the representative firm and perfect competition provide a good example. The classical economists recognized that the production process required specific capabilities to complete the array of tasks needed to produce the product and to coordinate this process. Originating from Marshall's efforts to preserve Smith's theoretical system, Edith Penrose (1959) developed a knowledge-based theory of the firm that fits well with the production-based theory developed by the classical economists without carrying the baggage of Marshall's representative firm.

Marshall (1961, p. 355) believed that 'the tendency to variation is a chief cause of progress', but he mentioned this to illustrate the principle of substitution. The breakdown of the principle of substitution was at the heart of the capital theoretical debates of the 1960s and is necessary for neo-classical theory (Kurz and Salvadori, 1995). Nonetheless, when quoting Marshall, Wesley Cohen and Franco Malerba (2001) emphasized that evolutionary models consider heterogeneity among firms as the engine of evolutionary dynamics, which may be closer in spirit to the biological analogies sometimes used by Marshall in his *Principles*. In contrast with Marshall's idea of the representative firm, however, these models adopt a population view, in which a heterogeneous population of firms (or agents) co-exist in the same economy. In these models, heterogeneity drives technical change and technological learning, which in turn drives the division of labour.

This chapter provided some examples of how some economists from the past have considered behavioural heterogeneity and technological diversity within the core of their theory. This heterogeneity can appear in a variety of different forms, including the labour and capital used in the production of a good or service, the products available on the market, different needs of consumers and producers, as well as the strategic and innovative behaviour of firms. Richard Nelson and Sidney Winter (1982) and J. Stanley Metcalfe (1998) have also emphasized the importance of behavioural heterogeneity and technological diversity to being competitive and for economic growth. These insights depend on standing on the shoulders of giants.

NOTES

1. Heinz Kurz (2006) points out, 'perhaps the most important problem which the economist dealing with the dynamism of the modern economy has to face is that of the growing heterogeneity and diversity of knowledge, commodities, services and labours performed as the economy evolves'.
2. The idea of the division of labour first appeared in Plato's *Republic*, where it is the main source of inequality. William Petty later integrated the idea into his political economy, but he also knew that people with no extensive training could perform many of the tasks in the division of labour.
3. Hayek (1945) also described the relationship between the division of labour and the division of knowledge.
4. For Smith, international trade widens the market and provides a vent to the resources that, in the absence of trade, would remain unemployed or underemployed. Specialization happens when a country has certain natural or acquired advantages that lead to lower absolute costs, and exchanges its surplus produce with the produce of other countries for which there is a demand in the home market.
5. Smith lays out the institutional framework necessary for a 'society of perfect liberty' in his *Theory of Moral Sentiments*, and then applies it in *The Wealth of Nations*. In this book he describes a society of perfect liberty in which socially desired outcomes depend on whether self-interested individuals can exercise self-control and adhere to the rules and codes of behaviour of the society.
6. The Lange–Hayek debates of the 1930s had a lot to do with competition. Lange adopted the method and content of marginalist theory as developed by Walras and Barone to show that socialism was feasible, provided functionaries obeyed the directives from the central planner. Hayek maintained that the price system resembled a communication network that the central planner could not duplicate.
7. However, Marshall (1890, p. 448) argued strongly that he did 'not assume that competition was perfect' because it 'requires a perfect knowledge of the state of the market'.
8. Kenneth Arrow (in Arthur, 1994) pointed out that Cournot (1838) developed his theories of monopoly and oligopoly because he knew that increasing returns were incompatible with free competition.

9. Sraffa's revival of the classical approach to value and distribution is found in his *Production of Commodities by Means of Commodities*, started in the late 1920s and published in 1960. Nevertheless, Sraffa was also concerned with economic growth and dynamics; ideas that were written in notebooks currently available at the Trinity College Library in Cambridge, but have not yet been published. A collected works of Sraffa is being planned by Cambridge University Press and will contain many of these notebooks.
10. Perfect competition requires that there is a sufficiently large number of independent, profit-maximizing business firms, with free access to all resources and complete knowledge of the market conditions so that no individual firm can influence the price. Any deviation from this ideal market condition leads to imperfect competition.
11. Gibrat's 'law of proportionate effect' implies that if the rates of growth of firms are identically and independently distributed, the distribution of the firms' size tends asymptotically to a lognormal. In other words, firm size follows a random walk.
12. Mansfield (1962) suggests three variations on the postulate: (1) include all firms, including those that exit during the period; (2) include all firms, except those that exit during the period; and (3) include only those firms that exceed the minimum efficient size in the industry.
13. There is a new and burgeoning literature on innovation and the productivity growth of firms based on Crépon et al. (1998). Peters (2005) found the persistence in innovation activities to be high between R&D and innovation survey data, whereas it tends to be lower with patent and major innovations.
14. Hopenhayn (1992) developed a model to explain the endogenous selection of heterogeneous firms in an industry and Melitz (2003) applied this model to trade theory. Pakes and Ericson (1998) developed a stochastic growth model with behavioural heterogeneity of firms that included passive and active learning strategies.

REFERENCES

Arthur, W.B. (1994), *Increasing Returns and Path Dependence in the Economy*, Ann Arbor: University of Michigan Press.

Axtell, R.L. (2001), 'Zipf distribution of U.S. firm sizes', *Science*, **293**, 1818–20.

Bottazzi, G. and A. Secchi (2002), 'On the Laplace distribution of firm growth rates', *LEM Working Paper Series, 2002/20*, Pisa.

Chamberlin, E.H. (1933 [1950]), *The Theory of Monopolistic Competition*, Cambridge, MA: Harvard University Press.

Chiaromonte, F. and G. Dosi (1993), 'Heterogeneity, competition and macroeconomic dynamics', *Structural Change and Economic Dynamics*, **4**, 39–63.

Coad, A. (2007), 'Firm growth: a survey', *Papers on Economics and Evolution, 0703*, Jena.

Cohen, W. and F. Malerba (2001), 'Is the tendency to variation a chief cause of progress?', *Industrial and Corporate Change*, **10**, 561–608.

Cournot, A.A. (1838 [1897]), 'Recherches sur les principes mathématiques de la théorie des richesses', trans. N.T. Bacon, *Researches into the Mathematical Principles of the Theory of Wealth*, New York: Macmillan.

Crépon, B., E. Duguet and J. Mairesse (1998), 'Research and development, innovation and productivity: an econometric analysis at the firm level', *Economics of Innovation and New Technology*, **7**, 115–58.

Debreu, G. (1959), *The Theory of Value*, New York: Wiley.

Dosi, G. (2005), 'Statistical regularities in the evolution of industries', *LEM Working Paper Series, 2005/17*, Pisa.

Dosi, G.A. Gambardella, M. Grazzi, and L. Orsenigo (2007), 'Technological revolutions and the evolution of industrial structures', *LEM Working Paper Series, 2007/12*, Pisa.
Eatwell, J. and M. Milgate (1994), 'Competition, prices and market order', in M. Colonna and H. Hagemann (eds), *Money and Business Cycles*, Aldershot, UK and Brookfield, USA: Edward Elgar.
Edgeworth, F.Y. (1881), *Mathematical Physics: An Essay on the Application of Mathematics to the Moral Sciences*, London: Kegan Paul.
Garegnani, P. (1984), 'Value and distribution in the classical economists and Marx', *Oxford Economic Papers*, **36**, 291–325.
Giammanco, M.D. (2002), 'Competition and technical progress in Marx: two different perspectives', *History of Economic Ideas*, **14**, 69–96.
Gibrat, R. (1931), *Les inégalités économiques*, Paris: Librairie du Recueil Sirey.
Hart, P.E. and S.J. Prais (1956), 'The analysis of business concentration', *Journal of the Royal Statistical Society*, **119**, 150–91.
Hayek, F.A. (1945), 'The use of knowledge in society', *American Economic Review*, **35**, 519–30.
Hayek, F.A. (1948), 'The meaning of competition', in F.A. Hayek, *Individualism and Economic Order*, Chicago: University of Chicago Press.
Hayek, F.A. (1978), 'Competition as a discovery procedure', in *New Studies in Philosophy, Politics, Economics, and the History of Ideas*, Chicago: University of Chicago Press, pp. 179–90.
Heckman, J.J. (2000), *Microdata, Heterogeneity and the Evaluation of Public Policy*, Nobel Prize Lecture, 8 December.
Hölzl, W. (2003), *Evolutionary Competition and the Evolutionary Theory of the Firm*, PhD thesis, University of Vienna.
Hopenhayn, H.A. (1992), 'Entry, exit, and firm dynamics in long run equilibrium', *Econometrica*, **60**, 1127–50.
Hymer, S. and P. Pashigian (1962), 'Firm size and rate of growth', *Journal of Political Economy*, **70**(6), 556–69.
Ijiri, Y. and H.A. Simon (1977), *Skew Distributions and the Sizes of Business Firms*, Amsterdam: North Holland.
Kalecki, M. (1945), 'On the Gibrat distribution', *Econometrica*, **13**, 161–70.
Klette, T.J. and A. Raknerud (2002), 'How and why do firms differ?', *Statistics Norway Discussion Paper 320*.
Knight, F.H. (1921), *Risk, Uncertainty and Profit*, Boston, MA: Houghton Mifflin.
Kurz, H.D. (2006), 'Whither history of economic thought? Going nowhere rather slowly?', *European Journal of the History of Economic Thought*, **13**, 463–88.
Kurz, H.D. and N. Salvadori (1995), *Theory of Production: A Long-Period Analysis*, Cambridge: Cambridge University Press.
Lavezzi, A. (2003), 'Division of labour and economic growth: Paul Romer's contribution in an historical perspective', in N. Salvadori (ed.), *Old and New Growth Theories. An Assessment*, Cheltenham, UK and Northampton, MA, USA: Edward Elgar.
Loasby, B.J. (1999), *Knowledge, Institutions and Evolution in Economics*, London: Routledge.
Lotti F., E. Santarelli, and M. Vivarelli (2003), 'Does Gibrat's Law hold in the case of young, small firms?', *Journal of Evolutionary Economics*, **13**, 213–35.
Löwe, A. (1965 [1977]), *On Economic Knowledge: Toward a Science of Political Economics*, enlarged edn, Armonk, NY: M.E. Sharpe.

Mansfield, E. (1962), 'Entry, Gibrat's Law, innovation, and the growth of firms', *American Economic Review*, **52**, 1023–51.
Marshall, A. (1890 [1961]), *Principles of Economics*, 9th variorum edn (ed. C.W. Guillebaud), London: Macmillan.
Marx, K. (1867 [1976]), *Capital: A Critique of Political Economy*, Vol. 1, London: Penguin.
Marx, K. (1894 [1981]), *Capital: A Critique of Political Economy*, Vol. 3, London: Penguin.
McNulty, P.J. (1968), 'Economic theory and the meaning of competition', *Quarterly Journal of Economics*, **82**, 639–56.
Melitz, M.J. (2003), 'The impact of trade on intra-industry reallocations and aggregate industry productivity', *Econometrica*, **71**, 1695–725.
Metcalfe, J.S. (1998), *Evolutionary Economics and Creative Destruction*, London: Routledge.
Nelson, R.R. and S.G. Winter (1982), *An Evolutionary Theory of Economic Change*, Cambridge, MA: Harvard University Press.
Opocher A. and I. Steedman (2006), 'The industry supply curve: two different traditions', *Discussion Papers in Economics 2006-02*, Manchester Metropolitan University.
Pakes, A. and R. Ericson (1998), 'Empirical implications of alternative models of firm dynamics', *Journal of Economic Theory*, **79**, 1–45.
Penrose, E. (1959), *The Theory of Growth of the Firm*, Oxford: Oxford University Press.
Peters, B. (2005), 'Persistence of innovation: stylised facts and panel data evidence', *ZEW Discussion Paper, 05-81*.
Pigou, A.C. (1928), 'An analysis of supply', *Economic Journal*, **38**, 238–57.
Ricardo, D. (1817 [1955]), *The Principles of Political Economy and Taxation*, Cambridge: Cambridge University Press.
Richardson, G.B. (1975), 'Adam Smith on competition and increasing returns', in A.S. Skinner, *Essays on Adam Smith*, Oxford: Oxford University Press.
Robertson, D.H., P. Sraffa, and G.F. Shove (1930), 'Increasing returns and the representative firm', *The Economic Journal*, **40**, 79–116.
Robinson, J. (1933), *Economics of Imperfect Competition*, London: Macmillian.
Robinson, J. (1980), *What are the Questions? And Other Essays*, Armonk, NY: M.E. Sharpe.
Romer, P. (1987), 'Growth based on increasing returns due to specialization', *American Economic Review*, **77**, 56–62.
Romer, P. (1990), 'Endogenous technological progress', *Journal of Political Economy*, **98**, S71–S102.
Rosenberg, N. (1974), 'Karl Marx on the economic role of science', *Journal of Political Economy*, **82**, 713–28.
Schumpeter, J.A. (1912 [1934]), *The Theory of Economic Development*, Cambridge, MA: Harvard University Press.
Schumpeter, J.A. (1942), *Capitalism, Socialism and Democracy*, New York: Harper and Brothers.
Simon, H.A. and C.P. Bonini (1958), 'The size distribution of business firms', *American Economic Review*, **48**, 607–17.
Smith, A. (1776 [1976]), *An Inquiry into the Nature and Causes of the Wealth of Nations*, eds R.H. Campbell, A.S. Skinner and W.B. Todd, Oxford: Oxford University Press.

Sraffa, P. (1925), 'Sulle ralazioni fra costo e quantita prodotta', *Annali di economia*, **2**.
Sraffa, P. (1926), 'The laws of returns under competitive conditions', *Economic Journal*, **36**, 535–50.
Sraffa, P. (1960), *Production of Commodities by Means of Commodities*, Cambridge: Cambridge University Press.
Steindl, J. (1945), *Small and Big Business*, Oxford: Basil Blackwell.
Steindl, J. (1952), *Maturity and Stagnation in American Capitalism*, Oxford: Basil Blackwell.
Steindl, J. (1965), *Random Processes and the Growth of Firms: A Study of the Pareto Law*, London: Griffen.
Stigler, G.J. (1957), 'Perfect competition, historically contemplated', *Journal of Political Economy*, **65**, 1–17.
Sutton, J. (1997), 'Gibrat's legacy', *Journal of Economic Literature*, **35**, 40–59.
Sutton, J. (1998), *Technology and Market Structure*, Cambridge, MA: MIT Press.
Targetti, E. and A.P. Thirlwall (eds) (1989), *The Essential Kaldor*, New York: Holmes & Meier.
Walras, L. (1874 [1954]), *Éléments d'Économie Pure*, Lausanne: Corbaz, definitive edn 1926, translated as *Elements of Economics*, London: George Allen & Unwin.
Young, A. (1928), 'Increasing returns and economic progress', *Economic Journal*, **38**, 527–42.

4. Heterogeneity, rationality and institutions

Tore Sandven

INTRODUCTION

Approaches to economics that are critical of the dominant neo-classical tradition often hold *heterogeneity* and *diversity* to be central characteristics of economic life. Conversely, the dominant tradition is claimed to neglect, misrepresent or deny heterogeneity and diversity. For instance, in his book *Les trous noirs de la science économique*,[1] Jacques Sapir proposes the concept of heterogeneity as a fundamental concept for an alternative economic theory. Diversity, he claims, should not be treated as accidental, as the product of imperfections, bound to be wiped out with time. On the contrary, 'diversity, from that of the actors to that of the trajectories of concrete economies, is the norm and should be conceptualised as such'.[2]

This chapter approaches the issue of heterogeneity or diversity by focusing on the question of institutional diversity in the economy. More specifically, different ways of understanding and evaluating this institutional diversity are discussed and related to the ideas one has of human rationality and human action. Or, conversely, an exploration is made of the implications of different conceptions of human rationality for the way institutional diversity may be understood.

INSTITUTIONAL DIVERSITY

As an issue distinguishing different approaches to an understanding of the economy, that of institutional diversity first and foremost concerns the relationship between the market or the market mechanism and other institutions or mechanisms in driving and coordinating economic action. For instance, in a discussion of the coordination mechanisms of modern capitalist economies, J. Rogers Hollingsworth, Philippe C. Schmitter and Wolfgang Streeck take institutional diversity as their point of departure.[3] They argue that economic coordination is not only accomplished through the market mechanism, but through a set of institutions that together make up the economy. In addition to markets, they identify corporate hierarchies (firms), the state, informal networks and associations as mechanisms of economic coordination.[4] More generally, they claim that economic arrangements are always embedded in and modified by the broader institutional context,[5] where we should also include such phenomena as social norms and values (for instance, standards concerning what counts as doing well in life in a given society or in a given social class).[6] Thus, their basic premise is that economic action is a special case of social action and should accordingly be understood as coordinated by institutional arrangements. This leads them to the notion of capitalism as an institutional order.[7] Hence, according to this approach, economic coordination is conceived as occurring through a set of institutions that together constitute the economy in question. Moreover, not only is diversity a characteristic of each economy in that each economy consists of a complex structure of diverse institutions, there is also diversity in the sense that there is considerable variation in these institutional structures across economies. This variation concerns both the institutions themselves, how they are coordinated with each other, and their relative importance in the total institutional structure.[8] Furthermore, we find this kind of variation in institutional structures not only across 'whole' economies, like national economies, but also across local economies and, notably, across economic sectors within these larger economies. Not least is there a variation across time: at all levels these institutional structures constantly change and develop. And here it is a central premise of the approach of Hollingsworth et al. that there are no good reasons for assuming that the development of different economies will imply a convergence of institutional structures towards one single type of institutional structure that in some sense should be 'best' or 'optimal'. Thus a basic objective of their approach becomes to 'capture this diversity of capitalism and render it manageable'.[9]

THE MARKET AND OTHER INSTITUTIONS

This focus on institutional diversity is very different from the approach of mainstream economic thought, as Hollingsworth et al. also emphasize.[10] In the mainstream neo-classical approach, the economy is basically thought of in terms of the market. In this way of thinking the market economy constitutes a unitary, self-regulating system. As we shall return to below, the economic order is here pictured rather as a natural order than as an institutional or social order. Typically, institutions other than the market are understood as logically secondary to the market to the extent that they are not neglected altogether. In the more orthodox, simplistic version of this approach other social institutions are defined in a purely negative way, as interfering with and distorting the workings of the market mechanism and thereby simply as sources of deviation from efficiency. In more sophisticated versions, these other kinds of institutions are understood as solutions to problems created by market imperfections or market failure such as, for instance, the absence of perfect information. But although institutions other than the market may be accorded a legitimate place in the economic order, they are still treated as logically secondary to the market as modifications to the perfect market solutions brought about by the presence of market imperfections, and where the solutions of the hypothetical perfect market still count as the standard of economic efficiency. It is how our understanding of these issues of institutional heterogeneity is associated with the way we understand human rationality that is the main focus of this chapter.

To get a grip on what is involved in this issue of heterogeneity, let us first see what other kinds of conceptual distinctions typically tend to be brought into the discussion when it is argued that understanding heterogeneity or diversity is essential to understanding economic processes.

It should be emphasized at the outset that no distinction is made here between heterogeneity and diversity and other related terms such as 'variety'. It is an objective of the discussion to clarify precisely what is meant by these terms.

HETEROGENEITY, CREATION AND UNCERTAINTY

In his book referred to above, Jacques Sapir treats heterogeneity as closely related to *creation* and *uncertainty*.[11] Heterogeneity, creation and uncertainty are central to what he terms *subjectivist* economics: conversely, he claims, neo-classical economic theory is characterized by its refusal to confront – and its inability to grasp – these types of phenomena.[12] By *subjectivist* we should here understand the idea that economic action is mediated by the manner in which the actors understand the situation they act within, for instance with regard to their expectations about the future. Thus they do not simply respond optimally to a situation whose traits are fully transparent and unambiguous. Furthermore, an important point here is that different actors understand the situation in different ways. This should not be understood as simply meaning that actors differ by being more or less right or wrong in their beliefs about the situation, so that the only freedom one has relative to doing what the situation dictates is to make mistakes. Rather, in a fundamental sense, the future is open and human beings are constantly creating new practices and products.[13] Thus by *subjectivist* we should also understand the idea that the world within which human beings act is not one that is simply pre-given and predetermined but one which, to a large extent, is being continually created and re-created through human practice, through human subjectivity. This creative and transforming aspect of human action has been emphasized by Anthony Giddens, who claims that this creative element is potentially present in even the most routine action: 'Every act which contributes to the reproduction of a structure is also an act of production, a novel enterprise, and as such may initiate change by altering that structure at the same time as it reproduces it.'[14]

In an article on differences among business firms, Richard R. Nelson also emphasizes the relationship between diversity and the creative aspect of human action.[15] Nelson argues strongly for the economic significance of what he calls *discretionary* firm differences, by which he means differences relating to how firms perceive the situation, and to their ambitions and strategies, and not simply differences that reflect differences in the (objective) conditions in which these firms operate. This is in explicit opposition to mainstream, neo-classical thinking, which he claims 'downplays or even denies the importance of such differences'.[16] Central to this denial of the importance of differences among firms, Nelson finds, is a theory of firm behaviour that portrays firms as facing given and known choice sets, and as having no difficulty in choosing the best action within these sets. Thus, in this approach the 'economic problem' is basically understood as 'getting private incentives right', and not as finding out what to do, 'which is assumed to be no problem'.[17] This assumption of given and known choice sets that the firms maximize on is particularly ill-suited for grasping what is involved in innovation. In his argument against models that simply apply the assumptions of standard neo-classical theory to innovation, Nelson emphasizes 'the fundamental uncertainty, the differences of opinion, the differences in perceptions about feasible paths, that tend to stand out in any detailed study of technical advance'.[18] At issue here is the very assumption of maximization: 'Does the assumption that "actors maximise" help one to analyse situations where some actors are not even aware of a possibility being considered by others?'[19]

Herbert Simon also argues against the view that what firms do can be reasonably depicted as 'choosing among a fixed set of alternative actions'. This, he claims, 'constitutes only a small, and relatively uninteresting part of the decision-making task within firms'. Rather, Simon sees business managements as mainly occupied with what he calls 'design tasks' – the 'design of products and of strategies for marketing, finance, manufacturing, and so on', claiming that 'the success of a business depends largely on how well these tasks are performed'.[20]

The question of creativity is stated very explicitly by G.L.S. Shackle, who asks if decisions are creative or predictable.[21] If, he claims, one thinks of the human actor as having a complete list of all actions open to him and complete knowledge of the results of all these actions, and likewise as being able to order all these results 'according to his own desires', then the actor's decision to do one thing rather than another would be 'as it were empty, the mere registering of a formal solution to a purely formal problem'.[22] Shackle holds that this approach fundamentally misses the logic of human action. He asks if 'the range of results seems to a man to be given to him objectively from without or created by him subjectively'. If we take the latter position; that is, if we ascribe to the individual 'an ultimate creative originality', then it becomes meaningless to speak of a complete list of such results. Thus Shackle defends the position that 'decision involves essential novelty', and consequently that 'prediction of human conduct is logically impossible'.[23]

HETEROGENEITY AMONG BUSINESS FIRMS

If we therefore reject the assumption that what the best thing to do for a firm is obvious, and perhaps actually assert that even in principle this is not something that can be known beforehand (which is what Shackle claims), and thereby accord a significant role for human creativity, we would expect there to be substantial heterogeneity among firms, even firms facing approximately the same conditions. As Nelson makes clear when making his case for the significance of discretionary firm differences, he implies 'a certain looseness of constraints, both in the short and long run, that gives room so that firms that differ in certain important respects can be viable in the same economic environment'.[24] This seems also to be what, in fact, we observe empirically. For instance, in a summary of recent empirical studies of firm behaviour and performance, J. Bradford Jensen and Robert H. McGuckin report that the behaviour of firms, even within the same geographical location and the same industry, differs dramatically, and that 'heterogeneity in the distribution of business units is pervasive along a wide variety of dimensions'.[25]

HETEROGENEITY AND THE IDEA OF NATURAL SELECTION

Against the position that discretionary firm differences matter, one might invoke a functionalist type of argument in terms of 'natural selection'.

Even though one might claim that firms cannot know beforehand what will be the best way to act, the heterogeneity of firms will only be a transient phenomenon. Those firms that fail to behave in the optimal way will not be able to survive the competition, while those whose behaviour happens, in fact, to coincide with the informed maximization of returns will prosper. Thus natural selection will ensure that only those firms adopting best practice will survive, and heterogeneity is consequently only a secondary phenomenon, little more than noise in the process towards equilibrium.[26]

However, apart from the fact that this still assumes that there is one 'best way' to act in a given situation, it is quite problematic to think of social evolution in the same terms as biological evolution. As Jon Elster reminds us, in biological evolution 'the organic environment changes slowly relative to the process of adaptation'.[27] The selection mechanism thus has time to do the work it is assumed to perform, namely to select the organisms that are best adapted to the given environment. The situation is quite the opposite in economic competition where 'the environment changes very fast – much faster than the process whereby unsuccessful firms go bankrupt and successful firms expand'.[28] Thus the selection mechanism does not have time to select those firms best adapted to the given environment. Long before any such equilibrium is reached, the environment is changed in important respects.

Here, one might object that selection will not only work through the disappearance of the non-optimal performers, but that best practice will also prevail through the poorer performers imitating the best. However, as Richard Nelson points out, it is often the case that what makes a firm strong or weak at any time is not well understood, not even within the firm itself.[29] Indeed, the widespread concern among firms with measuring performance, and the widespread dissatisfaction with existing performance measures, suggests that the very question of which firms do well, and which do less well, is not necessarily an easy one.[30] We may note here a phenomenon to which Marshall W. Meyer has drawn attention, namely the fact that 'while performance measures and measurement activity have proliferated over time, performance measures tend to be very weakly correlated with one another'.[31]

This suggests a further problem with the idea that natural selection should provide a mechanism that ensures only optimally performing firms will survive in competition. The fact that a firm does not perform well enough to survive is not simply a matter of registering an unambiguous fact. First, survival is not only a matter of how the firm has performed up until now, but also how it can be expected to perform in the future. As Meyer puts it, 'performance is ultimately future cash flows', and 'future cash flows cannot, by definition, be measured'.[32] But second, in accordance with the arguments of Shackle referred to above, this is not simply a matter of prediction either. Basically, whether a firm survives or not is a question of whether a sufficient number of people still believe this is an undertaking worth supporting, of their willingness to commit time, energy and resources to try to make it work. Often, there are conflicts and sharply differing points of view concerning these issues.

Thus at any one time there will be heterogeneity in the distribution of business firms, along a wide variety of dimensions, as Jensen and McGuckin say.[33] When circumstances then change, some firms will have greater potential than others for doing well, and those firms with a potential for doing well under the new conditions may not be the same as those that were doing well under the former conditions. Some of the heterogeneity among firms may be expressed by saying that firms have different strengths and weaknesses, and some conditions may favour certain kinds of strengths, other conditions other kinds of strengths. Thus some firms that were previously struggling may find themselves in a more favourable position when circumstances change. An implication of this is that a certain amount of heterogeneity among firms may be important for an economy's capacity for change and growth. Variation among firms heightens the probability that there will be firms ready to take advantage of the new opportunities arising as conditions constantly change. More generally, variation among firms may mean a wide range of different initiatives for change. These are issues that become important if one thinks of the future in terms of real uncertainty, as essentially open, where creativity, experimenting, trial and error, and learning by doing, are essential. Thus, from an evolutionary perspective one might say that heterogeneity among firms means that there is more variety for the selection mechanism to choose from.

THE VARIETY AND FLEXIBILITY OF INSTITUTIONAL ARRANGEMENTS

Hitherto we have focused on heterogeneity among business firms. We now turn explicitly to the issue of the heterogeneity of institutional arrangements. However, as is discussed below, the issue of the heterogeneity of economic actors is highly relevant for an understanding of the heterogeneity of institutional arrangements.

Some of the issues involved in the question of institutional heterogeneity are brought out by Robert M. Solow in an article on the relationship between economic history and economic theory.[34] A fundamental point for Solow is his disagreement with 'the attempt to construct economics as an axiomatically-based hard science',[35] an attempt that entails an assumption that there is 'a single universally valid model of the world'.[36] Solow argues against this belief that 'when it comes to studying the real world there is only one useful system of axioms and we already know what it is'.[37] Instead, he argues for 'a sense of the variety and flexibility of social arrangements' and for a better understanding of 'the interaction of economic behaviour and other social institutions'.[38] We should not expect economics to come up with 'a single monolithic model for all seasons', but instead 'a collection of models contingent on society's circumstances',[39] models that should be 'directly applicable in organising a historical narrative'.[40] In that way, economic theory would be able to enrich economic history, and economic theory would be 'taught something about the range of possibilities in human societies'.[41]

THE IDEA OF THE PERFECT MARKET

Solow thus argues for the importance of variation in economic institutions, and accordingly proposes the construction of different kinds of economic models to try to grasp different institutional circumstances. This is not easily reconcilable with neo-classical economic thinking, which is very much built around the theory of the perfect market economy as a point of departure and reference model. Bernard Guerrien emphasizes the normative character of this theoretical approach. For 'the great majority' of neo-classical economists, he claims, if there is an unacceptable gap between the theory and the real world, this does not mean that there is something wrong with the theory, but with the real world. 'The theory, they believe, has determined the conditions of the best of all possible worlds; if the real world differs from these, it is because it contains "imperfections" which one has to suppress in order to come closer to the theoretical conditions which allow one to obtain optimum'.[42]

Guerrien also notes, however, a somewhat different approach within the neo-classical tradition – what he calls an approach in terms of 'axiomization'. Instead of aiming to show that the market mechanism leads to an optimum, this approach turns the question the other way round and asks which assumptions one has to make in order to show that the market mechanism leads to an optimum. However, it has turned out that the assumptions necessary to show this are very strong in the sense that the conditions they describe are not very likely to be observed in real life. One has then been led to modify certain of the assumptions, but the price for this re-introduction of a minimum of realism has been a very important weakening of the generality of the conclusions.[43]

In any case the application of the theory of the perfect market or perfect competition as a reference model is central to neo-classical thinking, and as Mary O'Sullivan proposes, may simply be thought of as a defining characteristic of this tradition. She refers to neo-classical theory as 'that body of economic thought that uses the theory of the perfect market economy, whether explicitly or implicitly, as the benchmark for economic efficiency'.[44] She notes that 'most of contemporary economics' falls within this definition, and notably 'most of the so-called "market imperfections" literature'.[45]

Thus, in neo-classical thinking the perfect market economy constitutes a perfect or optimal economic system: if the market mechanism is allowed to work freely, without interference from other forces, this will result in an allocation of resources that is optimal in the sense that no better use can be made of available resources (given the 'initial' distribution of resources).[46] Interference with the free working of the market mechanism, and notably interference from other social institutions, will result in deviation from the optimum and loss of efficiency. This suggests that social institutions, other than the market, are defined purely negatively in relation to the market, as an intrusion of irrationality into the rationality of the perfect market. For instance, government regulations or trade unions are simply regarded as interfering with the employers' right to hire and fire, thereby causing rigidities, obstructing the free mobility of resources and thus hindering resources from being allocated to their most productive use. In this perspective social institutions are simply external hindrances to the workings of the market mechanism – interference from outside the market.

INSTITUTIONAL DIVERSITY AS A MANIFESTATION OF IMPERFECTION

We now turn to a trend within the neo-classical tradition that offers a somewhat alternative way of looking at institutions. This is the case for the so-called 'new institutional economics' that, according to Oliver E. Williamson, is now 'being incorporated within orthodoxy'.[47] Here institutions are not regarded simply as external hindrances to the workings of the market mechanism. They are nevertheless conceptualized as logically secondary to the perfect market. More specifically, they are seen as being derived from hindrances to the workings of the perfect market economy that are basically intrinsic to market exchange itself. Central to this approach is the acknowledgement that the assumptions necessary to sustain the idea of the perfect market economy as an optimal economic system 'not only are strong but are simply not observed in the real world'.[48]

The logic is well illustrated by Douglass C. North's exposition of the theoretical foundations of institutions. North's point of departure is that 'no institutions are necessary in a world of complete information'.[49] However, in real transactions each partner involved will not know if the other partner will live up to the agreements. Thus, with incomplete information 'cooperative solutions will break down unless institutions are created that provide sufficient information for individuals to police deviations'.[50] To fulfil this function an institution must contain both 'a communications mechanism that provides the information necessary to know when punishment is required', and a mechanism for actually carrying out the required punishment.[51] In essence, North's argument is that 'the players may devise an institutional framework to improve measurement and enforcement and thereby make possible exchange, but the resultant transaction costs raise the costs above the neoclassical level. The more resources that must be devoted to transacting to assure cooperative outcomes, the more diluted are the gains from trade of the neoclassical model.'[52]

Thus the neo-classical model of the perfect market economy is still the theoretical point of departure and reference. Institutions are derived by introducing modifications to the assumptions underlying the perfect market economy model, notably the assumption of perfect information. Relaxing the latter assumption means introducing uncertainty in exchange; institutions are then conceptualized as humanly devised constraints set up to reduce uncertainty in exchange. Defining institutions as constraints amounts to treating institutions in exactly the same way as 'the standard constraints of economics'; that is to say as defining the costs of alternative courses of action and thereby as structuring economic action. It is precisely these transaction costs that account for the deviation from the optimum defined by the perfect competition model of neo-classical economics, an optimum that can be approached to the extent that the transaction costs are low.[53] This kind of argument is an illustration of what Williamson has in mind when he refers to the proposition that 'the determinants of institutions are susceptible to analysis by the tools of economics'[54] as being fundamental to the new institutional economics.

We may note here that a significant element in the new institutional economics approach consists precisely of introducing heterogeneity into the uni-dimensionality of standard neo-classical theory. In the 'standard neo-classical Walrasian model', North says, 'commodities are identical, the market is concentrated at a single point in space, and the exchange is instantaneous. Moreover, individuals are fully informed about the exchange commodity and the terms of trade are known to both parties.'[55] By way of contrast, in the transaction cost perspective, products are multi-dimensional, and 'we get utility from the diverse attributes of a good or a service'.[56] Thus the value of what is exchanged will depend on the quality of these different attributes. Frequently, one will not know the quality of these attributes very precisely, and will have to use resources if one wants to find out more about them. Thus exchanges will 'entail costs that result from both parties attempting to determine what the valued attributes of these assets are'.[57] Moreover, there will be asymmetries of information among the players, since 'the seller of oranges' will know 'much more about the valuable attributes of the oranges than the buyer', and 'the used car dealer' will know 'much more about the valued attributes of the car than the buyer'.[58] This has further consequences, because 'not only does one party know more about some valued attribute than the other party, but he or she may stand to gain by concealing that information'.[59] Thus North compares the standard neo-classical perspective with his own in terms of the contrast between the simple and the complex, the uni-dimensional and the multi-dimensional: 'The contract specified by economic theory is simple, complete, and straightforward. It involves an exchange of a unidimensional product at an instant of time', whereas 'the contract in modern complex economies both is multidimensional and extends over time', and, 'the contract will typically be incomplete, in the sense that there are so many unknowns over the life of contracts extending over time that the parties will (deliberately) leave to the courts or to some third party the settlement of disputes that arise over the life of the contract'.[60]

However, it should be emphasized that although heterogeneity here is acknowledged and accorded an important role in the theoretical structure, it is at the same time conceived as a secondary phenomenon, being introduced precisely as a modification to the basic reference model, where the reference model is still seen as the standard against which actual situations are measured. Institutions are conceptualized as solutions to problems that arise with the introduction of imperfections (e.g. imperfect information) in the perfect competition framework. These institutions entail transaction costs that make outcomes deviate from the optimum specified by the reference model; but to the extent that these institutions are efficient in terms of facilitating exchange, transaction costs will be low and the optimum outcome claimed by standard neo-classical theory will be approached.[61] This perspective on heterogeneity and institutions reflects a fundamental acceptance of the conception of rationality of neo-classical economics. Although, as Williamson claims, there is 'close to unanimity' within the new institutional economics on the idea of *bounded rationality*,[62] this is understood simply as imperfection or limitation in relation to the perfect rationality of the neo-classical actor. Williamson simply equates bounded rationality with 'limited cognitive competence',[63] while North speaks of 'subjective and incomplete processing of information' playing 'a critical role in decision making'.[64] A fundamental premise here is the idea that, in any given situation, objectively there is one unique course of action that is the best (given the actor's preferences). Had the actor possessed the correct model of the world, he or she would have chosen this specific course of action. And this situation is also what is assumed to prevail if markets are competitive. Thus North subscribes to the idea that if markets are competitive, it will be the case that 'even though the actors may initially have diverse and erroneous models, the informational feedback process (and arbitraging actors) will correct initially incorrect models, punish deviant behaviour, and lead surviving players to the correct models'.[65] However, if there are market imperfections, for instance if markets are incomplete and 'the information feedback is fragmentary at best', then 'the historically derived perceptions of the actors shape the choices that they make'.[66] It is to the extent that these kinds of imperfections are present that 'the subjective models of actors modified both by very imperfect feedback and by ideology'[67] become important. Thus, according to North, ideology, which he characterizes as being 'based upon subjective perceptions of reality [plays] a major part in human beings' choices' precisely on account of the prevalence of 'incomplete processing of information' in decision-making.[68]

Hence heterogeneity and subjectivity are conceived here, in a fundamental sense, as *error*: if nobody made mistakes, everyone would have the same model of the world, namely the correct one. Heterogeneity thus expresses a regrettable imperfection or limitation, a secondary phenomenon in the sense of pertaining to the mere subjective representation of an essentially given, unambiguous reality. There is not much room here for a notion of subjectivity and heterogeneity as creativity. Introducing the idea of bounded rationality does not contribute to grasping this if bounded rationality is understood simply as being synonymous with 'limited cognitive competence'. Bounded rationality should be understood as the rationality of situated actors, acting in a world that to a large extent is open, not as an imperfect version of the rationality of an omniscient being contemplating a given static or predetermined world. Bounded rationality thus relates to a particular point of view, and in many cases different points of view cannot simply be compared in terms of distance from some hypothetically correct point of view, but rather represent different perspectives, different visions, different projects and possibilities, expressing different kinds of skills and abilities, different inclinations, and so on. These different visions and projects may turn out to be more or less viable, more or less successful, but this is precisely something that cannot in any objective way be predicted beforehand. Human action constantly creates something new, and this cannot be understood simply in terms of carrying out a preconceived plan. Rather, we have to do what J.S. Metcalfe calls an 'unfolding process', where plans and objectives instead have the status of giving a general direction and where 'the realisation of possibilities makes possible the specification of new possibilities'.[69]

THE AGENCY PROBLEMS AND THE FIRM

The logic of this type of attempt to understand institutions on the basis of neo-classical economic theory is well illustrated by how the institution of the firm is treated within the dominant economic tradition. A prominent example of this is the discussion of *corporate governance*, of how firms are governed and function.[70] In the traditional neo-classical approach there should be no problem here, and economists have often dismissed the issue as not worthy of consideration.[71] As Joseph E. Stiglitz has put it: 'Many economists argued that there was no need to look carefully into the black box called the firm: firms maximised profits (stock market value); if managers didn't, they would be replaced; and firms that didn't maximise wouldn't survive.' Accordingly, what went on inside the black box was mere detail. The behaviour of the firm could be described completely without knowledge of those details'.[72] In a perfect market economy there is no room for institutions; for institutional arrangements to have any role to play, there must be some disturbing element making the situation deviate from optimum of the perfect market. And what is seen as necessitating modifications to the perfect market framework is the presence of what within the neo-classical tradition is termed *agency problems*. Thus, in the words of Oliver Hart: 'Corporate governance does not matter in the absence of agency problems.'[73] 'This', he explains, 'is because in the absence of agency problems, all individuals associated with an organisation can be instructed to maximise profit or net market value or to minimise costs'.[74] What the agency problem is all about is that one cannot take precisely for granted compliance with the instruction to maximize the organization's profits. An essential reason for this is that 'information is imperfect and costly'.[75] Whether a manager in fact acts in the interests of the owner or not will often not be immediately apparent. To find out about this, the owner would typically have to collect and process much information, and this is costly. It follows that managers have 'considerable discretion', notably 'discretion to pursue their own interests'[76] and there is no guarantee that the particular interests of managers will coincide with the owners' interests in maximizing the profits of the firm. Thus: 'The fundamental problem of owners of firms is how to motivate their managers to act in the interests of the owners.'[77] This gives rise to a system of efforts to control managers through incentive schemes, monitoring, sanctions, and so on, and it is this apparatus of supervision and control of managers to which the term 'corporate governance' refers in the neo-classical perspective.

We should note that in principle this type of agency problem is not restricted to the relationship between owners and managers. Also, other categories of people associated with an organization, such as ordinary employees or workers, may have some discretion in pursuing their own interests, which again may not coincide with the owners' interests in maximizing the profits of the firm. Thus, within the overlapping traditions of neo-classical economics and rational choice theory there is now a substantial literature on how to develop incentive systems so that employees will find it in their interest to act in the interests of the owners.[78] However, it seems that the term 'corporate governance' has largely been reserved for agency problems in the relations between owners and managers. Especially, corporate governance is considered a significant issue in large, public companies where share ownership is widely diffused over a large number of small owners. This creates particularly difficult problems from the point of view of controlling managers so that they act in the interests of the owners, because 'the owners, that is, the shareholders, even though they typically have (ultimate) residual control rights in the form of votes, are too small and numerous to exercise this control on a day-to-day basis'.[79] Thus the issue of corporate governance in particular arises in the context of 'the separation of ownership and control', typical of the modern corporation.

For the mainstream economic approach, corporate governance thus becomes an issue of getting managers (and other people employed in the organization) to act in the interests of the owners where, in the context of the modern economy, owners are typically equated with shareholders.[80] The main reason that shareholders are considered to be owners is that they are *residual claimants*.[81] Whereas the other parties involved, like employees, lenders of money, suppliers, and others have their remuneration contractually specified in advance in the form of wages and salaries, interest on loans, payment for goods supplied, and so on, shareholders are the only party whose remuneration is directly tied to the economic returns of the firm. It follows that they are the direct carriers of the interest to maximize the firm's profits. Now, in the neo-classical way of thinking, to achieve the optimal outcome of the perfect market economy requires that firms maximize profits. It follows that 'the "maximisation of shareholder value" will result in superior economic performance, not only for the particular corporation but also for the economy as a whole'.[82] What stands in the way of this ideal of maximization of the profits of the firm and thus making corporate governance an issue is precisely the presence of an 'agency problem, or conflict of interest', as Hart puts it.[83]

CONFLICTS OF INTEREST

Now, we should note how the nature of this conflict of interests is conceived. The interests in question are thought of as purely egoistic interests; the behaviour is purely self-interested. It is a basic premise that individuals associated with an organization 'do not care *per se* about the outcome of the organisation's activities',[84] but only in so far as they imply costs and benefits for themselves. The conflict of interests is thus constructed as a conflict between the purely private interests of the managers and the general interest of profit maximization. Managers are supposed to act against the interests of the organizations they are running whenever the private benefits to them of so doing exceed their private costs; that is, whenever they can benefit from this and get away with it. Likewise, employees will act against the interests of the organization whenever this is to their own personal advantage. Notably, employees are thought to have an interest in *shirking*; they will collect their wages and 'work as little as they can get away with'.[85] Hence, the strong preoccupation in this theory with devising elaborate incentive schemes to align the private interests of managers and employees more closely with the goal of profit maximization.

That the economic actors in this conception are thought to be exclusively self-interested in their motivation for action also applies to the shareholders, but since the interests of the shareholders are in the maximization of the profits of the corporation, their particular personal interests happen to coincide with the general interest. Accordingly, in this way of thinking, conflict within organizations is exclusively conceived as a conflict between actors, all of whom try to further their own particular private interests, but where the interests of one category of actors happen to coincide with the general interest. The picture we thus get is one where the general interest in the optimal functioning of the economy in accordance with the model of the perfect market economy is obstructed by particular groups of actors like managers and employees furthering their own specific self-interests.

What is ruled out by assumption here is the possibility that conflicts of interest might be a multi-dimensional phenomenon, where one dimension of such conflicts might instead represent disagreement over the different perspectives on what will further the common good of the organization, or even of what will further the organization's contribution to society at large, and where particular self-interests are mediated through particular perspectives on the common good, and vice versa. Different groups will have different experiences; each will have different notions of what the problems are, what are strengths and weaknesses, of what is important and less important, and so on. If these different perspectives of the general good are recognized as such, and not exclusively as expressions of particular egoistic interests, it might be reasonable to view the organization's objectives as emerging through some kind of process of reconciliation of different perspectives, like in the reciprocal critical examination of arguments in a discussion, where the dominant perspective is transformed or modified through going into a dialogue with other perspectives, rather than seeing the organization's objectives as something that is unambiguously given in advance.[86]

IDENTITIES, IDENTIFICATION AND INTEGRATION

A weighty argument against the neo-classical agency problem conception of corporate governance is that there seems to be little empirical evidence for its central premise that individuals associated with an organization do not care *per se* about the outcome of the organization's activities, and will work for organizational goals exclusively to the extent that this contributes directly to their own economic self-interest. In a discussion of this view that only economic incentives or rewards will make people work for organizational goals, Herbert Simon claims that 'organisations would be far less effective systems than they actually are if such rewards were the only means, or even the principal means, of motivation available. In fact, observation of behaviour in organisations reveals other powerful motivations that induce employees to accept organisational goals and authority as bases for their actions',[87] the most important of these mechanisms being 'organisational identification'.[88]

On this point Stiglitz states: 'Simon argues persuasively that in successful organisations, workers *identify* with the organisation's objective; that they take on the organisation's objective as their own.'[89] What is said here about the motivation of employees or workers to act in accordance with organizational goals applies equally, or perhaps even more so, to managers.[90] Thus, in Stiglitz's view, 'the incentives paradigm can neither explain the structure of observed incentive schemes nor what it is that motivates managers and workers'. Consequently, 'if economists want to understand what makes managers work and what differentiates successful from unsuccessful organizations, we may need to look beyond the compensation schemes and the economists standard incentives paradigm'.[91] Simon is even more explicit about the limitations of the neo-classical approach in this area: 'The attempts of the new institutional economics to explain organisational behaviour solely in terms of agency, asymmetric information, transaction costs, opportunism, and other concepts drawn from neoclassical economics ignore key organisational mechanisms like authority, identification, and coordination, and hence are seriously incomplete.'[92]

Thus Simon challenges the agency problems approach to organizational functioning, where managers and employees are assumed not to care *per se* about the performance of the organization and will work for organizational goals only to the extent that this furthers their own self-interests. Simon argues to the contrary, that successful organizations are characterized precisely by managers and employees *identifying* with the goals of the organization, taking on these goals as their own. A central question then becomes what the conditions are for this kind of identification to occur. One answer may run along the lines that this requires some measure of integration in some kind of collective or community: one participates in, contributes to, some kind of common project. For this form of participation, so as not simply to be a matter of rhetoric and manipulation, membership must have some kind of real foundation, through some measure of commitment, some measure of communality of destiny, or the like. This is closely connected to the issue of *organizational integration*:[93] How are different types of actors integrated into the organization? Which ones are integrated into the organization, and which remain outsiders, treated as ordinary 'factors of production'? Under what conditions will different types of actors be integrated into the organization; in what ways, and what are the consequences for organizational learning, innovation and economic performance?

LEARNING, KNOWLEDGE AND DISCOVERY

Let us here further illustrate how the assumption of an unambiguously best action, somehow given in advance, structures the whole agency problems approach to corporate governance. This reflects the basic assumption of perfect information on the part of economic actors in the neo-classical model of the perfect market economy. To the extent that it is acknowledged that perfect information is not a fruitful assumption in most real-world situations, imperfect information is introduced precisely as a deviation from an ideal of perfect information, as a secondary complication brought into the picture, making adaptations deviate from the ideal situation of the perfect market economy. Notably, the relaxation of the assumption of perfect information and the consequent introduction of imperfect information is made through introducing an additional type of costs – information costs, the costs of gathering and processing information. As Israel M. Kirzner says of the dominant economic tradition's approach to knowledge, learning and discovery: 'For the mainstream, imperfect information is primarily a circumstance constraining the pattern of attained equilibrium.'[94] The ideal of perfect information, that is to say perfect prediction, with its underlying assumption of a given, finite world, still structures the whole approach.

The position where the ideal of perfect information structures the whole approach to corporate governance emerges very clearly in Oliver Hart's discussion of 'optimal principal–agent contracts'.[95] (The *principal* hires an *agent* to perform some task for him or her, for instance to run a firm. In this case, the owner is the principal, the manager is the agent.) Hart defines optimal principal–agent contracts as ones which 'are "comprehensive" in the sense that a contract specifies all parties' obligations in all future states of the world, to the fullest extent possible (i.e. to the extent that these obligations are observable and verifiable)'.[96] The notion of optimal principal–agent contracts for Hart is central to an understanding of corporate governance because if contracts are comprehensive, governance structure will not matter: 'In a comprehensive contracting world, everything has been specified in advance, i.e. there are no "residual" decisions.'[97] Thus, a second necessary condition for corporate governance to matter, in addition to the presence of agency problems, is that contracts are incomplete: 'Governance structure matters when some actions have to be decided in the future that have not been specified in an initial contract.'[98]

To explain that there are incomplete contracts Hart invokes the *costs* of writing comprehensive contracts. He refers to the transaction cost literature, which he claims has identified three costs that are particularly important:

> First there is the cost of thinking about all the different eventualities that can occur during the course of the contractual relationship, and planning how to deal with them. Second, there is the cost of negotiating with others about these plans. Third, there is the cost of writing down the plans in such a way that they can be enforced by a third party – such as a judge – in the event of a dispute.[99]

It is because of these costs that 'the parties will not write a comprehensive contract', but instead 'write a contract that is incomplete'.[100]

Thus the underlying idea here is that a comprehensive contract, where everything has been specified in advance, constitutes the ideal, but that consideration of the costs of establishing such a contract necessitates a deviation from this ideal. However, in accordance with what has been argued above, there is reason to have serious doubts about this whole notion that in principle it is possible to specify all eventualities in advance. Rather, one may claim that this is intrinsically impossible, that the whole idea is incompatible with basic traits of human action, such as its largely open-ended character, creativity, and initiative. Even for simple, standardized work operations the idea of specifying everything in advance is problematic.[101] More generally, instructions are typically not given as detailed commands to undertake specific actions, but, for instance, as more general instructions concerning the result to be produced.[102] As Herbert Simon reminds us: 'Obeying operating rules literally is a favourite method of work slowdown during labour–management disputes.' Instead, 'What is required is that employees take initiative and apply all their skill and knowledge to advance the achievement of the organisation's objectives.'[103]

PREDICTION, CONTROL AND TRUST

The idea that a comprehensive contract where everything has been specified in advance constitutes the ideal, but where considerations of cost necessitate a deviation from this ideal, is also highly problematic as a practical understanding with regard to the extent that the actors themselves have this understanding and are guided by it in their actions. Commenting on these issues in the case of cooperation between separate business units, Stewart Macaulay observes that 'carefully planned arrangements may create undesirable exchange relationships between business units. Some businessmen object that in such a carefully worked-out relationship one gets performance only to the letter of the contract. Such planning indicates a lack of trust and blunts the demands of friendship, turning a cooperative venture into an antagonistic horse trade.'[104] If one really has the setting up of comprehensive contracts where all possible eventualities have been specified in advance as an ideal, this looks more like an obsession with control than a rational way of collaborating to get things done. This applies whether we are dealing with relationships between separate business units, between managers and workers, or between owners and managers.

More generally, accounting for trust is a fundamental problem for theories that conceive human actors as exclusively opportunistic and self-interested. If actors are exclusively interested in costs and benefits to themselves, they will break their commitments whenever they find that this is to their advantage. Charles F. Sabel aptly characterizes this perspective on human action as 'a science of suspicion'.[105] As he observes, the corollary to the exclusive concern with own interests is a fear of being deceived: 'It makes the pursuit of self-interest and the fear of deception (because the others are pursuing their own interests, too) the spring of individual action and the guiding motive of institutional construction.'[106] As Mark Granovetter observes in this perspective, the way to hinder people from cheating and breaking commitments is to set up clever institutional arrangements that make these practices too costly to engage in. But as Granovetter emphasizes, these arrangements 'do not produce trust but instead are a functional substitute for it'.[107] Hence the distorted view of trust typically found in rational choice theory and related approaches: whether one trusts another person or not becomes exclusively a question of one's assessment of what the other person sees as being to his or her advantage.[108]

Again we may inquire into the practical implications of having this understanding as a practical understanding guiding one's own actions. Would this not imply that one does not trust the other person, but expects him to cheat and break commitments whenever he finds this to his advantage? And does one not also admit simultaneously that one is not to be trusted oneself but will likewise always be ready to cheat the other if one finds that one can profit from it? But of course, it would not be smart to say this openly, so one pretends instead that one is not exclusively self-interested and pretends not to believe the other to be exclusively self-interested, while at the same time expecting the other to put up the same pretence. To the extent that the actors themselves have this understanding this would seem precisely to imply a social world dominated by suspicion and deceit, as Sabel indicates.[109] It is doubtful whether this kind of social world would function very well!

We thus see how the neo-classical agency problems approach to the running of firms in a fundamental sense reduces the heterogeneity of interests and perspectives relating to the firm to one single, uni-dimensional, unambiguous perspective. According to this view, firms are to be run in the exclusive interests of the owners or shareholders. This is so because the latter are held to be residual claimants, and it is precisely the residual claimants who are thought to have a direct interest in maximizing the profits of the firm. Moreover, what course of action will in fact be best to further the interests of the owners (i.e. what will in fact maximize the profits of the firm) is assumed in principle to be clear and obvious, an objectively given fact. That different actors have different models of the world, plays a central role in the agency problems approach, but this is conceived strictly within the confines of the idea of 'asymmetric information'. The correct model of the aspect of the world in question is assumed to be given, and then the differences among actors are thought of as a difference between those who have the correct model and those who have an incorrect model, or perhaps somewhere between those with a model close to the correct one and those with a model further away from the correct one. Specifically, managers (agents) will typically know more about the situation confronting the firm than shareholders (principals). As Richard Nelson observes, for the agency problems approach, the question of the running of firms is basically seen as a question of getting private incentives right, not of finding out what to do.[110] The issue becomes one of aligning the interests of managers, and also other people employed by the firm, with the interests of shareholders or owners, so that managers and other employees do not use their advantaged position concerning knowledge of the firm's operations to further their own interests at the expense of shareholders or owners.

DIFFERENT PERSPECTIVES OF THE FIRM

Let us look more explicitly at the view of the firm as an institution here and contrast this with other ways of looking at it. The firm in the neo-classical tradition is quite clearly conceived as an *object of property*, specifically the property of shareholders. This corresponds to what Ronald Dore has called the *property view* of the firm, which he compares with three other ways of looking at the firm.[111] According to Dore, the 'key assumption' of the property view, is that 'the legal situation . . . represents social reality'. In this view 'a company is an entity set up by its capital-providing members to further their own material interests', and 'the managers are their agents with a duty to give priority to that shareholder interest'. The treatment of employees is purely instrumental in relation to the goal of furthering shareholder interests. Hence, 'policies of "worker involvement", paternalistic welfare policies and premium wages above market "going rates" may all be permissible tactics, provided that they are "manipulations" designed to yield better value for money in the purchase of labour'.[112]

The property view closely resembles the *arena view*, in which the firm is seen as an arena where different kinds of actors (such as managers, skilled workers, suppliers, etc.), 'motivated exclusively by self-interest', make contracts and bargains of various kinds. Productive activities are seen as the fulfilment of such contracts. In this view, 'the organisation of a firm can be "dissolved" into a network of contracts' between exclusively self-interested actors.[113] Dore comments that 'principal–agent theory' is compatible with both the property view and the arena view.[114]

Dore contrasts the property view and the arena view with the *entity/community view*, where he distinguishes two sub-versions: the *managerial community view* and the *employee community view*. In the entity/community view the firm is seen not merely as a set of instrumental arrangements, but more as a social entity analogous to a nation. This involves, on the one hand, that the firm is seen as an entity that transcends the group of individuals involved in it at any one time. The individuals involved may come and go, but the firm persists. On the other hand, the group of people who at any one time embody the firm 'constitute a community, tied together by bonds of interest in the community's fate, obligations of cooperation and trust, the sharing of similar risks'.[115] What motivates the distinction between two sub-versions of this view is the fact that there is variation concerning the definition of the boundaries of this community. The community may 'be limited to the group of senior managers, or senior and middle managers, or it may be extended to the whole body of people who work in the firm, blue collar and white collar', although in the latter case only core workers are usually included; that is, not part-time or temporary workers.[116] To the extent that the community is limited to managers, either at the senior level or including middle managers, we are concerned with the managerial community view, while to the extent that all people working in the firm are included, we have the employee community view. Dore claims that principal–agent theory is never compatible with the employee community view, although it may sometimes be 'not incompatible with' the managerial community view.[117] Correspondingly, from the perspective of the entity/community view, and especially in its employee community version, the interests of shareholders are not a central concern. On the contrary, in this view 'the shareholders are just one of the groups of outsiders who have to be taken into account for the community to survive and prosper'.[118]

PROCESSES OF CONSTRUCTION OF A COLLECTIVE INTEREST

We see that the entity/community view of the firm, especially in its employee community sub-version, differs in important respects from views based on neo-classical theory and its derivatives. Notably, in the neo-classical approach, what the interests of different kinds of actors are, whose interests are to prevail and how best to pursue these interests, are assumed to be objectively given facts, so that problems of governance reduce to an issue of incentives, of how to align the interests of agents/managers/employees with principals/owners/shareholders. This is not a question of seeing governance as one of reconciliation of different, and partly contradictory, interests and perspectives and initiatives.

By contrast, this is precisely the kind of question that is likely to become central in perspectives based on an entity/community view of the firm. For instance, in their book *Dérives du capitalisme financier*,[119] Michel Aglietta and Antoine Rebérioux, argue strongly against the view that firms are the property of shareholders and that they should be run exclusively in the interests of the latter. According to Aglietta and Rebérioux, by opting for liquidity, shareholders cannot at the same time opt for control. Choosing liquidity means precisely to distance oneself from the enterprise; liquidity is 'synonymous with exteriority'.[120] And 'the more financial markets are liquid', they claim, 'the more exterior to the enterprises are shareholders'.[121] By contrast, Aglietta and Rebérioux emphasize the collective character of the enterprise, and claim that both in its 'technological, financial, cognitive and organisational dimensions' the development of capitalism has reinforced this collective character.[122] Thus they see the issue of governance not primarily as having to do with control, but with 'the formation of a collective interest which expresses itself in an objective which is acknowledged and accepted by the parties involved in the activities of the enterprise'.[123] They also emphasize that the collective interest is not something that is pre-given; it does not exist prior to the process of deliberation that brings it about. In this process 'the private interests of the parties are transformed; the collective interest is neither the aggregation nor the confrontation of particular interests'.[124]

SOCIAL DIVISIONS

The arguments of Aglietta and Rebérioux connect with Simon's points about the identification of employees with the goals of the organization. The integration of employees in the organization and their identification with its goals implies a transformation of the private interests of the parties, a process through which the goals of the organization themselves are formed. However, we should emphasize that the transformation of private interests in the formation of a collective interest in no way means that the interests of different types of actors simply become identical, and that conflicts of interest disappear. This is not simply a question of divergence between individual interests and collective interests, but also of the individual belonging to different, overlapping collective entities. People working for a given firm will also be more or less integrated in other kinds of collective entities, internalizing in various manners and degrees the perspectives, orientations and objectives of these other entities. These different orientations will partly reinforce one another, partly be in conflict, and so forth. People's identity, loyalties and commitments refer only in part to the firm where they are employed. To a large extent, and in varying degrees, people also have their identities defined elsewhere. The Japanese case is perhaps extreme in that employees' identities and whole lives are to a very high degree tied to the company where they are employed. Discussing so-called *lean production* in Japanese firms, Wolfgang Streeck claims that 'it depends on workers making themselves freely available to the organisation in a way that cannot be expected in a pluralist social structure'.[125] Streeck here specifically draws a contrast with Germany:

> In a society such as Germany, where workers do not primarily identify with their place of employment, involvement of individuals in secondary organizations of work must be *negotiated*, protecting workers' 'privacy' from being consumed by requirements of organizational efficiency. Reliable and enforceable demarcations

> of rights and responsibilities, in particular of managerial power, is felt to be indispensable as a brake on what in a pluralist environment is perceived as a totalitarian tendency of organizations under competitive pressures to absorb their members.[126]

This indicates that there is nothing absolute about the boundaries delimiting a firm or an organization. People working in an organization will not only be more or less integrated into the organization in question, but in different ways and degrees in other kinds of social entities, carrying with them commitments to and orientations of these other types of entities into the organization. This will influence how they interpret the goals of the organization, how they see their own role, how they see conflicts of interests, and so forth.

THE FORMATION OF IDENTITIES AND SELF-INTERESTS

When questions of the identification of employees with organizational goals and of the integration of employees in the organization are made central, one in effect makes the issue of the identity of actors, of the formation of identities, including the formation, mediation and transformation of self-interests, a central concern. By contrast, it is an absolutely fundamental premise of the neo-classical approach, including the transaction cost approach, new institutional economics, principal–agent theory, and so on, that the identities of actors, their wants and preferences, are magnitudes that are given in advance.

We can see this very clearly for instance in Douglass North's treatment of institutions and of culture more generally in terms of *constraints*, referred to above. Thus institutions are conceived in the same way as 'the standard constraints of economic theory'. Together with the latter 'they define the choice set and therefore determine transaction and production costs, and hence the profitability of engaging in economic activity'.[127] Furthermore, according to North, 'culture defines the way individuals process and utilise information', and 'the way by which the mind processes information is not only the basis for the existence of institutions, but is a key to understanding the way informal constraints play an important role in the make-up of the choice set'.[128] In other words, institutions and culture are thought to influence what the alternatives facing the actors are, including what the actors *believe* the alternatives to be. But institutions and culture are *not* assumed to affect the actors' *valuation* of these alternatives; that is to say, they are not assumed to affect the actors' preferences. There is no consideration whatsoever of culture as involved in the formation of individuals, in influencing their wants, needs, preferences and aspirations.

COMMENSURABILITY AND PREFERENCES

Aglietta and Rebérioux get to the fundamentals of the implicit conception of personal identity underlying neo-classical theory when they point out that the neo-classical project of a 'pure' science of economics presupposes 'that it is possible to build a science of the efficiency of means for goals which are pre-given and expressed in separate individual utilities'.[129] We should also emphasize an essential aspect of this conception of the identity of actors as defined by separate pre-given preferences, namely the assumption that each actor's preference structure consists of a complete ranking of all possible present and future 'consumption bundles', or more generally 'outcomes' or states of the world.[130] This assumption is essential because it assures *ex ante* the commensurability of all goods. The heterogeneity of goods is supposed already at the outset to be reduced to a common standard through the pre-given preference structures of separate individuals. As Jacques Sapir argues, this homogeneity hypothesis is central because it is this hypothesis that 'justifies hypotheses concerning the preferences which permit the standardisation of behaviour *ex ante*'.[131] This assumption of the commensurability of all possible goods through the pre-given complete preference order of separate economic actors is thus a basic presupposition for the neo-classical notion of the perfect market economy as an optimal economic system that serves as a benchmark against which all existing economic systems are to be measured.

However, this idea of each actor having a complete preference ordering, ranking all possible present and future states of the world, does not make much sense from a perspective emphasizing the open-ended character of human action, the constant possibility of creating something new, where the realization of possibilities makes it possible to see still further possibilities.[132] This is not only a question of the impossibility of knowing beforehand what the future alternatives will be. There is also reason to question the more general conception of rational action that sees actors as approaching each and every situation with pre-given, precise preferences and rational action then as consisting in the actor choosing among all available alternatives the one that is highest ranked in the actor's preference ordering. Human beings are not static entities: they change, learn, grow and develop, and an integral part of this learning, growth and development is that preferences change and develop. Moreover, actors are normally aware of this and, perhaps, at least we could say that this type of awareness is to be expected that of a *rational* actor. Hence approaching each and every situation with precise, preconceived notions of what one wants should not be considered as characteristic of rationality. Rather, if – in every situation – one tries to control what happens in accordance with a preconceived conception of how one wants things to be, one's behaviour becomes rigid, inefficient, static. In effect, if this type of instrumental approach becomes exclusive, this would amount not to an efficiently functioning rationality, but rather to an obsession with control.[133] Although the instrumental mode of relating to the world is essential to being rational, it is equally essential that this instrumental mode is balanced by other modes of relating to the world, where we are more receptive and open and let things happen – let ourselves be carried along. As Herbert Simon said of his own career: 'My career was settled at least as much by drift as by choice.'[134] Where one wants to go is to a large extent something one gradually finds out.

THE FORMATION AND TRANSFORMATION OF PREFERENCES

We have seen that it is normal that preferences change, and accordingly learning and exploration, development and growth are central phenomena also in this regard.[135] We learn from other people, are open to influence and suggestions, respond to initiatives from – and engage into dialogue with – other people. Notably, we may in particular be prepared to learn from specific people whom we somehow acknowledge as having a special competence concerning, for instance, taste or style; we acknowledge that certain people are good at dressing, making food, decorating their homes, and so on. Moreover, books, journals, magazines, TV programmes, are full of advice and discussions on such matters of taste and style, life style, and so forth. However, it should be emphasized that these kinds of processes of preference change are not only, and unambiguously, processes that can be characterized as rational, in the sense of having to do with exploration, trying out, discussion, exchange of points of view, learning and growth, and so on. Quite the contrary, this is also a field where manipulation, power, domination and struggle for distinction are central phenomena. Advertising is perhaps the most obvious example. Producers and creators of different products are typically persons who have a special competence in matters concerning their types of product, and thus are among the people we are particularly prepared to learn from. No doubt, many producers genuinely believe in the products they make, but at the same time they have an interest in selling these products and thus in getting people to believe that their products are exceptionally good and useful, and their discourse about their products may easily degenerate into manipulation. But issues of manipulation and domination are not only relevant in relation to advertising. Issues of consumption styles and standards of consumption are central in the relations of symbolic power and struggles for distinction that, notably, have been the centre of attention of Pierre Bourdieu.[136] According to Michel Aglietta and Anton Brender, in our type of society, characterized by the tendency to generalization of the wage relationship,[137] the mode of consumption is central to the relations within and between social strata.[138] Following Bourdieu, they propose to place consumption 'at the heart of social competition', and claim that 'this interpretation has the merit of establishing an endogenous dynamic to account for the extraordinary diversification of industrial objects which have invaded the mode of life of all social categories. The normalisation of the tension between imitation and distinction is decisive for the rhythm of change in the mode of consumption.'[139]

INSTITUTIONS AND PERSONAL IDENTITIES

In the foregoing, we have tried to show that there are complex processes at work in the formation and transformation of preferences, and that there will often be an ambiguous relationship between learning from others and being dominated by others. An implication of this is that the neo-classical project of a pure economics of the efficiency of means relative to pre-given preferences of isolated individuals is untenable. More generally, this also applies to the whole idea of explaining the origin of institutions taking fully formed individual actors defined by pre-given preferences as an absolute point of departure, as does North, among others. Institutions and culture cannot be understood simply as *constraints* on the actions of given individuals with pre-given preferences, as North would have it. On the contrary, institutions and culture are fundamentally involved in the formation of individuals and their preferences.

This means that preferences, and more generally the identities of the actors, cannot simply be treated as pre-given entities. Preferences change as persons learn, grow and develop. Notably, they change through learning from other people, through engaging in discussion and exchange of points of view with other people. Accordingly, there is a significant collective dimension to the formation and transformation of preferences. An important background to this is that, as Samuel Scheffler observes: 'in general, it is in one's interest to be comfortably integrated into the society in which one lives', which, among other things, notably also includes 'to find the society's culture and practices congenial, to experience one's own participation in them as natural, and to be regarded by others as a full-fledged fellow participant'.[140] An implication of the foregoing is a notion of the *basically rational* character of these processes of formation and transformation of preferences, even though these processes often, in varying degrees, also have significant elements of rivalry, conflict, manipulation, relationships of power and domination, and so on.

THE BASIC RATIONALITY OF PREFERENCE CHANGE

Characterizing these processes of preference-change as basically rational stands in sharp contrast to the approach of neo-classical theory and of rational choice theory more generally, where, to the extent that the phenomenon of preference change is acknowledged at all, this is treated as a problem – an anomaly. Specifically, preference-change is typically treated as *irrationality*, and this in particular applies where the preference change is brought about through the influence of other people. Although it may be acknowledged that actors' preferences often change, *normatively* the neo-classical model is nevertheless still held to be valid. If actors always were rational, they would not let their preferences be changed and, furthermore, to the extent that they in fact are rational, their preferences do not change.

Thus normatively, as a standard, the neo-classical model of the perfect market economy that presupposes pre-given preferences of separate individuals, is still held to be valid. Here, preference-change would appear as yet another element of deviation from optimality. We cannot go further into these issues in the present context; they will have to be taken up on a later occasion.[141] On such an occasion we should also discuss the tendency within the tradition of neo-classical economics and rational choice theory to try to account for behaviour that apparently does not conform to the assumption of utility maximization by postulating maximization at a *meta*-level. Preference change could, for instance, then be understood as the result of a decision taken in accordance with the principle of utility maximization on the basis of meta-preferences, where these meta-preferences are then assumed to be pre-given and unchanging.[142]

IMPLICATIONS FOR THE UNDERSTANDING OF COMMENSURABILITY

The foregoing implies that commensurability cannot simply be taken as an unproblematic given, as simply expressing the complete and unchanging preference structures of pre-given individual actors. Commensurability is something that is being continuously created and re-created through diverse social processes, and which is never achieved once and for all but is constantly open to contention. As Jacques Sapir argues, commensurability cannot be understood as perfect and *ex ante*, but as existing only as a tendency and *ex post*.[143] Among other things, one's perspectives on this issue have important implications for how one understands *money*. According to Sapir, neo-classical theory sees money simply as one commodity among others, raised to the status of *numéraire*, whereas traditions that are critical of the neo-classical perspective in general understand money as an institution.[144] He claims that 'the issue of money is thus a central point if one starts from the hypothesis of a considerably heterogeneous society where centralisation has to be realised *ex post*, through institutions which have their own history and dynamic'.[145] In this connection he quotes Max Weber, for whom 'money prices are the product of conflicts of interest and of compromises; they thus result from power constellations'.[146]

CONCLUSION: CHALLENGING THE IDEA OF OPTIMALITY

In conclusion, institutions and culture more generally cannot simply be understood as constraints on the maximizing actions of static human actors with pre-given, complete preference structures. On the contrary, institutions and culture more generally are crucially involved in the formation and transformation of actors and identities and preferences. Moreover, it was argued above, from the perspective of the human actor emphasizing personal learning, development and growth, that these processes of the formation and transformation of preferences should be understood as basically rational. At the same time, it is important to emphasize that in varying degrees these processes are typically mixed up with relationships and processes of power, domination, manipulation and rivalries of all sorts. However, understanding these processes of formation and transformation of preferences as nevertheless *basically* rational, implies a claim to the effect that not even *normatively* does the neo-classical notion of the perfect market economy have validity as a standard of efficiency somehow representing the optimal way of organizing economic activity. Not even as an ideal or Utopian view of how things could have been if everyone was rational and had unlimited cognitive competence, does this notion have any validity. The static individual with given and constant preferences forming the basis of neo-classical theory is not so much paradigmatic of rationality as of stagnation and an obsession with control.

As should be clear from the above discussion, taking heterogeneity seriously, rejecting the idea of a complete and unproblematic commensurability *ex ante* residing in the given preference structures of static individuals, also implies a rejection of the claims to the validity of central neo-classical ideas like maximization and optimality. In their book *An Evolutionary Theory of Economic Change*, Richard Nelson and Sidney Winter argue against the proposition that firms maximize profits, a basic assumption of neo-classical economic theory. They agree that striving for profit is an important objective for business firms, but they disagree that this is fruitfully dealt with through the neo-classical approach. Hence, they endorse an assumption of 'profit seeking' or 'profit-motivated striving', but not an assumption of profit maximization.[147] They note that this distinction is especially important in the context of *change*:

> In a sufficiently calm and repetitive decision context, the distinction between striving for profit and profit maximisation may be of little moment, but in a context of substantial change it matters a great deal. Strict adherence to optimisation notions either requires or strongly encourages the disregard of essential features of change – the prevalence of Knightian uncertainty, the diversities of viewpoint, the difficulties of the decision process itself, the importance of highly sequential 'groping' and of diffuse alertness for acquiring relevant information, the value of problem-solving heuristics, the likely scale and scope of actions recognised *ex post* as mistaken, and so forth.[148]

The neo-classical theory of profit maximization, on the other hand, treats 'a firm's choice sets as obvious to it and the best choice similarly clear and obvious'.[149] Basically, in this theory profit maximization is something that has a precise, unique solution.[150] Thus Nelson and Winter contend that neo-classical theory gives a static and distorted picture of profit-motivated striving, depicting economic actors as 'automaton maximizers' merely picking the best alternative from a given choice set, a given set of alternatives.[151] Rejection of these notions of maximization as something that has an obvious, precise, unique solution, also implies a rejection of the neo-classical claims for the optimality of the perfect market economy. And, as Richard Nelson emphasizes: 'Once the optimality argument falls away, there would seem to be no way to avoid comparing how market organisation performs against alternative structures.'[152] Moreover, as Nelson also observes: 'Modern market capitalism (or any plausible alternative system) is very complex and variegated',[153] and market organization thus comes in many different forms. So rather, once arguments based on the optimality of the model of the perfect market economy are rejected as invalid or irrelevant, there is no way to avoid comparative empirical investigation of different institutional arrangements, implying a comparison of different forms of market organizations, different ways markets are embedded in wider social structures, different norms and standards guiding the orientation of market participants, and so on. Modern capitalist economies are highly complex institutional structures, and these institutional structures also vary considerably *across* economies.[154] Different institutional arrangements will have different strengths and weaknesses, for instance in different regards, in different kinds of situations, in different kinds of production, and so forth.

An implication of these points is also that it becomes an open question to what extent it is appropriate to characterize modern capitalist economies as 'market economies'. William Lazonick has been very critical of this label. Although he acknowledges that 'market exchange remains a distinctive feature of advanced capitalist economies', he also claims that 'the historical experience of capitalist development demonstrates the growing importance of *organisational coordination* relative to *market coordination* in the value-creation process'.[155] This, in essence, is also a point made by Herbert Simon. He questions why we should call our economies 'market economies' at all, and suggests that 'organizational economy' would be a better term, 'with market relations among organizations'.[156] Here, we should emphasize that this is not simply an issue of the respective importance of markets in relation to other types of institutions and arrangements, but also of the very conception of what markets themselves are all about and how they function.

In any case, once the optimality argument falls away, one cannot treat the institutional heterogeneity or diversity of modern capitalist economies as accidental or secondary, as an expression of 'imperfections'. This also means that economic coordination cannot fruitfully be conceptualized in terms of the notion of *equilibrium*, a state that would automatically be attained and maintained had it not been for the existence of different kinds of imperfections that make the system deviate from optimality. Taking heterogeneity seriously, especially within the context of a decentralized economy, means taking as a point of departure the constant and permanent possibility of incompatibility of plans and actions among the different economic actors. Jacques Sapir expresses this by saying that we should take *crisis* as the point of departure. In this perspective, he claims, we should regard actually existing economies not as imperfect deviations from equilibrium, but as a set of solutions to the problem of crisis, where some solutions are 'relatively long-lasting and others not, some relatively successful and others not'. This set of solutions is embodied in a hierarchically structured complex of institutions that Sapir sees as a construction, partly non-intentional, but also partly intentional, that expresses and contains the conflicts and divergences of perspectives that run through the society in question. The coherence of this set of institutions has thus nothing inevitable about it, but is always only a temporary and local coherence.[157] Also J.S. Metcalfe rejects the notion of equilibrium as appropriate for analysing economic coordination, and argues that one gets a far better grasp of economic processes through the notion of *order*: 'From the evolutionary point of view of the evolutionary economist, it is far better to refer to the consistency of actions as a temporary *order*, for coordination leads to order, not to equilibrium.'[158] A central issue then becomes how order is created and re-created, and an important point here is precisely that 'every position of temporary economic order creates within itself the conditions to change that order'.[159]

NOTES

1. Jacques Sapir, *Les trous noirs de la science économique: Essai sur l'impossibilité de penser le temps et l'argent*, Paris: Albin Michel, 2000.
2. Ibid., p. 275.
3. J. Rogers Hollingsworth, Philippe C. Schmitter and Wolfgang Streeck, 'Capitalism, sectors, institutions, and performance', in J. Rogers Hollingsworth, Philippe C. Schmitter and Wolfgang Streeck (eds), *Governing Capitalist Economies: Performance and Control of Economic Sectors*, New York: Oxford University Press, 1994, pp. 3–16.
4. Ibid. pp. 4–8.
5. Ibid. p. 3.
6. See Douglass C. North, who defines institutions as 'the humanly devised constraints that structure political, economic and social interaction'. He distinguishes between two kinds of institutions: informal constraint, of which he mentions sanctions, taboos, customs, traditions and codes of conduct; and formal rules, of which he mentions constitutions, laws and property rights. See Douglass C. North, 'Institutions', *Journal of Economic Perspectives*, Volume 5, Number 1, Winter 1991, pp. 97–112. Reference, with quote, is to p. 97. The choice of the word *constraint* in defining institutions will be commented on below.
7. Hollingsworth, Schmitter and Streeck, op. cit., p. 4.
8. Ibid., p. 8.
9. Ibid.
10. Ibid., pp. 3–4.
11. Sapir, op. cit., p. 42.
12. Ibid.
13. Ibid.
14. Anthony Giddens, *New Rules of Sociological Method: A Positive Critique of Interpretative Sociologies*, London: Hutchinson, 1976, p. 128.
15. Richard R. Nelson, 'Why do firms differ, and how does it matter?', in Richard R. Nelson, *The Sources of Economic Growth*, Cambridge, MA: Harvard University Press, 1996, pp. 100–19. Originally published in 1991 (in *Strategic Management Journal*).
16. Ibid., p. 101.
17. Ibid., p. 105.
18. Ibid., p. 108.
19. Ibid.
20. Herbert A. Simon, *An Empirically Based Microeconomics*, Cambridge: Cambridge University Press, 1997, p. 51.
21. G.L.S. Shackle, *Time in Economics*, Amsterdam: North Holland Publishing Company, 1958, pp. 21–3.

22. Ibid., p. 21.
23. Ibid.
24. Nelson, op. cit., p. 101.
25. J. Bradford Jensen and Robert H. McGuckin, 'Firm Performance and Evolution: Empirical Regularities in the US Microdata', *Industrial and Corporate Change*, Volume 6, Number 1, 1997, pp. 25–47. Quotation is from p. 27.
26. See Sidney G. Winter's summary of what he terms the 'classic defense of the rationality-as-optimization paradigm in economic theory', in Sidney G. Winter, 'Comments on Arrow and on Lucas', in Robert M. Hogarth and Melvin W. Reder (eds), *Rational Choice: The Contrast between Economics and Psychology*, Chicago: The University of Chicago Press, 1987, pp. 243–50, especially p. 245. The natural selection argument figures prominently in Milton Friedman, 'The methodology of positive economics', in Daniel M. Hausman (ed.), *The Philosophy of Economics: An Anthology*, Cambridge: Cambridge University Press, 1984, pp. 210–44, see especially p. 223. (Originally published in 1953 in Milton Friedman, *Essays in Positive Economics*.)
27. Jon Elster, *Nuts and Bolts for the Social Sciences*, Cambridge: Cambridge University Press, 1989, p. 80.
28. Ibid.
29. Nelson, op. cit., p. 101.
30. See Marshall W. Meyer, 'Finding performance: The new discipline in management', in Andy Neely (ed.), *Business Performance Measurement: Theory and Practice*, Cambridge: Cambridge University Press, 2002, pp. 51–62.
31. Marshall W. Meyer, with Kenneth C. O'Shaughnessy, 'Organizational design and the performance paradox', in Richard Swedberg (ed.), *Explorations in Economic Sociology*, New York: Russell Sage Foundation, 1993, pp. 249–78. Quotation from p. 249. Meyer refers to this as the 'performance paradox'.
32. Meyer, 'Finding performance', op. cit., p. 54.
33. Jensen and McGuckin, op. cit., p. 27.
34. Robert M. Solow, 'Economics: is something missing?', in William N. Parker (ed.), *Economic History and the Modern Economist*, Oxford: Basil Blackwell, 1986, pp. 21–29.
35. Ibid., p. 21.
36. Ibid., p. 25.
37. Ibid., p. 22.
38. Ibid., p. 29.
39. Ibid., p. 23.
40. Ibid., p. 24.
41. Ibid.
42. Bernard Guerrien, *L'économie néo-classique*, Paris: Éditions La Découverte, 1989, p. 11.
43. Ibid., pp. 10–11.
44. Mary O'Sullivan, *Contests for Corporate Control: Corporate Governance and Economic Performance in the United States and Germany*, Oxford: Oxford University Press, 2000, p. 3.
45. Ibid.
46. See Guerrien, op. cit., pp. 46–54.
47. Oliver E. Williamson, 'The new institutional economics: taking stock, looking ahead', *Journal of Economic Literature*, Volume XXXVIII, Number 3, September 2000, pp. 595–613. Quote is from p. 596.
48. Douglass C. North, *Institutions, Institutional Change and Economic Performance*, Cambridge: Cambridge University Press, 1990, p. 57.
49. Ibid.
50. Ibid.
51. Ibid.
52. Ibid., p. 58.
53. For this argument, see in particular ibid., pp. 3–6 and pp. 57–58, and North, 'Institutions', op. cit., pp. 97–8.

54. Williamson, op. cit., p. 595. Williamson here quotes from R.C.O. Matthews, 'The economics of institutions and the sources of economic growth', *Economic Journal*, Vol. 96, No. 4, 1986, pp. 903–18. Quote is from p. 903.
55. North, *Institutions, Institutional Change and Economic Performance*, op. cit., p. 30.
56. Ibid., p. 28.
57. Ibid., p. 29.
58. Ibid., p. 30.
59. Ibid.
60. Ibid., p. 52.
61. See for instance ibid., p. 58.
62. Williamson, op. cit., p. 600.
63. Ibid.
64. North, *Institutions, Institutional Change and Economic Performance*, op. cit., p. 23.
65. Ibid., p. 16.
66. Ibid., pp. 95–6.
67. Ibid., p. 95.
68. Ibid., p. 23.
69. J.S. Metcalfe, 'Institutions and progress', *Industrial and Corporate Change*, Volume 10, Number 3, 2001, pp. 561–86. Quote is from p. 570.
70. For a thorough treatment of this issue, see for instance Mary O'Sullivan, *Contests for Corporate Control: Corporate Governance and Economic Performance in the United States and Germany*, op. cit.
71. This position is very explicit in Friedman, 'The methodology of positive economics', op. cit.
72. Joseph E. Stiglitz, 'Symposium on organizations and economics', *Journal of Economic Perspectives*, Volume 5, Number 2, Spring 1991, pp. 15–24. The quote is from p. 15.
73. Oliver Hart, 'Corporate governance: some theory and implications', *The Economic Journal*, Volume 105, Number 430, May 1995, pp. 678–89. The quote is from p. 678.
74. Ibid.
75. Stiglitz, op. cit., p. 16.
76. Ibid.
77. Ibid.
78. See, for instance, Gary J. Miller, 'Managerial dilemmas: political leadership in hierarchies', in Karen Schweers Cook and Margaret Levi (eds), *The Limits of Rationality*, Chicago: The University of Chicago Press, 1990, pp. 324–48.
79. Hart, op. cit., pp. 680–81.
80. Indeed, corporate governance is often simply defined as the processes of supervision and control 'intended to ensure that the company's management acts in accordance with the interests of the shareholders'. See J.E. Parkinson, *Corporate Power and Responsibility: Issues in the Theory of Company Law*, Oxford: Clarendon Press, 1993, p. 159. See also Xavier Vives, 'Corporate governance: does it matter?', in Xavier Vives (ed.), *Corporate Governance: Theoretical and Empirical Perspectives*, Cambridge: Cambridge University Press, 2000, pp. 1–21, especially p. 1.
81. For a critical exposition of these views, see for instance Mary O'Sullivan, *Contests for Corporate Control: Corporate Governance and Economic Performance in the United States and Germany*, op. cit.; also Michel Aglietta and Antoine Rebérioux, *Dérives du capitalisme financier*, Paris: Albin Michel, 2004, especially pp. 39–72.
82. O'Sullivan, *Contests for Corporate Control*, op. cit., p. 43.
83. Hart, op. cit., p. 678.
84. Ibid.
85. Miller, op. cit., p. 341, see also pp. 327–8.
86. For related ideas, see the treatment of organizational learning in William Lazonick and Mary O'Sullivan, *Perspectives on Corporate Governance, Innovation, and Economic Performance*, report to 'Corporate Governance, Innovation, and Economic Performance in the EU', a research project funded by the Targeted Socio-Economic Research (TSER) Programme of the European Commission (DGXII) under the Fourth Framework

Programme, Fontainebleau: INSEAD, 2000, pp. 73–9, where they discuss ideas of authors like Peter Senge on 'shared vision', 'co-creating' and 'generative learning'. According to D. Flood, to whom Lazonick and O'Sullivan refer (op. cit., pp. 76–7), creating a 'shared vision' is 'the antithesis of visioning exclusively from the top of a management hierarchy'. The concern here is with 'how to move visioning from the top of the management hierarchy to a widespread intensive dialogue – from "telling" to "co-creating."' By 'co-creating' is meant 'a widespread and collaborative process where a shared vision is built in a mood of generative learning'. The quotes here are from D. Flood, *Rethinking the Fifth Discipline: Learning Within the Unknowable*, Routledge, 1999, p. 24, which, in turn, discusses P. Senge, C. Roberts, R. Ross and B. Smith, *The Fifth Discipline Fieldbook*, 1994.

87. Herbert A. Simon, 'Organizations and markets', *Journal of Economic Perspectives*, Volume 5, Number 2, Spring 1991, pp. 25–44. The quote is from p. 34.
88. Ibid.
89. Stiglitz, op. cit., p. 22. Italics in the original.
90. See Simon, op. cit., p. 30.
91. Stiglitz, op. cit., p. 22.
92. Simon, op. cit., p. 42.
93. On organizational integration, see William Lazonick, *Business Organization and the Myth of the Market Economy*, Cambridge: Cambridge University Press, 1991, especially pp. 23–58 (the chapter entitled 'Institutional foundations of industrial dominance and decline').
94. Israel M. Kirzner, 'Entrepreneurial discovery and the competitive market process: an Austrian approach', *Journal of Economic Literature*, Volume XXXV, March 1997, pp. 60–85. The quote is from p. 62.
95. Hart, op. cit., pp. 679–80.
96. Ibid., p. 679.
97. Ibid.
98. Ibid.
99. Ibid., p. 680.
100. Ibid.
101. See for instance, Alain Lipietz, *Towards a New Economic Order: Postfordism, Ecology and Democracy*, Cambridge: Polity Press, 1992, p. 4 (originally published in French in 1989), on the separation of conception and execution of work operations in 'Taylorism'.
102. See Simon, op. cit., p. 31.
103. Ibid., p. 32.
104. Stewart Macaulay, 'Non-contractual relations in business: a preliminary study', *American Sociological Review*, Volume 28, Number 1, 1963, pp. 55–67. The quote is from p. 64. This passage is quoted and discussed in Mark Granovetter, 'Economic action and social structure: the problem of embeddedness', *American Journal of Sociology*, Volume 91, Number 3, November 1985, pp. 481–510. Here the quote is on p. 496.
105. Charles F. Sabel, 'Constitutional ordering in historical context', in Fritz W. Scharpf (ed.), *Games in Hierarchies and Networks: Analytical and Empirical Approaches to the Study of Governance Institutions*, Frankfurt am Main: Campus Verlag, 1993, pp. 65–123. The quote is from p. 65.
106. Ibid.
107. Granovetter, op. cit., p. 489.
108. This view is very explicit in one of the most authoritative statements of rational choice theory, James S. Coleman's *Foundations of Social Theory*, Cambridge, MA: The Belknap Press of Harvard University Press, 1990. See especially the chapter entitled 'Relations of Trust', pp. 91–116.
109. Sabel, op. cit., p. 65.
110. Nelson, 'Why do firms differ, and how does it matter?', op. cit., p. 105, p. 107.
111. Ronald Dore, 'The distinctiveness of Japan', in Colin Crouch and Wolfgang Streeck (eds), *Political Economy of Modern Capitalism: Mapping Convergence and Diversity*, London: Sage Publications, 1997, pp. 19–32, especially pp. 19–20.

112. Ibid., p. 19.
113. Ibid., p. 20.
114. Ibid.
115. Ibid., p. 19.
116. Ibid., p. 20. For William Lazonick where exactly the boundaries of this community lie constitutes precisely one of the basic characteristics distinguishing British, American and Japanese capitalism, where in British capitalism the community is typically limited to senior managers, in American capitalism it is extended to include also middle managers, and in Japanese capitalism includes all the people who work in the firm, again apart from 'peripheral', i.e. part-time or temporary, workers. See Lazonick, *Business Organization and the Myth of the Market Economy*, op. cit., especially pp. 23–58 (the chapter entitled 'Institutional foundations of industrial dominance and decline').
117. Dore, op. cit., p. 20.
118. Ibid.
119. Aglietta and Rebérioux, *Dérives du capitalisme financier*, op. cit.
120. Ibid., p. 346.
121. Ibid., p. 349.
122. Ibid., p. 333.
123. Ibid.
124. Ibid., p. 339.
125. Wolfgang Streeck, 'Lean production in the German automobile industry: a test for convergence theory', in Suzanne Berger and Ronald Dore (eds), *National Diversity and Global Capitalism*, Ithaca: Cornell University Press, 1996, pp. 138–70, quoted from p. 153.
126. Ibid., p. 152. Italics in the original.
127. North, 'Institutions', op. cit., p. 97.
128. North, *Institutions, Institutional Change and Economic Performance*, op. cit., p. 42.
129. Aglietta and Rebérioux, op. cit., p. 339.
130. See, for instance, the definition of a *preference ordering* in Shaun Hargreaves Heap, Martin Hollis, Bruce Lyons, Robert Sugden and Albert Weale, *The Theory of Choice: A Critical Guide*, Oxford: Blackwell, 1992, p. 346: 'A preference ordering is a ranking of all possible outcomes in accordance with one's preferences.' This is assured when an actor's preferences meet the requirement of *completeness*, together with the requirement of *transitivity* and the purely formal requirement of *reflexivity*. See ibid., pp. 5–6.
131. Sapir, *Les trous noirs de la science économique*, op. cit., p. 89.
132. Sapir characterizes this idea of a complete ordering of all possible present and future states of the world as meaningless, see ibid., p. 91.
133. These issues have been discussed at some length in a previous work by the author, see Tore Sandven, 'Autonomy, adaptation, and rationality – a critical discussion of Jon Elster's concept of "Sour Grapes"', *Philosophy of the Social Sciences*, Volume 29, Number 1, March 1999, pp. 3–31 (Part I), and Volume 29, Number 2, June 1999, pp. 173–205 (Part II).
134. Herbert A. Simon, *An Empirically Based Microeconomics*, op. cit., p. 192 (from the section entitled 'Herbert A. Simon's autobiographical sketch', pp. 191–6).
135. See James G. March, 'The technology of foolishness', in James G. March, *Decisions and Organizations*, Oxford: Basil Blackwell, 1988, pp. 253–65, originally published in 1971 (in *Civiløkonomen*), and James G. March, 'Bounded rationality, ambiguity, and the engineering of choice', in March, *Decisions and Organizations*, op. cit., pp. 266–93, originally published in 1978 (in *Bell Journal of Economics*).
136. The basic reference here is Pierre Bourdieu, *La distinction: Critique sociale du jugement*, Paris: Éditions de Minuit, 1979.
137. Aglietta and Brender calls this type of society 'la société salariale'. See Michel Aglietta and Anton Brender, *Les métamorphoses de la société salariale: La France en projet*, Paris: Calmann-Lévy, 1984, especially pp. 7–22.
138. Ibid., p. 97. More generally on the dynamics of consumption standards, see the whole of the section entitled 'Solvabiliser et différencier la demande', ibid., pp. 97–109.

139. Ibid., p. 99.
140. Samuel Scheffler, *Human Morality*, Oxford: Oxford University Press, 1992, pp. 141–2.
141. Some of the aspects of the issue of preference-change and the question of the rationality or irrationality of this have been thoroughly dealt with in the context of a critical discussion of Jon Elster's treatment of the issues of personal autonomy and 'endogenous preference change' in the previous work of the author referred to above, see Sandven, 'Autonomy, adaptation, and rationality – a critical discussion of Jon Elster's concept of "Sour Grapes" ', op. cit.
142. For a critical discussion of this type of attempt to rescue the theory by removing maximization to a meta-level, see ibid., Part II, pp. 195–204. See also Sidney Winter's critical comments on the tendency among neo-classical economists to portray the economic actor as a 'superotimizer' in Winter, 'Comments on Arrow and on Lucas', op. cit., pp. 246–7.
143. Sapir, *Les trous noirs de la science économique*, op. cit., p. 95.
144. Ibid., p. 180.
145. Ibid., p. 188.
146. Max Weber, *Economy and Society: An Outline of Interpretive Sociology*, Berkeley: University of California Press, 1968, reissue 1978, originally published in German in 1920–22, p. 108. See Sapir, op. cit., p. 189.
147. Richard R. Nelson and Sidney G. Winter, *An Evolutionary Theory of Economic Change*, Cambridge, MA: The Belknap Press of Harvard University Press, 1982, pp. 30–1.
148. Ibid., p. 31.
149. Nelson, 'Why do firms differ, and how does it matter?', op. cit., p. 107.
150. On the importance attached to unique predictions both in neo-classical economics and in rational choice theory more generally, see, for instance, Jon Elster, 'When rationality fails', in Cook and Levi (eds), op. cit., pp. 19–51, especially pp. 24–5; or Jon Elster, *Nuts and Bolts for the Social Sciences*, Cambridge: Cambridge University Press, 1989, especially pp. 30–2. According to Elster, the theory gives unique predictions in standard cases. Elster then discusses some special, non-standard cases where the theory does not give unique predictions, where thus 'rationality fails'. The claim that the theory in standard cases predicts one unique course of action as the one that maximizes utility, and which thus in this perspective is the one that rationality dictates, presupposes, of course, that the preferences and the beliefs (concerning the alternatives available) of the actor be known in advance. The implication of the argument above is that this is not a very fruitful assumption, perhaps not even very meaningful. At a more fundamental level one may question whether the description of human action as following from decisions where actors compare alternatives and then pick out the best of them (see, for instance, Jon Elster, *Making Sense of Marx*, Cambridge: Cambridge University Press, 1985, p. 13) is very fruitful as a general description of the structure of human action. We cannot go further into these questions here.
151. Nelson and Winter, *An Evolutionary Theory of Economic Change*, op. cit., p. 32. The term 'automaton maximizers' they take from W.J. Baumol, 'Entrepreneurship in economic theory', *American Economic Review*, Volume 58, 1968, pp. 64–71.
152. Richard R. Nelson, 'The problem of market bias in modern capitalist economies', *Industrial and Corporate Change*, Volume 11, Number 2, 2002, pp. 207–44, quoted from p. 221.
153. Ibid., 222.
154. See, for instance, Colin Crouch and Walfgang Streeck, 'Introduction: the future of capitalist diversity', in Crouch and Streeck (eds), *Political Economy of Modern Capitalism: Mapping Convergence and Diversity*, op. cit., pp. 1–18, and, more generally, all the contributions to this volume.
155. Lazonick, *Business Organization and the Myth of the Market Economy*, op. cit., p. 59.
156. Simon, 'Organizations and markets', op. cit., p. 28–9.
157. Sapir, *Les trous noirs de la science économimique*, op. cit., pp. 270–1.
158. Metcalfe, 'Institutions and progress', op. cit., p. 572.
159. Ibid., p. 575.

REFERENCES

Aglietta, M. and A. Brender (1984), *Les métamorphoses de la société salariale: La France en projet*, Paris: Calmann-Lévy.

Aglietta, M and A. Rebérioux (2004), *Dérives du capitalisme financier*, Paris: Albin Michel.

Bourdieu, P. (1979), *La distinction: Critique sociale du jugement*, Paris: Éditions de Minuit.

Coleman, J.S. (1990), *Foundations of Social Theory*, Cambridge, MA: The Belknap Press of Harvard University Press.

Crouch, C. and W. Streeck (1997), 'Introduction: the future of capitalist diversity', in C. Crouch and W. Streeck (eds), *Political Economy of Modern Capitalism: Mapping Convergence and Diversity*, London: Sage Publications, pp. 1–18.

Dore, R. (1997), 'The distinctiveness of Japan', in C. Crouch and W. Streeck (eds), *Political Economy of Modern Capitalism: Mapping Convergence and Diversity*, London: Sage Publications, pp. 19–32.

Elster, J. (1985), *Making Sense of Marx*, Cambridge: Cambridge University Press.

Elster, J. (1989), *Nuts and Bolts for the Social Sciences*, Cambridge: Cambridge University Press.

Elster, J. (1990), 'When rationality fails', in K.S. Cook and M. Levi (eds), *The Limits of Rationality*, Chicago: University of Chicago Press, pp. 19–51.

Friedman, M. (1984), 'The methodology of positive economics', in D.M. Hausman (ed.), *The Philosophy of Economics: An Anthology*, Cambridge: Cambridge University Press, pp. 210–44. (Article originally published in 1953.)

Giddens, A. (1976), *New Rules of Sociological Method: A Positive Critique of Interpretative Sociologies*, London: Hutchinson.

Granovetter, M. (1985), 'Economic action and social structure: the problem of embeddedness', *American Journal of Sociology*, **91** (3), 481–510.

Guerrien, B. (1989), *L'économie néo-classique*, Paris: Éditions La Découverte.

Hart, O. (1995), 'Corporate governance: some theory and implications', *The Economic Journal*, **105** (430), 678–89.

Heap, S.H., M. Hollis, B. Lyons, R. Sugder and A. Weale (1992), *The Theory of Choice: A Critical Guide*, Oxford: Blackwell.

Hollingsworth, J.R., P.C. Schmitter and W. Streeck (1994), 'Capitalism, sectors, institutions, and performance', in J.R. Hollingsworth, P.C. Schmitter and W. Streeck (eds), *Governing Capitalist Economies: Performance and Control of Economic Sectors*, New York: Oxford University Press.

Jensen, J.B. and R.H. McGuckin (1997), 'Firm performance and evolution: empirical regularities in the US microdata', *Industrial and Corporate Change*, **6** (1), 25–47.

Kirzner, I.M. (1997), 'Entrepreneurial discovery and the competitive market process: an Austrian approach', *Journal of Economic Literature*, **XXXV**, 60–85.

Lazonick, W. (1991), *Business Organization and the Myth of the Market Economy*, Cambridge: Cambridge University Press.

Lazonick, W. and M. O'Sullivan (2000), *Perspectives on Corporate Governance, Innovation, and Economic Performance*, report to 'Corporate Governance, Innovation, and Economic Performance in the EU', a research project funded by the Targeted Socio-Economic Research (TSER) Programme of the European Commission (DGXII) under the Fourth Framework Programme, Fontainebleau: INSEAD.

Lipietz, A. (1992), *Towards a New Economic Order: Postfordism, Ecology and Democracy*, Cambridge: Polity Press. (Originally published in French in 1989.)
Macaulay, S. (1963), 'Non-contractual relations in business: a preliminary study', *American Sociological Review*, **28** (1), 55–67.
March, J.G. (1988a), 'The technology of foolishness', in J.G. March, *Decisions and Organizations*, Oxford: Basil Blackwell, pp. 253–65. (Article originally published in 1971.)
March, J.G. (1988b), 'Bounded rationality, ambiguity, and the engineering of choice', in J.G. March, *Decisions and Organizations*, Oxford: Basil Blackwell, pp. 266–93. (Article originally published in 1978.)
Metcalfe, J.S. (2001), 'Institutions and progress', *Industrial and Corporate Change*, **10** (3), 561–86.
Meyer, M.W. (2002), 'Finding performance: the new discipline in management', in A. Neely (ed.), *Business Performance Measurement: Theory and Practice*, Cambridge: Cambridge University Press, pp. 51–62.
Meyer, M.W. with K.C. O'Shaughnessy (1993), 'Organizational design and the performance paradox', in R. Swedberg (ed.), *Explorations in Economic Sociology*, New York: Russell Sage Foundation, pp. 249–78.
Miller, G.J. (1990), 'Managerial dilemmas: political leadership in hierarchies', in K.S. Cook and M. Levi (eds), *The Limits of Rationality*, Chicago: University of Chicago Press, pp. 324–48.
Nelson, R.R. (1996), 'Why do firms differ, and how does it matter?', in R.R. Nelson, *The Sources of Economic Growth*, Cambridge, MA: Harvard University Press, pp. 100–19. (Article originally published in 1991.)
Nelson, R.R. (2002), 'The problem of market bias in modern capitalist economies', *Industrial and Corporate Change*, **11** (2), 207–44.
Nelson, R.R and S.G. Winter (1982), *An Evolutionary Theory of Economic Change*, Cambridge, MA: The Belknap Press of Harvard University Press.
North, D.C. (1990), *Institutions, Institutional Change and Economic Performance*, Cambridge: Cambridge University Press.
North, D.C. (1991), 'Institutions', *Journal of Economic Perspectives*, **5** (1), 97–112.
O'Sullivan, M. (2000), *Contests for Corporate Control: Corporate Governance and Economic Performance in the United States and Germany*, Oxford: Oxford University Press.
Parkinson, J.E. (1993), *Corporate Power and Responsibility: Issues in the Theory of Company Law*, Oxford: Clarendon Press.
Sabel, C.F. (1993), 'Constitutional ordering in historical context', in F.W. Scharpf (ed.), *Games in Hierarchies and Networks: Analytical and Empirical Approaches to the Study of Governance Institutions*, Frankfurt am Main: Campus Verlag, pp. 65–123.
Sandven, T. (1999), 'Autonomy, adaptation, and rationality – a critical discussion of Jon Elster's concept of "Sour Grapes"', *Philosophy of the Social Sciences*, **29** (1), 3–31 (Part I) and **29** (2), 173–205 (Part II).
Sapir, J. (2000), *Les trous noirs de la science économique: Essai sur l'impossibilité de penser le temps et l'argent*, Paris: Albin Michel.
Scheffler, S. (1992), *Human Morality*, Oxford: Oxford University Press.
Shackle, G.L.S. (1958), *Time in Economics*, Amsterdam: North Holland Publishing Company.
Simon, H.A. (1991), 'Organizations and markets', *Journal of Economic Perspectives*, **5** (2), 25–44.

Simon, H.A. (1997), *An Empirically Based Microeconomics*, Cambridge: Cambridge University Press.
Solow, R.M. (1986), 'Economics: is something missing?', in W.N. Parker (ed.), *Economic History and the Modern Economist*, Oxford: Basil Blackwell, pp. 21–9.
Stiglitz, J.E. (1991), 'Symposium on organizations and economics', *Journal of Economic Perspectives*, **5** (2), 15–24.
Streeck, W. (1996), 'Lean production in the German automobile industry: a test for convergence theory', in S. Berger and R. Dore (eds), *National Diversity and Global Capitalism*, Ithaca, NY: Cornell University Press, pp. 138–70.
Vives, X. (2000), 'Corporate governance: does it matter?', in X. Vives (ed.), *Corporate Governance: Theoretical and Empirical Perspectives*, Cambridge: Cambridge University Press, pp. 1–21.
Weber, M. (1968), *Economy and Society: An Outline of Interpretive Sociology*, Berkeley: University of California Press. (Reissue 1978, originally published in German in 1920–22.)
Williamson, O.E. (2000), 'The new institutional economics: taking stock, looking ahead', *Journal of Economic Literature*, **XXXVIII** (3), 595–613.
Winter, S.G. (1987), 'Comments on Arrow and on Lucas', in R.M. Hogarth and M.W. Reder (eds), *Rational Choice: The Contrast between Economics and Psychology*, Chicago: The University of Chicago Press, pp. 243–50.

5. Conceptual framework for an analysis of diversity and heterogeneity in the knowledge economy and society

Elias G. Carayannis

INTRODUCTION

The emerging glo*C*alizing (i.e. simultaneously globalizing *and* localizing) (Carayannis and von Zedwitz, 2005; Carayannis and Alexander, 2006) frontier of converging systems, networks and sectors of innovation that is driven by increasingly complex, non-linear and dynamic processes of knowledge creation, diffusion and use, confronts us with the need to reconceptualize – if not reinvent – the ways and means by which knowledge production, utilization and renewal take place in the context of the knowledge economy and society (gloCal knowledge economy and society).

Perspectives from and about different parts of the world and diverse human, socioeconomic, technological and cultural contexts are interwoven to produce an emerging new worldview on how specialized knowledge, that is embedded in a particular sociotechnical context, can serve as the unit of reference for stocks and flows of a hybrid, public/private, tacit/codified, tangible/virtual good that represents the building block of the knowledge economy, society and polity.

We postulate that one approach to such a reconceptualization is what we call the 'Mode 3' system consisting of 'Innovation Networks' and 'Knowledge Clusters' (see definitions below) for knowledge creation, diffusion and use (Carayannis and Campbell, 2005). This is a *multi-layered, multi-modal, multi-nodal* and *multi-lateral* system, encompassing mutually complementary and reinforcing innovation networks and knowledge clusters consisting of human and intellectual capital, shaped by social capital and underpinned by financial capital.

The 'Mode 3 Innovation Ecosystem' is in short the nexus or hub of the emerging 21st-century Innovation Ecosystem,[1] where people,[2] culture[3] and technology[4] (Carayannis and Gonzalez, 2003 – forming the essential 'Mode 3 Innovation Ecosystem' building block or 'knowledge nugget' – Carayannis, 2004) meet and interact to catalyze creativity, trigger invention and accelerate innovation across scientific and technological disciplines, public and private sectors (government, university, industry and non-governmental knowledge production, utilization and renewal entities) and in a top-down, policy-driven as well as bottom-up, entrepreneurship-empowered fashion.

We address the concept of heterogeneity, in the socioeconomic sense, in a multi-disciplinary fashion including perspectives from innovation, entrepreneurship and competitiveness concepts combined with systems theory to management, economics and sociology. As opposed to mainstream economics, evolutionary and systemic approaches in innovation studies place the issue of firm heterogeneity at the center of economic development. This is because *entrepreneurship* and *innovation* activities are understood as *heterogeneity-inducing microprocesses*. Firm heterogeneity is continuously reshaped as a result of market selection forces. Institutional arrangements (such as patents, trade marks) regulations, competition policies, sectoral, national and regional economic knowledge infrastructures shape both market selection and diversity creation.

In this chapter, we depart from the biological view of heterogeneity linking it to variety and selection mechanisms in a closed system context, and attempt to weave a series of perspectives at the macro-, meso- and micro-levels to form a new view of heterogeneity that builds on a perspective of mutually driving, complementary and reinforcing processes of co-opetition (a hybrid form of collaboration and competition), co-specialization and co-evolution (C3) (Carayannis and Gonzalez, 2003; Carayannis, 2004; Carayannis et al., 2005; Carayannis and Campbell, 2005) (see Figures 5.2, 5.3 and 5.4). In particular, we view heterogeneity as both a cause and an effect of the input, process and output (IPO) innovation stages that in turn are catalyzed and shaped by C3.

In other words, the main thesis and motivation of this chapter is that heterogeneity (from the Greek etymological roots of the word to mean literally 'possessing different genes') has been serving as the trigger, catalyst and accelerator of sociotechnical change that generates positive socioeconomic development.

Nowadays, it is often perceived that globalization serves as both a catalyst of accelerated development as well as an agent of chaotic disruption resulting in socioeconomic and political dislocations. In light of this, the key message of the chapter is that heterogeneity could be understood as a *mindset* and a *practice* where complexity and diversity are leveraged strategically in a manner that promotes *sustainable entrepreneurship*.

We define *sustainable entrepreneurship* as the creation of viable, profitable and scalable firms that engender the formation of self-replicating and mutually enhancing innovation networks and knowledge clusters leading towards what we call robust competitiveness.

We define *robust competitiveness* as a state of economic being and becoming that avails systematic and defensible 'unfair advantages' to the entities that are part of the economy and is built on mutually complementary and reinforcing low, medium and high technology, public and private sector entities (government agencies, private firms, universities, and non-governmental organizations).

In other words, robust competitiveness results from an emerging 21st-century *innovation ecosystem* (also called 'Mode 3' Innovation Ecosystem) (Carayannis and Campbell, 2005) which is defined as follows:

> *A 21st-Century Innovation Ecosystem* is a multi-level, multi-modal, multi-nodal and multi-agent system of systems. The constituent systems consist of innovation meta-networks (networks of innovation networks and knowledge clusters) and knowledge meta-clusters (clusters of innovation networks and knowledge clusters) as building blocks and organized in a self-referential or chaotic, fractal, knowledge and innovation architecture, which in turn constitute agglomerations of human, social, intellectual and financial capital stocks and flows as well as cultural and technological artifacts and modalities, continually co-evolving, co-specializing, and co-opeting. These innovation networks and knowledge clusters also form, re-form and dissolve within diverse institutional, political, technological and socio-economic domains including Government, University, Industry, Non-governmental Organizations and involving Information and Communication Technologies, Biotechnologies, Advanced Materials, Nanotechnologies and Next Generation Energy Technologies. (Carayannis and Campbell, 2006)

Heterogeneity is the quality of being diverse and not comparable in kind (Webster's Dictionary). Why are we concerned with this concept? Heterogeneity may indeed be one of the central reasons why and how the human species survived, evolved and prospered both biologically as well as socioeconomically. In this chapter we focus on the socioeconomic aspects of heterogeneity and, in particular, whereas '*adaptation*' has been the biological manifestation of the forces of variety and selection acting on the natural gene pool, we postulate that 'innovation' has been the socioeconomic manifestation of the forces of *co-opetition*, *co-specialization* and *co-evolution* (C3) acting on the knowledge economy and society (see Figures 5.1 and 5.2).

Specifically, we see in Figure 5.1 the juxtaposition and differentiation of the conceptual context of heterogeneity in the socioeconomic versus the natural domains, and the ensuing linkages with innovation versus

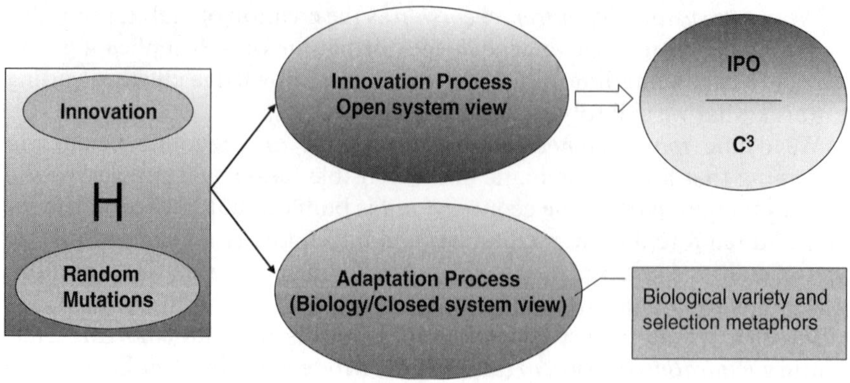

Note: IPO: Input – Process – Output; C^3: co-opetition, co-specialization, co-evolution

Figure 5.1 The concept of heterogeneity (H)

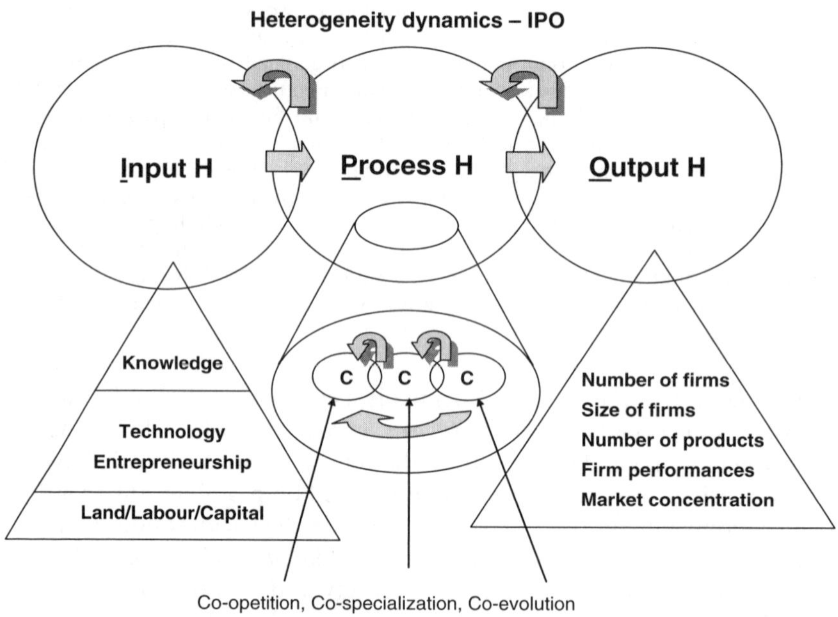

Figure 5.2 Heterogeneity dynamics – input/process/output

adaptation respectively. Moreover, in the context of heterogeneity dynamics, we distinguish across three classes of heterogeneity (input, process and output) (see Figure 5.2). Specifically, in Figure 5.2 we depict in summary form a socioeconomic system for adding value via cascaded, interconnected

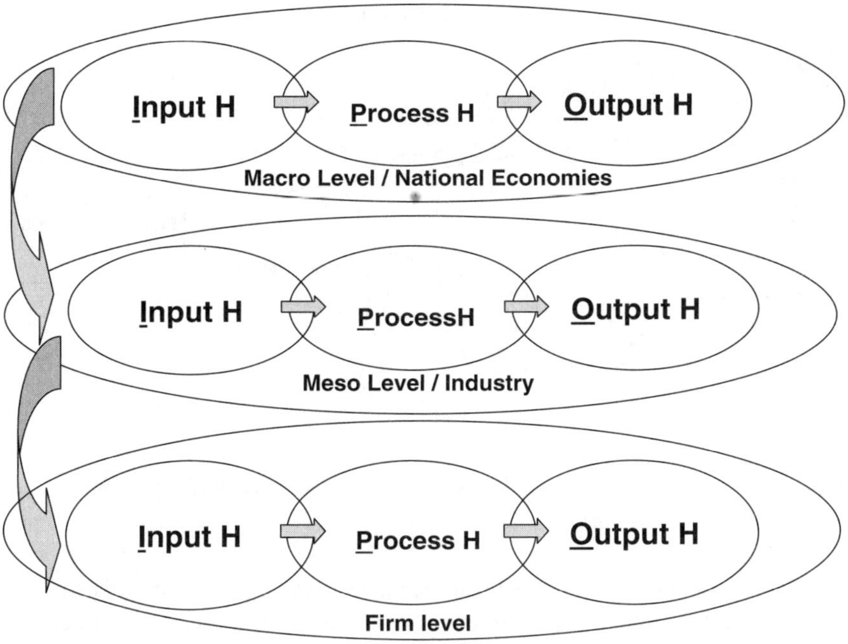

Figure 5.3 Heterogeneity dynamics – micro/meso/macro levels

and interacting stages of socioeconomic being and becoming where the concept of C3 as discussed above drives the creation, diffusion and use of knowledge.

This system consists of multiple layers (macro, meso, micro as shown in Figure 5.3) as well as modules (*input*, *process*, *output*). The presence of heterogeneity in those layers and modules acts as both cause and effect for driving the value-creating, diffusing and potentially destroying, processes of *co-opetition, co-specialization* and *co-evolution* (C3) as shown in Figures 5.2 and 5.3 (Carayannis and Campbell, 2005).

Input, process and output heterogeneity deals with the issue of value creation in a socioeconomic context as later discussed briefly in the chapter summaries included in this proposal:

- *Input heterogeneity* refers to the variety and diversity of the key inputs to economic activity, namely, land, labor, capital, technology and entrepreneurship as identified by Adam Smith, Ricardo, and Joseph Schumpeter among others. Intrinsic in all these inputs is knowledge, which has been increasingly the key source of value adding of most human endeavors.

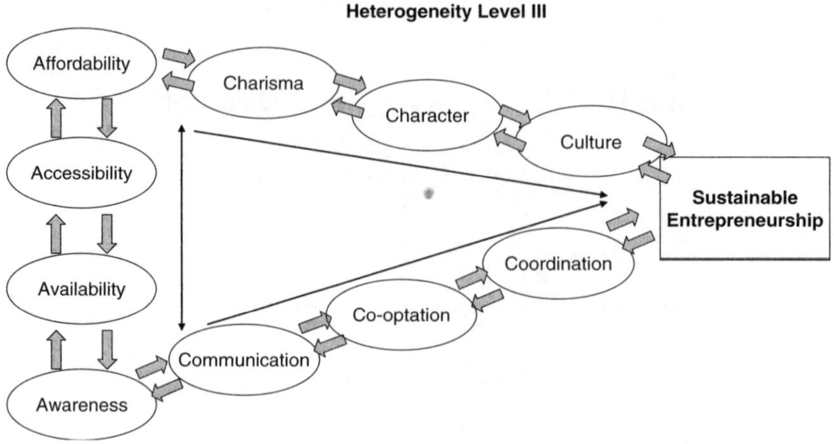

Figure 5.4 Heterogeneity dynamics – micro level stages, drivers and determinants

- *Process heterogeneity* reflects the variety and diversity intrinsic in the ways that the key inputs to economic activity are leveraged, allocated, re-combined and re-created as part of the processes of technology innovation and entrepreneurship aiming at the maximization of value added.
- *Output heterogeneity* reflects the diverse ways and means that the value added of economic activity combining and leveraging the key inputs discussed earlier, is captured and exploited, namely, number and size of firms, firm performance, market concentration, number and rate of renewal of products and services, as well as public–private sector partnerships structure and performance, to name a few.

Again, these issues are further delineated, profiled and discussed below and also depicted schematically in Figures 5.3 and 5.4. In Figure 5.3, we show the macro-, meso-, and micro-levels of socioeconomic analysis where heterogeneity manifests itself and shapes as well as is impacted by socio-economic processes. In Figure 5.4, we show at the micro-level, the key success factors for *sustainable entrepreneurship* that is one of the major pillars of *robust competitiveness* as discussed earlier. In Figure 5.5, we show the Schumpeterian process of '*Creative Destruction*' combined with its complements ('*Creative Creation*', '*Destructive Creation*', and '*Destructive Destruction*') mapped onto the technology lifecycle curve (S-curve) also providing the context for the 'Horizon' and 'Memory' dimensions of a system's lifecycle discussed later on.

MODE 3 INNOVATION ECOSYSTEM: PERSPECTIVES FROM SOCIOECONOMICS AND BIOLOGY

Most contemporary theories of heterogeneity emphasize the *'learning potential'* (Carayannis, 1993, 1994, 2000, 2004; Carayannis and Alexander, 1999, 2006; Carayannis and Laget, 2004; Carayannis et al., 2005) of belonging to a heterogeneous and complex super-system, which integrates several sub-systems, and accordingly enables coordination. The upshot is that heterogeneity may – under conditions yet to be specified – enable combinations of different forms of knowledge – different specializations. By comparison, the evolution of a species in nature is established within the development of the geology of the earth, global climate change, and several other naturally given macro factors, which shape and transform ecologies. Within these ecologies, species compete, struggle for survival, and try to adapt to changes, or fail.

In the context of socioeconomics, an emerging 'species' is the form of knowledge creation, diffusion and use paradigm called the 'Mode 3' Innovation Ecosystem[5] by Carayannis and Campbell. The 'Mode 3' Innovation Ecosystem is based on a number of distinct building blocks of a 21st-century innovation ecosystem as earlier defined, namely innovation networks[6] and knowledge clusters[7] (Carayannis and Campbell, 2005):

> The 'Mode 3 INNOVECO' is in short the nexus or hub of the emerging 21st-century Innovation Ecosystem, where *people*, *culture*[8] and *technology*[9] (forming the essential 'Mode 3 INNOVECO' building block or 'knowledge nugget'[10]) meet and interact to catalyze creativity, trigger invention and accelerate innovation across scientific and technological disciplines, public and private sectors (government, university, industry and non-governmental knowledge production, utilization and renewal entities) and in a top-down, policy-driven as well as bottom-up, entrepreneurship-empowered fashion. One of the basic ideas is: *co-existence*, *co-evolution* and *co-specialization* of different knowledge paradigms and different knowledge modes of knowledge production and knowledge use as well as their co-specialization as a result. We can postulate a dominance of knowledge heterogeneity at the systems (national, trans-national) level. Only at the sub-system (sub-national) level we should expect more homogeneity. This understanding we can paraphrase with the term 'Mode 3' [emphasis added] (Carayannis et al., 2005; Carayannis and Campbell, 2005) (see Figures 5.2, 5.3 and 5.5).

As a point of departure, the interest in the concept refers to its successful use in biology, where the heterogeneity of an interbreeding population is an important element in the theory of the evolution of the species, founded by Darwin. Accordingly, let us by way of passing, consider the relation between socioeconomics as explored in the social sciences and economics, and biology.

	CREATION	DESTRUCTION
CREATIVE	I: The beginning of the S-curve	II: The end of an S-curve
DESTRUCTIVE	III: The beginning of a new S-curve	IV: The end of a technological regime or business cycle

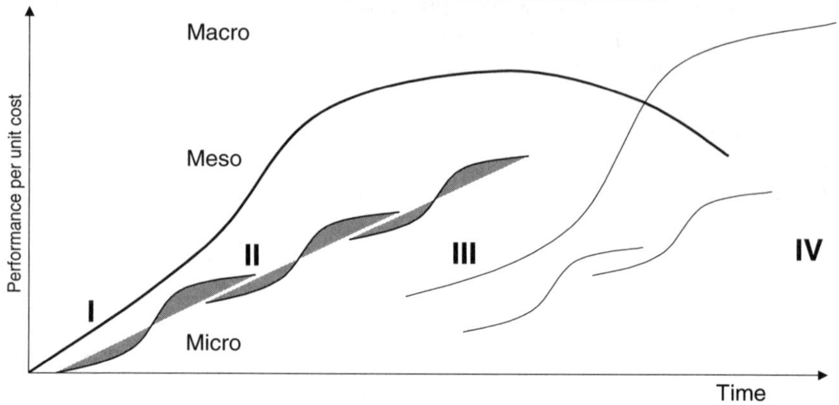

Figure 5.5 Heterogeneity dynamics – creative destruction

Heterogeneity is a collective property of *the population*. It follows from the previous discussion that heterogeneity as a concept refers to at least three levels of analysis:

- The *context* of the interbreeding population, in biology often referred to as ecologies, in the learning economy we call this context *Embedding* of knowledge. This context determines the borders of the interbreeding population.
- The *size* of the interbreeding population – where heterogeneity is measured.
- The variation between and among *individuals* making up the population.

The definition of the population in biology partly may be conditioned by natural barriers (mountains, glaciers, oceans) leading to conditioned *loss of reproductive capacity*. In the learning economy, we refer to specialized knowledge. Similarly, we have seen that *in a learning economy*, barriers around a specialized path-dependent innovation system, reducing the capacity for absorption of new forms of knowledge, may be changed or removed through various forms of corporate or business strategies (horizon scanning), through innovation policy interventions, like 'foresight', and through entrepreneurship – crossing the institutionalized borders between forms of knowledge. Heterogeneity may be seen as created through various combinations of:

- Variations *among individual members of the population*. In our case, individual members are specialized forms of knowledge and products in the national economy.
- Variations in *relations between individual forms of knowledge*. Variations in relations, we would like to argue, may create heterogeneity even though the individuals are pretty similar .

In the simple case, there is a 1:1 relation between knowledge and market product. The function of the knowledge is the direct product on the market, with no intervening networks or market relations. This is the case with a craft-based industry, based on harvesting of natural products, where the skilled craftsman knows pretty much everything there is to know in the production of the product – and the tastes of the customers. In other cases, the knowledge we are referring to may be input into larger and more complex value chains, combining several forms of knowledge through various forms of economic relations, like markets, employment contracts, projects, and so on.

Our basic unit of analysis is *heterogeneity* in terms of types of shared and living *knowledge organized as structured innovation networks and knowledge clusters (being part of existing and emerging innovation systems)*. These networks have two core properties:

- They have a *structure*, which is a fixed set of relations between the forms of knowledge involved in the network.
- The structure enables the system to generate a *function*. The function may be some form of economically useful *performance*, like producing one or more competitive products.

EVOLUTIONARY ECONOMICS

Turning to many of the basic works within the tradition of evolutionary economics and population ecology, one hardly finds a mention of the concept of heterogeneity (Nelson and Winter, 1982; Metcalfe, 1998; Aldrich, 1999), although the principle of heterogeneity may be regarded as an underlying precondition for many of the evolutionary approaches (Andersen, 1994).

While a key focus in analyzing evolution is on mechanisms related to variation and selection as the two main issues, the principle of heterogeneity may be more implicit in the analyses; that is, mechanisms inducing variation contribute to increased heterogeneity, while selection mechanisms contribute to the reduction of heterogeneity.

Schumpeter is a natural starting point for discussing heterogeneity. His perception of innovation is seeing the production process as the process of combining various resources in specific ways. The combination of various resources, or combining 'materials and forces' as he phrased it, is a key to understand his concept of development. According to Schumpeter, development is something that basically comes 'from within' the economic system, and is related to changes in the way production is organized; that is, resources are combined in new ways. Based on this, Schumpeter introduced the concept of *new combination* (1934/96, pp. 65–6) and defined development by the introduction of a new combination. His most famous definition of entrepreneurship is related to five different ways in which new combinations may be organized: (1) introduction of a new good or quality of good; (2) introduction of a new method of production; (3) opening up a new market; (4) the conquest of a new source of supply or raw materials or half-manufactured goods, and (5) the carrying out of a new organization of industry.

The concept of *context* is fundamental for our analysis of entrepreneurship and diversity. Although Schumpeter was not very explicit in defining the context in which he analyzed entrepreneurship, and for instance did not address various and specific environmental factors that might have an influence on the entrepreneur, he very clearly related the role of the entrepreneur to the economic system. When discussing entrepreneurship, he did this in the context of the capitalist system, and his major concern was how the intervention of entrepreneurs contributed to disturbing the existing system and the current 'circular flow'.

Schumpeter's theory on economic development was originally written during the early parts of the twentieth century.[11] In this work, the role of the entrepreneur was regarded as much more important for economic development than was the case in Schumpeter's later works, in particular his work on capitalism, socialism and democracy (Schumpeter, 1943/1996). This shift in Schumpeter's view on the role of the entrepreneur demonstrates how important the systemic context is for understanding the phenomenon of entrepreneurship, and this has been an important background for developing the concept of technological regime and applying this concept in a number of evolutionary analyses. In the following we summarize some of these differences, and relate them to the concept of *technological regime* (TR).

A number of authors have discussed these principles and summarized the mechanisms of evolution. The seminal work of Nelson and Winter (1982) may represent the starting point of the evolutionary approach (Saviotti, 1997), and their approach was based on the three building blocks of organizational routines, search behavior and selection environment (Van den Bergh, 2004). A number of authors have followed up on this and summarized the principles in different ways (see for instance Aldrich, 1999; Carlsson and Stankiewicz, 1991; McKelvey, 1997; Peneder, 2001; Saviotti, 1997). Among them Edquist (see Saviotti, 1997, p. 6) has stated that evolutionary theories often include the following elements:

- The point of departure is the existence of reproduction of entities like genotypes in biology or a certain set-up of technologies and organizational forms in innovation studies.
- There are mechanisms that introduce novelties in the system (i.e. mechanisms that create diversity). They include significant random elements, but may also produce predictable novelties (e.g. purpose-oriented development work). In biology the novelties are mutations and in our context they are innovations.
- There are mechanisms that select among the entities present in the system. This increases the relative importance of some and diminishes that of others. The selection process reduces diversity and the mechanisms operation may be the 'natural selection' of biology or the 'market selection' of competition as regards technical change. Together the selection mechanisms constitute a filtering system that functions in several stages and leads to a new set-up of, for example technologies and organizational forms. There might also be feedback from the selection to the generation of new innovations.

The simple version of this is that a population, system or technology develops through mechanisms of variation and selection. Diversity is a necessary condition for change; that is, evolutionary change depends on the existence of diversity in economic actions (Peneder, 2001). Not only does this mean that an initial variety is required, the continuing creation of variety, for instance through new information, is also required, as the initial diversity will be 'consumed' during the process of selection (Peneder, 2001). This means that the variation and selection processes are working continuously and in some kind of interaction, it is no stage model or sequential model, but rather 'a sort of harmonica movement' (Van den Bergh, 2004).

Diversity is a necessary condition for evolution. The formation of new firms may be regarded as the manifestation of diversity, and it is the variety in the system that governs the pace and direction of change (Metcalfe,

2004). However, this does not mean that it will make sense to focus on diversity in isolation. Diversity per se will not create evolution, it is how the 'system' reacts to diversity that is important, and these reactions may be described through the type of selection processes that are at play. This means that evolution is determined by efforts to exploit diversity, and these efforts will depend on available competences and the system's learning processes (Cohendet and Llerena, 1997). *The continuous interplay between variety creating and selection mechanisms governs evolution.*

AN OPEN SYSTEM PERSPECTIVE

In focusing on system learning dynamics, we have to pay attention to the relation between '*Creative Accumulation*', with a low level of heterogeneity, and '*Creative Destruction*', with a high level of heterogeneity (see Figure 5.5). One could consider 'Creative Accumulation' (see Schumpeter's Mark I and II comments, 1943/1996) as a manifestation of co-specialization, an interim point between 'Creative Destruction' (Schumpeter, 1943/1996) and what we call '*Destructive Creation*' (see Figure 5.5). This move is possible through opening the issue of the specific *heterogeneities of human knowledge*. Heterogeneities of *human knowledge* – as opposed to forms of biological life, are feeding into a particular form of *knowledge 'mating'* or *knowledge 'osmosis'* (Carayannis, 2000–2005; Carayannis and von Zedwitz, 2005), which, as described by Niklas Luhmann (1984, 1994), is the act whereby humans share knowledge with each other – through shared *understanding* of what they are doing, enabling expectations and, hence, learning and adapting.

In saying this, however, we must at the same time remember that *knowledge* in this context is not an abstract substance. We are not interested in knowledge from the perspective of *stored symbols and texts*, whether it is found carved into a stone, in a library or in a hard-drive. Our interest is in *living knowledge*, used in practice for policy making and economic useful purposes, in contexts of shared understanding. Niklas Luhmann (1984, 1994) argues that shared understanding is a basic form of autonomous human self-organization, which may be seen as a form of life. This should not be regarded as allegorical. Quite the contrary, shared understanding – or social systems – is created interactively; the systems evolve through reproduction – and they die when they go out of practice.

Living knowledge is *embedded* in bodies, brains, spaces, institutions, organizations, communities of practice, as well as communities sharing formalized forms of knowledge. This is why, in analyzing the heterogeneities of knowledge, we must relate to the deeper layers of organizational, regional and institutional arrangements within which different species of interbreeding knowledge is embedded. As argued by Lam (2000), these forms of institutional embedding of knowledge are crucial in *structuring* the relations between different forms of knowledge in the processes of interactive learning leading to innovation. We wish to better understand the tension between *heterogeneity* – which opens up the way for major changes generated by random events (radical innovations) – and *structure*, which tends to give evolution a specific direction. This question is accordingly similar to another problem, that of the relation between an open and a closed system. Whereas a closed system follows its own, internally defined path – until it eventually is struck by some unforeseen disaster – an open system may adjust to changes in the environment. This, again, is a part of an even wider debate.

CO-OPETITION, CO-SPECIALIZATION AND CO-EVOLUTION OF TECHNOLOGY AND INSTITUTIONS

The discussion of the relation between *random processes* and *structures* in innovation system theory[12] was opened by Nelson and Winter (1982). They were concerned with *differences in productivity between national economies*. In their attempt to explain contemporary *differences*, they emphasized the dualism between two phenomena to be located at two different points in time: 'variation', followed by 'selection'. The corollary was '*co-evolution of technology and institutions*' (Nelson, 1991): 'Technology and the structure of industry co-evolve, and this process leads to growth in productivity, which is a statistical property of the system as a whole' (Nelson, 1991, p. 21).

His basic model of evolution took as a point of departure 'systematic selection where somewhat random variation plays a central role' (Dosi et al., 2000, p. 22). These more or less fixed structures, which Nelson and Winter – based on Abernathy and Utterback, referred to as dominant *technological paradigms*. A dominant technological paradigm may be considered as a special case, a sectorial innovation system, which selects only what fits into the paradigm from the beginning. Hence the technological paradigm starts to close. Abernathy and Utterback were criticized by Blauwhof who, based on Hughes and Latour, pointed out that in the invention process, prior to the phase of closing in on mature products described by Abernathy and Utterback, the networks of the entrepreneurs were wide open.

By including the process of invention or innovation, Blauwhof argues, Hughes and Latour identified a communicative process (*interactive learning*) where different forms of knowledge were integrated through processes of 'translation', which enabled new knowledge combinations. In this early phase of inventions, the market product (the innovation) was an abstract idea, which was redefined and transformed by actors trying to find some way into the market. The 'invention system' was *open*, as different options were tested (see Figure 5.5). Unlike the phase of invention, the mature product within an existing technological paradigm emerges with a fixed set of different forms of knowledge, linked within a specific structure. In this case the internal complexity of the innovation system may be larger than in the phase of invention, but the 'strengths of loose ties' are not as prominent as before. This is the parallel in the learning economy of forming a new species. In nature, part of a population that is specializing in a new direction may sooner or later discover that it has lost the capacity to mate with the population from which it once came. This loss of mating capacity is the birth of a new species.

CLOSED SYSTEMS OF HUMAN KNOWLEDGE?

There is some merit in claiming that there are distinct and closed knowledge species among the humans. For instance, anthropologists like Bourdieu (1977), argue strongly for the position that communities of practice, sharing tacit knowledge, are likely to disintegrate and disappear if their knowledge is integrated into modern, codified knowledge systems. The reasons for this are the radical differences between the logic of practice and the logics of codified discourse. The elements structuring systems of tacit knowledge – 'Doxa' (ancient Greek for shared tribal beliefs in a closed system view) – are likely to be discredited. This, according to Bourdieu, unleashes processes of orthodoxy – defense of Doxa – which destroy the traditional order from within.

This position may be questioned, as several authors, such as Polanyi, Granovetter and others, have identified forms of 'traditional' organization, like reciprocity, which not only prosper but also succeed in competition in modern markets, as illustrated in the case of industrial districts. Moreover, closing in completely is a special case. In the history of science, there may be paradigmatic revolutions (Kuhn, 1971), which may make old forms of scientific knowledge obsolete. Beyond this, the market as a selection mechanism is destroying forms of knowledge all the time. Among other things, knowledge destruction has to do with the speed of transformations that are generated by the market. Even though there might *in principle* be 'mating options', the speed of destruction may be too fast. Closing in too much is a mortal sin.

OPENING UP THROUGH INCREASING SYSTEM-INTERNAL HETEROGENEITY

In looking for species of specialized human knowledge, then, we must regard strategies of specialization in the context of the external threats of the market, and the capacity for agility, adaptation and turnaround. Given this rapid speed of destruction, even highly integrated clusters may – at least if they are forced to by the market – open up for new forms of knowledge. What enables this strategy is the *multi-dimensionality* of human knowledge systems. Humans may embed their various forms of specialized knowledge *in layers*. These layers may be interrelated through points of dense interactivity, where interactive learning is possible, such as organizations, regional clusters, or single humans. Here, complex processes of interactive translation and communication across different specializations are possible. For instance, two widely diverse knowledge systems may be mating inside a single human body, resulting in an entrepreneurial achievement.

Indeed, the drive towards increased specialization has resulted in the evolution of specialists who reap the benefits of crossing borders and initiating unusual or unheard of acts of mating – the *Schumpeterian entrepreneur*. This peculiar form of socializing of humans – and their entrepreneurialism – enhances the adaptability of human knowledge to changes. In nature, this option is lost – once the new species is formed, and the barriers against mating are established, there is no return. For systems that are contextualized by the global market, staying specialized should always – as pointed out by Blauwhof – be balanced by diversity – creating buffers, enabling rapid mating with new forms of knowledge if times are changing.

Opening and Closing as Strategic Options

To *open or close* may be seen as optional strategies. The standard argument for closing is:

1. Reducing external transaction costs. Instead of investing in the transaction costs involved in contacting external sources of knowledge, the system may focus on its own specialization, thus . . .
2. Avoiding internal complexity. Closing in and specializing may be seen as a strategy to avoid the internal complexity that is necessary to be able to relate to and integrate external knowledge. It is better and more efficient to make it simple, and stay specialized within a narrow niche.
3. Specialization may be profitable. Specialization in the function of the system may prove to be a profitable strategy that is rewarded by the

market, as the specialist may avoid price competition from other, less sophisticated competitors.

Strategies of specialization have their downsides as well.

1. Exposure to random events. A closed system is a system with no knowledge of its environment. It may be exposed to random and rapid destruction triggered by the market.
2. Loss of 'adaptive capacity'. A too narrow specialization may run the risk of turning the system into a unique species, which loses the mating option that may be necessary when the niche is made obsolete.

The normal case is a 'semi-open' system, which combines a *border* with a *horizon*. The border is protecting the inner core from the complexities of the environment, allowing internal specialization strategies to proceed. *By monitoring the horizon, the system is able to see what is coming, and adjust before it is too late*.

THE HORIZON: REVIEWING POTENTIAL FUTURE PARTNERS AND THREATS

The '*horizon*' is the knowledge the system has internally of its own environment. The horizon includes *potential* partners for co-opetition. It identifies relevant others, as well as some form of expectation of their functions, based on 'some form of' understanding of what they are supplying, including a memory of past experiences. Since these are external relations, these expectations are based on simplifications. Actions based on simplifications may, accordingly, lead to surprises and frustrations. A system that is capable of learning manages to combine four elements:

1. Knowledge of the environment (a horizon), including potential partners, as well as threats and opportunities not exploited at the moment.
2. *Expectations* of the environment, simplified assumptions, based on:
3. A *memory of past experiences*. The memory includes experiences of past events. It generates the ability to become frustrated in the future, if expectations are not met.
4. Ability to recognize and act on *frustration* and correct the course.

In extreme cases of simplification, the external relation may be regarded as just another standardized item to be bought in the market. All the knowledge available will be a name, classifying the service of product – and an expectation as to the standard. This is the case of a pure market relation, which does not involve any interactive learning. On the other hand, the system may relate to the external agency as another system; that is, knowledge may include some form of knowledge of the inner complexities of the other system. Then it may be possible to foresee a more interactive relation, involving a deeper sequence of co-opetition, co-specialization and co-evolution.

A *learning* system may relate flexibly to its environment. It may, for instance, decide to open up for other systems and integrate their internal complexities, or it may find out that this is just a disruption.

CONCLUSION

This chapter has been about diversity and heterogeneity in knowledge systems. In Figure 5.1 we acknowledge the knowledge system involves the presence and interactions of input, process, and output factors in the knowledge society and economy manifested via co-existence, co-opetition, co-evolution, and co-specialization processes. We have further studied and discussed the ways and means that diversity and heterogeneity – two key properties of the knowledge system – influence how knowledge is created, diffused and used. Our discussion of knowledge systems has been open-ended. We have thus attempted to provide an emerging conceptual framework to serve as the 'intellectual sandbox' and 'creative whiteboard space' of the mind's eyes of 'knowledge weavers' (*Wissensweber*)[13] across disciplines and sectors as they strive to tackle the 21st-century challenges and opportunities for socioeconomic prosperity and cultural renaissance based on knowledge and innovation.

As a result of the glocalized nature and dynamics of state-of-the-art, specialized knowledge, one needs to cope with and leverage two mutually reinforcing and complementary trends:

a. Micro–macro – the symbiosis and co-evolution of top-down national and multi-national science, technology and innovation public policies, technological paradigms and institutional complementarities, as well as bottom-up technology development and knowledge acquisition, private initiatives; and
b. The leveling of the competitive field across regions of the world via technology diffusion and adoption, accompanied and complemented by the formation and exacerbation of multi-dimensional, multi-lateral, multi-modal and multi-nodal divides (cultural, technological, socio-economic, etc.).

Key findings and contributions of the chapter can be grouped in terms of theory, methodology and empirical results as follows:

a. In terms of enriching theory, this chapter postulates that heterogeneity and diversity constitute major drivers of sustainable and potentially more equitable economic development.
b. Moreover, in advancing theory, this chapter attempts to promote the understanding of the role of heterogeneity and diversity in the inter-linkages of rationalities identities, preferences and intentions to macro-level phenomena such as institutions, national systems, regional city economies, the evolution of technological paradigms and the ways in which multi-level innovation systems work.
c. In advancing theory, as well as methodology and empirical evidence, this chapter provides some insights as to how heterogeneity, diversity, entrepreneurship and innovation can act, interact and impact on each other and their environment, as well as on individual and institutional actors in social and/or economic settings.
d. In terms of theory again, this chapter strives to advance the understanding of patterns of co-evolution and co-specialization at the micro, meso and macro levels – and in particular how micro factors, events and processes impact those at the meso and macro levels and vice versa.
e. In terms of both theory, policy and practice, this chapter in essence provides a more coherent and congruent framework for understanding and potentially anticipating the phenomena of innovation and entrepreneurship through a more profound understanding of the nature, dynamics and impacts of heterogeneity and diversity. In this manner, one could thus more effectively trigger and catalyze sustainable entrepreneurship and robust competitiveness as they are outlined in the text.

Where opportunities for further advancement of our work lie include the following points, which constitute potentially a roadmap for our future research:

a. One could better and more fully integrate our findings in the form of a more robust innovation ecosystem at the macro, meso and micro levels that can serve as a normative operationalization rather than just an interpretive framework, and thus better leverage the insights on heterogeneity's role and implications discussed in this chapter.
b. One could also better integrate aspects of public and private sector institutions and governance, promoting more efficacious policies in

support of diversity that enhances entrepreneurship- and innovation-based competitiveness.

c. One could more fully research when, how and why heterogeneity and diversity enhance rather than undermine the open innovation and entrepreneurship ecosystem.

All in all, heterogeneity and diversity remain powerful concepts that retain their enchanting capacity via their versatility, complexity and potential.

NOTES

1. Furthermore, see Milbergs (2005).
2. See discussion on democracy in the conclusion of this article.
3. '*Culture* is the invisible force behind the tangibles and observables in any organization, a social energy that moves people to act. Culture is to the organization what personality is to the individual – a hidden, yet unifying theme that provides meaning, direction, and mobilization' (Killman, 1985).
4. *Technology* is defined as that 'which allows one to engage in a certain activity . . . with consistent quality of output', the '*art of science and the science of art*' (Carayannis, 2001) or '*the science of crafts*' (Braun, 1997). We consider the following quote useful for elucidating the meaning and role of a '*knowledge nugget*' as a building block of the 'Mode 3 Innovation Ecosystem': 'People, culture, and technology serve as the institutional, market, and socio-economic "glue" that binds, catalyzes, and accelerates interactions and manifestations between creativity and innovation as shown in Figure 3, along with public–private partnerships, international Research & Development (R&D) consortia, technical/business/legal standards such as intellectual property rights as well as human nature and the "creative demon". The relationship is highly non-linear, complex and dynamic, evolving over time and driven by both external and internal stimuli and factors such as firm strategy, structure, and performance as well as top-down policies and bottom-up initiatives that act as enablers, catalysts, and accelerators for creativity and innovation that leads to competitiveness' (Carayannis and Gonzalez, 2003, p. 593).
5. The 'MODE 3' Systems Approach for knowledge creation, diffusion and use: 'Mode 3' is a multi-lateral, multi-nodal, multi-modal, and multi-level systems approach to the conceptualization, design, and management of real and virtual 'knowledge-stock' and 'knowledge-flow' modalities that catalyze, accelerate, and support the creation, diffusion, sharing, absorption, and use of co-specialized knowledge assets. 'Mode 3' is based on a system-theoretic perspective of socioeconomic, political, technological, and cultural trends and conditions that shape the co-evolution of knowledge with the 'knowledge-based and knowledge-driven, gloCal economy and society' (Carayannis and Campbell, 2006).
6. Innovation Networks are real and virtual infrastructures and infra-technologies that serve to nurture creativity, trigger invention and catalyze innovation in a public and/or private domain context (for instance, government–university–industry public–private research and technology development co-opetitive partnerships) (Carayannis and Campbell, 2006).
7. Knowledge clusters are agglomerations of co-specialized, mutually complementary and reinforcing knowledge assets in the form of 'knowledge stocks' and 'knowledge flows' that exhibit self-organizing, learning-driven, dynamically adaptive competences and trends in the context of an open systems perspective (Carayannis and Campbell, 2006).
8. '*Culture* is the invisible force behind the tangibles and observables in any organization, a social energy that moves people to act. Culture is to the organization what personality

is to the individual – a hidden, yet unifying theme that provides meaning, direction, and mobilization' (Killman, 1985).

9. *Technology* is defined as that 'which allows one to engage in a certain activity . . . with consistent quality of output', the '*art of science and the science of art*' (Carayannis, 2001) or '*the science of crafts*' (von Braun, 1997).

   We consider the following quote useful for elucidating the meaning and role of a '*knowledge nugget*' as a building block of the 'Mode 3 INNOVECO': 'People, culture, and technology serve as the institutional, market, and socio-economic "glue" that binds, catalyzes, and accelerates interactions and manifestations between creativity and innovation as shown in Figure 3, along with public–private partnerships, international Research & Development (R&D) consortia, technical/business/legal standards such as intellectual property rights as well as human nature and the "creative demon". The relationship is highly non-linear, complex and dynamic, evolving over time and driven by both external and internal stimuli and factors such as firm strategy, structure, and performance as well as top-down policies and bottom-up initiatives that act as enablers, catalysts, and accelerators for creativity and innovation that leads to competitiveness' (Carayannis and Gonzalez, 2003, p. 593).

10. See Carayannis (2004).
11. The original German version was published in 1911, while the English version was published for the first time in 1934.
12. Innovation system theory emerged out of micro-level studies of technological systems as well as middle (meso-) and macro-level studies of *innovation systems* and innovation through interactive learning (Freeman, 1987; Lundvall, 1992; Nelson, 1993; Edquist, 1997), and through several EU and OECD publications (OECD, 1999), where the NIS perspective was promoted. A critical and comprehensive review of this literature is presented in a recent publication by Miettinen (Miettinen 2002). Despite the fact that 'innovation system' is a phrase that has been on everyone's lips for the last ten years, 'system' is more often than not used as a heuristic device in the literature. To Lundvall, 'system' simply explained interactivity, seen in contrast to linear knowledge transfer. Schienstock and Hämäläinen (2001) define 'innovation system' with reference to the *function of knowledge conversion*, understood as *new knowledge creation, diffusion, and commercial utilization*, in short, *the knowledge process* (see also Carayannis and Campbell, 2005).
13. The term constitutes the brainchild or *conceptual branding* of the authors as part of this journey of discovery and ideation.

REFERENCES

Aldrich, H. (1999), *Organizations Evolving*, London: Sage Publications.

Andersen, E.S. (1994), *Evolutionary Economics. Post-Schumpeterian Contributions*, London and New York: Pinter.

Bourdieu, P. (1977), *Outline of a Theory of Practice*, Cambridge Studies in Social and Cultural Anthropology, Cambridge: Cambridge University Press.

Carayannis, E.G. (1993), *Incrementalisme Strategique*, Paris: Le Progrès Technique.

Carayannis, E.G. (1994), *Gestion Strategique de l'Apprentissage Technologique*, Paris: Le Progrès Technique.

Carayannis, E.G. (2000), 'Investigation and validation of technological learning versus market performance', *International Journal of Technovation*, **20**, 389–400.

Carayannis, E.G. (2000–2005), 'GWU lectures on technology, innovation and entrepreneurship', unpublished lectures 2000–2005, Graduate School of Business, George Washington University.

Carayannis, E.G. (2001), *The Strategic Management of Technological Learning*, Boca Raton, FL: CRC Press.

Carayannis, E.G. (2004), 'Measuring intangibles: managing intangibles for tangible outcomes in research and innovation', *International Journal of Nuclear Knowledge Management*, **1**(1–2), 49–67.

Carayannis, E.G. and J. Alexander (1999), 'Winning by co-opeting in strategic government–university–industry (GUI) partnerships: the power of complex, dynamic knowledge networks', *Journal of Technology Transfer*, **24**(2/3), 197–210. Awarded 1999 Lang-Rosen Award for Best Paper by the Technology Transfer Society.

Carayannis, E.G. and J.M. Alexander (2006), *Global and Local Knowledge: Global Transatlantic Public–Private Partnerships for Research and Technology Development*, Basingstoke: Palgrave Macmillan.

Carayannis, E.G. and D.F.J. Campbell (eds) (2005), *Mode 3: Knowledge Creation, Diffusion and Use in Innovation Networks and Knowledge Clusters: A Comparative Systems Approach across the US, Europe and Asia*, Westport, CT: Quorum Books/Greenwood Press.

Carayannis, E.G. and E. Gonzalez (2003), 'Creativity and innovation = competitiveness? When, how, and why', in Larisa V. Shavinina (ed.), *The International Handpaper on Innovation*, Amsterdam: Pergamon, pp. 587–606.

Carayannis, E.G., and P. Laget (2004), 'Trans-Atlantic innovation infrastructure networks: public–private, EU–US R&D partnership', *Journal of R&D Management*, **34**(1), 17–31.

Carayannis, E.G. D. Popescu, C. Sipp and M. Stewart (2005), 'Technological learning for entrepreneurial development (TL4ED) in the knowledge economy (KE): case studies and lessons learned', *Technovation*, **26**(4) 419–43.

Carayannis, E.G. and M. von Zedwitz (2005), 'Architecting glocal (global–local), real–virtual incubator networks (G-RVINs) as catalysts and accelerators of entrepreneurship in transitioning and developing economies: lessons learned and best practices from current development and business in incubation practices, *International Journal of Technovation*, **25**(2), February, 95–110.

Carlsson, B. and R. Stankiewicz (1991), 'On the nature, function and composition of technological systems', *Journal of Evolutionary Economics*, **1**, 93–118.

Cohendet , P. and P. Llerena (1997), 'Learning, technical change, and public policy: how to create and exploit diversity', in C.H. Edquist (ed.), *Systems of Innovation: Technologies, Institutions, and Organizations*, London: Pinter, pp. 223–41.

Dosi, G., R.R. Nelson and S.G. Winter (2000), 'Introduction', in G. Dosi, R.R. Nelson and S.G. Winter (eds), *The Nature and Dynamics of Organizational Capabilities*, Oxford: Oxford University Press, pp. 1–22.

Edquist, C.H. (ed.) (1997), *Systems of Innovation: Technologies, Institutions, and Organizations*, London: Pinter.

Freeman, C. (1987), *Technology Policy and Economic Performance: Lessons from Japan*, London: Francis Pinter.

Killman, R. (1985), *Gaining Control of the Corporate Culture*, New York: McGraw-Hill.

Kuhn, T. (1971), *The Structure of Scientific Revolutions*, Chicago: University of Chicago Press.

Lam, A. (2000), 'Tacit knowledge, organizational learning and societal institutions: an integrated framework', *Organization Studies*, **21**(3), 487–513.

Luhmann, N. (1984), *Social Systems*, Stanford, CA: Stanford University Press.

Luhmann, N. (1994), *Die Wirtschaft der Gesellschaft*, Frankfurt am Main: Suhrkamp Tassenbuch Wissenschaft.
Lundvall, B.-Å. (ed.) (1992), *National Systems of Innovation: Towards a Theory of Innovation and Interactive Learning*. London: Pinter.
McKelvey, M. (1997), 'Using evolutionary economics to define systems of innovation', in C.H. Edquist (ed.), *Systems of innovation: Technologies, Institutions, and Organizations*, London: Pinter, pp. 200–22.
Metcalfe, J.S. (1998), *Evolutionary Economics and Creative Destruction*, London and New York: Routledge.
Metcalfe, J.S. (2004), 'The entrepreneur and the style of modern economics', in G. Corbetta, M. Huse and D. Ravasi (eds), *Crossroads of Entrepreneurship*, Boston/New York/Dordrecht: Kluwer Academic Publishers, pp. 33–51.
Miettinen, M.R. (2002), *National Innovation System: Scientific Concept or Political Rhetoric*, Helsinki: Edita.
Millbergs, E. (2005), 'Presentation to National Innovation Initiative', April.
Nelson, R.R. (1993), *National Systems of Innovation: A Comparative Study*, Oxford: Oxford University Press.
Nelson, R.R. and S.G. Winter (1982), *An Evolutionary Theory of Economic Change*, Cambridge, MA and London: The Belknap Press.
OECD (1999), *Managing National Innovation Systems*, Paris: OECD.
Peneder, M. (2001), 'Dynamics of initial cluster formation: the case of multimedia and cultural content', in OECD (ed.), *Innovative Clusters. Drivers of National Innovation Systems*, Paris: OECD, pp. 303–13.
Saviotti, P.P. (1997), 'Innovation systems and evolutionary theories', in C.H. Edquist (ed.), *Systems of Innovation: Technologies, Institutions, and Organizations*, London: Pinter, pp. 180–99.
Schienstock, J. and T. Hämäläinen (2001), *Transformation of the Finnish Innovation System: A Network Approach*, SITRA report series 7, Helsinki: SITRA.
Schumpeter, J.A. (1934, 1996), *The Theory of Economic Development*, London: Transaction Papers.
Schumpeter, J.A. (1943, 1996), *Capitalism, Socialism and Democracy*, London and New York: Routledge.
Van den Bergh, J.C.J.M. (2004), 'Firm behaviour and organisation from an evolutionary perspective', in H.L.F. de Groot, P. Nijkamp and R.R. Stough (eds), *Entrepreneurship and Regional Economic Development. A Spatial Perspective*, Cheltenham, UK, and Northampton, MA, USA: Edward Elgar, pp. 15–45.
Von Braun, C.F. (1997), *The Innovation War*, Upper Saddle River, NJ: Prentice Hall.

6. Towards a communicative theory of diverse innovation systems

Finn Orstavik

INTRODUCTION

The concept of *innovation systems* has gained wide currency in the scholarly analysis of innovation (Freeman and Pavitt, 2002), and has become integrated in recent years into core policy rhetoric in the OECD and the European Union (Mytelka and Smith, 2002). The systemic approach is important for several reasons. Theoretically, it allows for the explicit analysis of diverse rationalities (Sandven, Chapter 4, this volume), and it makes it possible to argue convincingly that even radical innovation can be the outcome of human intentions and not simply the result of serendipity (Godø, Chapter 2, this volume). In policy, the systemic approach is useful as a basis for developing specific and hands-on economic, industrial, and innovation policies. For this reason, the systemic approach has been received enthusiastically by policy makers, in spite of the fact that the theoretical foundations of the concept have yet to be sufficiently developed (Acha et al., 2004; Edquist, 1997; Miettinen, 2002).

The objective of this chapter is to contribute to the conceptual foundations of a new and more robust theory of innovation systems. Contrary to Edquist's idea that the concept of innovation systems may be exploited and can be rendered useful even when renouncing theoretical rigour (Edquist, 1997), it can be argued that such rigour is essential. It is obviously important scientifically, if innovation systems theory is to be taken seriously in the future. In the longer term, theoretical rigour is also vital in order to secure the legitimacy of systemic innovation policies.

The discussion in the present chapter is related to several recent efforts that have been done to advance innovation systems theory.[1] We propose to anchor the innovation system concept more firmly in sociological theory. Geels has already made an effort in this direction. He has used the concept 'sociotechnical system' in analyses of innovation (Geels, 2004), and has discussed how technological transitions can be seen as evolutionary reconfiguration processes in such systems (Geels, 2002).

Geels' effort is important. However, further progress can be made by investigating the use of the concept of 'system' in more detail, and the assumption that the coordination of complex systems has to be hierarchical in nature. In fact, the hierarchical modelling of innovation systems, for example as bureaucratic governance structures, does not reflect adequately some of the basic properties of innovation processes. Even in contexts that (Godø, 2000) has characterized as dominated by 'technological regimes', such processes are self-organized rather than hierarchically controlled. There is a need to make a change in innovation systems theory so that the main focus is removed from an assumed strategic apex of policy making (Freeman, 1987; Okimoto, 1989) or business strategy (Porter, 1980; Wolfrum, 1991), onto the multiple strategic loci of actual innovation processes. This is the key issue to be researched in this chapter. An important question is how to proceed in order to achieve this theoretical shift. The regrouping that is proposed here consists of relating innovation systems theory to the social systems theory of Niklas Luhmann (Luhmann, 1988, 1995). Rather than subsuming innovation systems under the category of sociotechnical systems, or under Callon's concept of heterogeneous networks (Hassard and Law, 1999; Callon et al., 1986), innovation systems should be defined as a specific type of social system. Such a theory of innovation systems as a kind of social system, however, has to be based on a systematic analysis of the process of innovation. Such an analysis must concern itself with reflexivity, with learning, and with what we here call 'developmental constellations'. In this chapter we consider such developmental constellations in some detail. The further elaboration of the theory of innovation systems is, however, outside the scope of our discussion here.

ORIGINS OF THE SYSTEMIC APPROACH IN INNOVATION STUDIES AND POLICY

Innovation entered centre stage in academic research and policy debates on growth and employment during the 1980s and 1990s. In this period, the unemployment–inflation crisis of the 1970s and failed policy attempts to counter the economic downturn had undermined the belief that macroeconomic governance was in a position to secure long-term stability and growth (Freeman et al., 1982). A co-development of policy and innovation theory commenced, and this continues to form the basis for new systems-oriented innovation theory, as well as for systemic innovation policies in industrialized countries (Mytelka and Smith, 2002).

The systemic approach emerged as attempts were made to synthesize antithetical positions in earlier debates. Is it supply or demand that is the most fundamental driver of technical change (Rosenberg, 1982; Schmookler, 1966)? Is it the individual entrepreneur or the big corporation that is the engine of growth in modern economies (Dosi, 1988)? Is it the support of basic science, or targeted support to specific industries and technologies that best promotes economic growth (Nelson, 1982)? The current literature on such issues is vast. It was, however, an important milestone when the concept of a 'national innovation system' entered the mainstream of innovation theory. This happened by way of two independent volumes, edited respectively by Bengt-Åke Lundvall (1992) and Richard R. Nelson (1993).

In their later effort to delineate a genealogy of the national innovation system concept, Freeman and Lundvall have pointed out that the idea of innovation systems was not new at the time (Freeman, 1995; Lundvall et al., 2002). Freeman (1987) also demonstrated how the origins of the concept of national innovation systems can be traced back to Adam Smith (1776/1981) and Friedrich List (List and Lloyd, 2001). Tracing the theoretical roots of the concept, the works of the Austrian economist Joseph Schumpeter cannot be avoided (Hagedoorn, 1996; Schumpeter, 1976, 1982, 1983, 2006), since Schumpeter discussed how, as an economic system, capitalism was essentially an organized system of economic development and change, giving the term 'innovation' a central place in his analysis. However, some of the seminal contributions to classical social science and sociology are also relevant. The writings of Max Weber (1978), Karl Marx (1962), Ferdinand Tönnies (1988), Georg Simmel (1983), and Emile Durkheim (1964), are important. All these authors shared an interest in explaining the relationship between institutional specificities in capitalism and economic, technological and social change. In a historical perspective, capitalism had proved itself as an unrivalled dynamic force of economic and social change; the question was, what produced this exceptional dynamism?

An underlying issue in all these efforts was the difficulty of building a unified theoretical model accounting both for structural mechanisms (of capitalism) as well as for the significance of individual action. Up until today this striving to combine the individualistic conceptions of freedom and action and the understanding of structures and institutions as basic formations of social life, has presented a challenge. Niklas Luhmann's work is no exception in this respect. His ambition to exploit systems theory in sociology was motivated by a perceived crisis in sociology caused by the inability to devise a general sociological theory that could build a bridge between individualistic and structuralist conceptions of social reality. Such a general theory would, he argued, be needed in order to establish sociology more firmly as an autonomous scientific field (Luhmann and Baecker, 2002).

In sociology, Talcott Parsons had made the most fundamental effort of the post-war period to fill the gap and build a general sociological theory bridging the ideas of system, structure and individual action (Parsons, 1968, 1979). His work had, and still has, considerable impact on theoretical efforts in sociology. It is in particular relevant for us here, as Parsons significantly influenced the thinking of Niklas Luhmann (Vanderstraeten, 2002).

From the 1960s, academic debate in the field of economic and technological change was preoccupied with similar issues. A growing, qualitatively oriented literature on economic behaviour and business ventures filled important knowledge gaps left by neo-classical economic modelling. Important histories were told about the emergence of technology and technical systems (Bijker and Law, 1992; Bijker et al., 1989; Hughes, 1983, 1986; Kidder, 1981). Focus was on the role of individuals and groups, the multidimensionality of motivations for entrepreneurial action, the significance of the fact that innovation is about bringing forth something new, and the significance of this for the way in which we understand innovation processes. It became obvious that specific efforts connected to innovation have much wider significance than the short-term profitability of firms. Innovation has to do with long-term business strategy and sustainability, but it also concerns fundamental human values and politics. What do we want the world to be and to become, and who should be the ones to decide (Noble, 1977, 1986; Winner, 1977, 1986)?

ADVANCES IN INNOVATION SYSTEM THEORY

The institutional structure of industrial societies and the dynamics of capitalism had been a key concern of sociologists for more than 100 years. Nevertheless, it was Nelson and his colleagues in the USA, and Lundvall and his colleagues in Europe, who put the concept of the innovation system on the research agenda in the early 1990s. In their pioneering volumes, the focus put on the national system of innovation reflected a policy concern with the institutionalization of the R&D system and the policy-making system (Lundvall et al., 2002). Nelson and his colleagues (Nelson, 1993) analysed the performance of the system as a whole in the light of its overall institutional structuring. They were interested in the lag manifest in the US economy, and the emergence of Japan as a technological powerhouse. Lundvall was more interested in the anatomy of the relationship between academia and business. He focused on the inherent limitations of the concept of a linear relationship between science, technology and business. He was motivated to model the learning processes of capitalism in terms of communicative action rather than in the traditional fashion where science is placed at the apex of societal development and commercial business activities make up the low-level operating core (Lundvall, 1988, 1992).

The idea that there is an institutional system of innovation was attractive to policy makers, for rather obvious reasons. If the structuring of the policy system has important consequences for the ability of business to innovate and for economies to grow, then the continued development of the political system must be at the heart of any effort to strengthen national economic performance. At the national level, the structuring of the institutional system of innovation, with its ministries, policy agencies, research institutions, business associations and enterprises, became a key object of study.

However, the specific attention to institutional structuring at the level of nation states was recognized as problematic from the outset. In an increasingly globalized world, innovation performance would have to be seen not only in the context of nation states. Most obviously, regions can themselves be distinct areas of economic activity. Geographers were soon to point out that in policy as well as in theory, regional innovation systems would be an even more important object of study than the national system. Cooke developed this perspective in a number of publications (Cooke, 2002; Cooke et al., 2004). In his work, the distinction between innovation systems and clusters fades. The innovation system is seen as a geographically bounded conglomerate of commercial and non-commercial (private as well as public) organizations, interacting, pursuing partly coordinated and complementary value creation objectives, in a dynamic system of production (Orstavik, 2004).

In his effort to advance the state of the art in innovation analysis, Geels chose sectoral innovation systems theory as his point of departure (Geels, 2002, 2004). Such theories focus on the actors, the relationships, and the actual efforts to generate novelty in a specific area. In this context, 'area' is neither geographical space, nor an institutional space such as a nation state. Geels, however, understated the distinctness of sectoral innovations systems theory by stating plainly that this is just another 'level of analysis' besides the national and regional levels (Geels, 2004, p. 897). In fact, sectoral innovation systems theory abstracts the systems concept completely from geographical space. A different type of criterion for defining the specific character of the system is chosen.

Drawing attention to notable contributions by Malerba, Carlsson, Hughes and others (Breschi and Malerba, 1997; Carlsson, 1995; Hughes, 1983; Malerba, 2002), Geels showed how different authors have defined 'sector' in similar, but not identical ways.

Franco Malerba and colleagues define the sectoral system based on industry. They aim to provide a multi-dimensional, integrated and dynamic view of sectors. The sector is a set of 'individuals and organizations at various levels of aggregation, with specific learning processes, competencies, organizational structure, beliefs, objectives and behaviours . . . [that] interact through processes of communication, exchange, co-operation, competition and command' (Malerba, 2002, p. 248).

Bo Carlsson defines sector as a technological system; that is, a network of agents interacting in a specific area of technology under a particular institutional infrastructure to generate, diffuse and utilize technology (Carlsson and Stankiewicz, 1991, cited from Geels, 2004, p. 898).

Thomas P. Hughes and others analyse the integration of social and technical aspects in the evolution of large and complex technical systems such as power systems, complex machinery, and so on (Bijker and Law, 1992; Bijker et al., 1989; Hughes, 1983, 1986).

Considering these contributions, Geels made the following observations: first, that Malerba in his work on sectoral innovation systems tends to ignore the importance of the user side, and in some cases focuses exclusively on firms, ignoring other types of significant actors. Second, that technological systems theory as formulated by Carlsson and Stankiewicz reduces the system to a social system or a network, and fails to take properly into consideration the material dimension of networks. Third, that although proponents of sectoral innovation systems theory all focus on linkages between elements and co-evolution, there is little agreement on what elements and what linkages are essential (Geels, 2004).

Geels wanted to advance the state of the art in sectoral innovation systems theory by promoting four changes:

- To move beyond the sectoral systems framework, by making sociotechnical systems the proper object of study.
- In terms of elements of systems, to set the focus on the three broad categories of systems, actors and institutions.
- In terms of analytical focus, to pay much more attention to institutions.
- To address explicitly the question how sociotechnical systems change.

Geels wished to present a coherent conceptual perspective in which these four developments are realized. He presented a broad conceptual framework in which insights from sociology are exploited, in particular with regard to the dynamic and mutual structuring of individual identities and the cultural and material aspects of societies. He looked on innovation as broad sociotechnical change processes, originating in niches, potentially propagating effects over time into the overall social structure ('sociotechnical landscapes').

Geels' basic idea is sound: it is important that innovation research should remove disciplinary blinkers and consider properly the relevant contributions that have already been made in sociological theory and in systems theory. This should be done in order to properly account for both the 'supply side' and the 'demand side' in change processes. It will help in avoiding ambiguity in the application of concepts such as institutions, and also avoid being stuck in structural analysis when what is called for is the analysis of systems dynamics. However, while Geels stated the problem and the challenge for innovation theory adequately, his proposed solution is inadequate.

Our key criticism of Geels is related to his way of conceptualizing systems, in particular his proposal to frame the discussion of systems in the particular language of sociotechnical systems. This approach does not alleviate the problems he himself pointed out with respect to 'some current literatures which group together too many heterogeneous elements' (Geels, 2004, p. 898). Quite the contrary. When Geels made sociotechnical systems a basic category in his conceptual scheme, he claimed that basic elements and resources of such systems are, among other things, scientific knowledge, natural resources, machines, cultural meaning, property rights, and so on (p. 900). However, it is far from clear that such an amorphous set of elements should be considered as basic elements of sociotechnical systems. Rather, it would appear that the categories are chosen rather haphazardly

Furthermore, Geels stated that sociotechnical systems are created by human beings, not as individuals but as social groups. These groups have relative autonomy, distinct jargons, and a common culture (2004, pp. 900–1). Rather trivially, one can object to his claim that natural resources are created by human beings in social groups, and propose instead that the defining feature of natural resources is that they are resources not produced by human beings at all. More important, however, is that the concept of 'social group' is unclear. Geels introduced a definition of sociotechnical systems as 'the linkages between elements necessary to fulfil societal functions' (2004, p. 900). Among these functions are transport, communication and nutrition. Technology is a crucial element, so that the production, use and diffusion of technology are sub-functions in the sociotechnical system. This conceptual move was intended to reflect a pragmatic analytical strategy. It was not based on a fundamental differentiation of actors, rules and systems as basic elements in the real world.

In this way, Geels interpreted the term 'system' as a 'functional complex of material artefacts'. He narrowed his discussion by introducing the terms 'resources' and 'materials' in order to specify what the key elements of systems are. It is self-defeating for Geels that he chose to separate 'systems' (the technical, material aspects) from acting human beings, as this undermines the sociotechnical system concept itself. The key property of sociotechnical systems is the integration of the technical and the social aspects of the human condition.

As a final point in our critical remarks, we will point out that 'rules' by Geels were taken to be an independent reality, a systemic entity that exists outside human beings. Furthermore, he claimed that rules 'do not exist as single, autonomous entities. Instead, they are linked together and organised into rule systems. Rule systems may be purely private rule or "personality systems" or they may be collectively shared systems' (Geels, 2004, p. 904). It is not reasonable to separate rules in general from human beings and their knowledge. It is not reasonable, because rules are of many kinds, and only a limited set of effective rules are codified and formally stated. Among the most consequential rules are those that are tacit, cultural specificities and norms, and which are part of what people know without necessarily being able to state them explicitly, using language (Polanyi, 1962). Such rules are cognitive and social phenomena, and are not external to the human mind.

The conclusion is that Geels' proposal to base future research on the concept of sociotechnical systems should be rejected. A different conceptual scheme ought to be established. Such a scheme should build more directly on Lundvall's insight that a theory of innovation and innovation systems has to place interactive learning at its core. It ought also to take into account that reality in its social and material aspects is constituted in reflexive, interactive processes. It is in these transformative efforts – in what Giddens calls processes of structuration – where people and groups take each other's efforts into consideration, that innovation materializes (Giddens, 1984). Clearly, the public system of rules, regulations, organizations and policy tools influences this interaction. However, as Coriat and Weinstein (2001) show, the concept of an innovation system is more general and cannot be reduced to the institutional setup alone.

PARADIGMS OF SYSTEMS THEORY

A system is commonly understood to be a group of interacting (interrelated, interdependent) elements forming a whole. Such a whole can be static, as when symbols are arranged in a certain order that makes sense, or simply makes an impression (for example, the black and white squares of a chess board). Very often, however, the complex is a 'functional whole'; parts are related such that the totality of parts produces an output based on inputs and on systematic interactions between the elements comprising the system. An engine can be used a metaphor for this understanding of what a system is. A structure of functionally related elements, in which some parts may have a degree of freedom to move with respect to the other elements, but where the functional relationships still depend completely on the structure as a whole remaining unchanged. Within such a framework of understanding, systems dynamics is limited to the interoperation of parts within the given structure.

As we have seen, Geels sees a system in this way, and he is not alone among innovation theorists. Edquist employs a definition of systems as 'complexes of elements or components, which mutually condition and constrain one another, so that the whole complex works together, with some reasonably clearly defined overall function' (Edquist, 1997, p. 13). He points out that all the important contributions to innovation systems theory face a huge challenge when asked to specify the borders of the systems. If an innovation system is 'all important economic, social, political, organizational, institutional, and other factors that influence the development, diffusion, and use of innovations', then, 'the crucial issue . . . becomes one of identifying all those important factors' (p. 14).

In his general analysis of social systems, Niklas Luhmann develops a theory that places dynamics at the heart of the theoretical edifice rather than the problem of specifying what 'the whole' is. He argues that there are two fundamental developments in systems theory that have to be taken into consideration. These changes increase its inherent complexity, but they nevertheless make systems theory a much more promising resource for efforts to develop theory in social science (Luhmann, 1988, pp. 15–29).

The first of these changes concerns a reformulation of the basic concern of systems theory. The system is no longer to be understood as all the elements that compose the 'whole' and the relationships between them. Rather, it is the interaction between system and environment that is essential. Theory based on the whole/part dichotomy conceives the system as a fixed structure or a machine, while theory based on the system/environment dichotomy takes as its premise that there is no fixed structure and no stable engine, but rather an 'emergent whole'. Rather than trying to take a snapshot picture of elements that at a certain point in time may be said to comprise the whole system, the theoretical focus is placed on the dynamic interactions of the system and the environment.

The second fundamental change in systems theory, Luhmann argues, and the change that introduces a distinct third paradigm, consists of introducing the idea of self-organization into second-paradigm systems theory. The essential point is that certain systems are able to monitor themselves and to use information about the internal state of affairs in the system to influence the evolution of the system itself. Such a self-organizing, or autopoietic system, is neither a hierarchical structure governed from the top, nor a mechanism governed by exogenous forces (as is the case with the engine and a machinist). It is an emergent complex of elements where intelligence is distributed and governance happens locally and interactively. This is so because very complex systems cannot but rely on local self-governance. No central coordination mechanism will be able to rival the information processing and exploitation capability of a non-hierarchical, distributed system (Luhmann, 1988, p. 27 and pp. 405–6).

KNOWLEDGE DYNAMICS AND INNOVATION

Knowledge is obviously important for innovation. How knowledge should be accounted for in theories of technical and economic change has long been recognized as an important issue. Following Arrow's seminal contributions in mainstream economics (Arrow, 1962a, 1962b), knowledge has largely been dealt with as a public good with particular characteristics. In Arrow's work, as well as in most of the subsequent economic literature, knowledge has been equated with information (Tunzelmann, 1995). That is, knowledge is seen as existing externally of human beings and as something that human beings are able to create, store and transfer.

Lundvall and colleagues point out how the traditional treatment of knowledge as information is problematic (Lundvall et al., 2002). However, although they do not equate knowledge with information, they continue to deal with knowledge as external to human beings, and emphasize that research on innovation systems should focus more on 'business services that specialise in producing, gathering, and selling knowledge' (p. 224). In addition, they state that 'innovation systems work through the introduction of knowledge into the economy' (p. 225). Such terminology concerning knowledge and learning is inadequate.

The problem with equating knowledge and information has been recognized by many authors. In their effort to formulate different perspectives to the rivalling Keynesian active-state policies on the one hand and neo-liberal tenets of laissez-faire on the other, Nelson and Winter positioned evolutionary theory as an alternative framework for the analysis of innovation and economic growth (Nelson and Winter, 1982). They showed that in trying to explain economic action and technological change, evolutionary theory cannot but concern itself with learning. However, the authors themselves avoided a thorough discussion of knowledge and learning. Their key concern was to argue a certain stability of individual economic action, within a framework of bounded rationality. In their version of evolutionary theory, it was routines that came to play the pivotal role of providing stability.

In the biological realm, one can claim that animals and plants do not learn, in the sense that genetic dispositions in individuals do not change in response to environmental impacts. Nelson and Winter argued that a similar lack of ability to learn is important in the economy. Routines form an indispensable glue in the economic system. The stability provided by routines allows selection mechanisms to do their work. In a competitive situation, better routines allow firms to survive, while firms with less powerful routines perish.

While it may be tempting to hypothesize the dominance of quasi-biological selection mechanisms in social and economic life, it is obvious that such mechanisms cannot be the only ones governing change processes in these domains. The ability of human beings and their organizations to learn has to be accounted for. Human beings reflect on experiences, evaluate situations, and use their cognitive and practical skills to adjust their behaviour. Not only do they adjust their own behaviour within a given definition of a situation, they also adapt cognitively and reflexively, and they improve the understanding of their own situation in interactions with other human beings. That human beings define their situations interactively and reflexively sets social systems apart from biological systems. Hence, simple biological models of evolution cannot be applied. In addition, this also creates conditions for systems created by human beings to be self-organizing.

Wrapping up this part of the discussion, we agree with Loasby that while reaping valuable insight from models of evolution, evolutionary economic theory must focus on the issues of knowledge and learning, in a different manner to that taken by neo-Darwinian biologists (Loasby, 2002). Learning becomes a key factor in the analysis of how economic systems and economic actions develop over time, and the way knowledge is understood becomes a critically important issue.

When knowledge is treated as if it was a concrete object existing outside and independently of the minds of human beings, the relationship between the 'knowledge objects' and human beings is taken to be binary. You can either have a specific knowledge, or not have it. Furthermore, with the help of language and code systems, knowledge can be stored as information, outside the human body and the human brain. Knowledge can be acquired by absorbing such information. Hence, knowledge can be produced, diffused and absorbed, in a similar way as any other commodity (good).[2]

The basic problem with this line of argument is the thesis that knowledge can be externalized from the minds of human beings, and that it can be transferred between human beings 'as is'. In his work on social systems, Luhmann (and a number of phenomenologists before him) has shown that the acquisition of knowledge cannot be reduced to the transfer of information (Joas, 1989; Luhmann, 1988; Mead, 1934). Communication concerns the transmission and receiving of information, of meaningful data. In an important sense, information sent is not the same as information received. This is because the meaningfulness of data has to be constructed by each individual in their social context. This reconstruction depends on a continuing recursive communication process in which assumptions about intended meaning are tested (Luhmann, 1988, p. 196). Communication is an interactive process that may converge on a shared point of view. However, in this interactive process *new knowledge is being constructed.*

It follows from this that much of what is usually considered to be 'knowledge' and 'knowledge production', is actually communication. For instance, the writing down of a text must be called 'communication' rather than 'knowledge production', whereas books and libraries as well as the World Wide Web with its contents, are communication devices rather than 'knowledge objects', or 'items of codified knowledge'. In the terminology that is proposed here, knowledge is not out there as objects external to living human beings: knowledge is in people's minds and in their problem-solving capabilities and skills only. This implies that in its essence knowledge is the *capacity to handle complexity* (Luhmann, 1988, pp. 156–60). Knowledge is the ability of individuals to deal with complex situations and to deal with a complex environment.

Schumpeter pointed out how innovators create something new on the basis of an old order, and how the new stands not only on the shoulders of the old, but on the ruins of some significant parts of this old order. Innovation is not only creative, it is also destructive (Schumpeter, 1983). What is important for an innovator is to develop abilities to cope with what is initially unfamiliar. The complexity of a new environment cannot initially be handled effectively.

It is important to emphasize the reflexive and cognitive content of this creative process. It is in the dialogue and the negotiation of meanings and decisions that the results find their form. Hence, communication has to be a crucial element in any attempt to construct a general model of innovation and innovations systems. However, communication, knowledge and learning are intimately connected to observable realities such as established work practices and routines, formal rules and regulations, infrastructures, technical systems, and so forth. To apply a well known and somewhat clichéd formulation, innovation does not happen, and knowledge is not created, in a vacuum. Learning takes place in interactions between people, and between people and artefacts. Learning is always influenced by established practices and the perception of normative elements in human culture (Lesser et al., 2000; Wenger, 1999).

DEVELOPMENTAL CONSTELLATIONS

The dynamics of innovation processes have been described in a compelling way by Van de Ven and his colleagues, and in an equally compelling way by social constructivists focusing on heterogeneous network building (Bijker et al., 1989; Bijker and Law, 1992; Latour, 1987; Latour and Woolgar, 1979; Van de Ven et al., 1989; Van de Ven et al., 1999). While Van de Ven and his colleagues have mapped processes in real time following innovation projects as they unfold, constructivist writers have as a rule employed a more historical research method. The studies show that individuals and groups have to involve actors around them in interactive processes of communication. In this interaction, perspectives and interests are negotiated, meaning is stabilized, and people and artefacts made to interoperate in novel ways.

Elements from different domains of reality (psychic, social, technical, etc.) are interrelated and can be said to contribute to each other's 'emerging as established reality' in the course of innovation processes. This idea is contained, for example, in Thomas Hughes' thesis about the Seamless Web (Hughes, 1986), and in Michel Callon's (and others') theory that innovation processes are exercises in heterogeneous network building (Rip et al., 1986).

Even though Hughes, Callon and many others may tend to conceive of innovation processes as coordinated and driven from a single strategic apex, this is not a necessary precondition for their network perspective to make good sense. The idea of a visionary scientist or entrepreneur as a driver of innovation may be enticing, since it makes it easy to explain why change processes take place in spite of all the obstacles that must be overcome. What drives the innovation processes, if not single individuals or firms with a particular entrepreneurial or Napoleonic spirit (Schumpeter, 1983, 2006)?

What is wrong with the idea of an entrepreneur or a visionary scientist forming a strategic apex for control and coordination is not the idea of the visionary scientist or the entrepreneurial spirit. The problem is the belief that there is a singular person or a monolithic system of hierarchical control that governs the innovation process. There are many actors and many drivers in innovation processes. Innovation is carried out in an interplay between people and organizations, in contexts where technologies, infrastructures and cultural conditions are both enabling resources, yet, at the same time, obstacles to change. Geels proposed distinguishing between *technical systems*, *actors* and *rules*. Instead, the proposal is made here that theorizing should be based on the five basic elements in innovation that we have highlighted in our analysis. These are: people, communication, routines, rules and materials.

- People – that is, individuals and groups of human beings – act as innovators, making use of their knowledge and their practical competences.
- Communication is the compound process in which data are exchanged, meaningful information is developed, and knowledge is interactively constructed. The communication we are interested in here emanates from people and groups that engage in innovation efforts.
- Routines are the relevant regularities in behaviour – contextually embedded procedures (actions and action sequences) that are repeated in a similar way over time.
- Rules are the codified descriptions of behavioural procedures that innovators (and others that are relevant for the innovation effort) live by. Rules are developed internally in organizations, or made effective for the functioning of individuals, groups, and organizations by external agents. However, non-codified, informal rules and norms are part of what people know. Here, this is taken to be part of the complexity-handling capacities of people; that is, knowledge, and not part of what we define as rules.
- Materials are relevant materials, artefacts, technologies and technical systems, infrastructure, and so on.

Of course, these elements can be seen as constituent parts of any society, and any economy. They are found in different configurations in different places. Configurations change over time. In the most general sense, innovation is the reconfiguration process in which these basic elements and the relationships between them are transformed. The core of innovation, then, is a co-production of change in which new configurations of the basic elements emerge. People, communication, rules, routines, and materials co-develop and shape each other in non-linear ways during the development process. Communication plays a pivotal role in these dynamics, as it forms the medium in which processes can be sustained.

Lacking more adequate means, we illustrate these dynamics in a somewhat primitive way, in Figure 6.1.[3]

Figure 6.1 can be read as an abstract rendering of the five basic elements of innovation that we have introduced. However, it may also be interpreted as depicting one actor in innovation, on the level of aggregation that we choose. In the following, we focus on business innovation. We assume that Figure 6.1 represents one organizational entity in the innovation process, such as a firm. Innovation in the firm involves changes in technology and materials, and in the routines and rules of work processes. In addition, innovation involves employees learning new skills and also having to face the fact that established skills become obsolete. Communication is the key to bringing forth and sustaining the change processes.

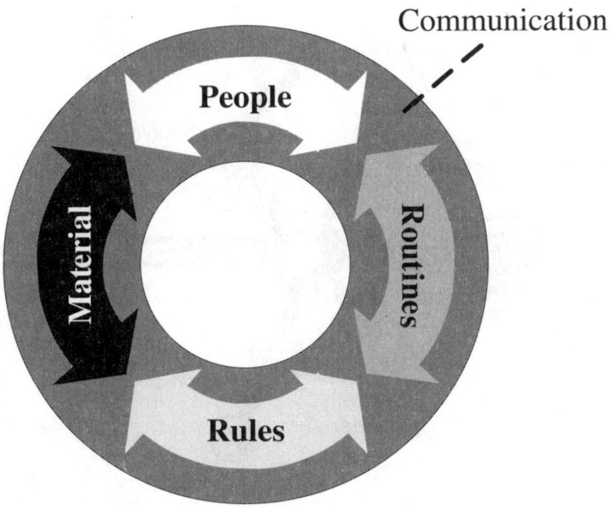

Figure 6.1 Elements of innovation

Whether the innovator initially is an individual, a business enterprise, or another type of organization, innovation cannot rely solely on intra-mural reconfigurations. In the same way as an individual is unable to innovate without involving others, so is any business firm (or another type of organization) dependent on others. The innovating actor is embedded in a specific context. The 'significant others' for an innovating firm are those other companies, government institutions, research milieus, NGOs, or other entities, that are relevant for effecting the specific changes that the innovator sets out to realize. These actors form a heterogeneous group. In order for innovation to be successful, communication must be effective in a way that allows all of them to adjust their efforts to be compatible with the changes that are generated elsewhere in the group. The structure is clearly also a network. However, in order to highlight the fundamental importance of the heterogeneity of actors and interfaces, and the dynamic nature of interrelationships, the term 'network' is not employed here. Instead, the emergent group is defined as the 'developmental constellation' of the innovation process. The term 'developmental' is used to emphasize the double dynamic that is present. The constellation is established in order to generate change, and at the same time the constellation itself changes in the process. Figure 6.2 is a simple illustration of such a constellation.

The number of relevant partners, interfaces and interactions taking place may obviously be very high in real innovation processes. Only a limited number of interface activities are shown in Figure 6.2, as anything

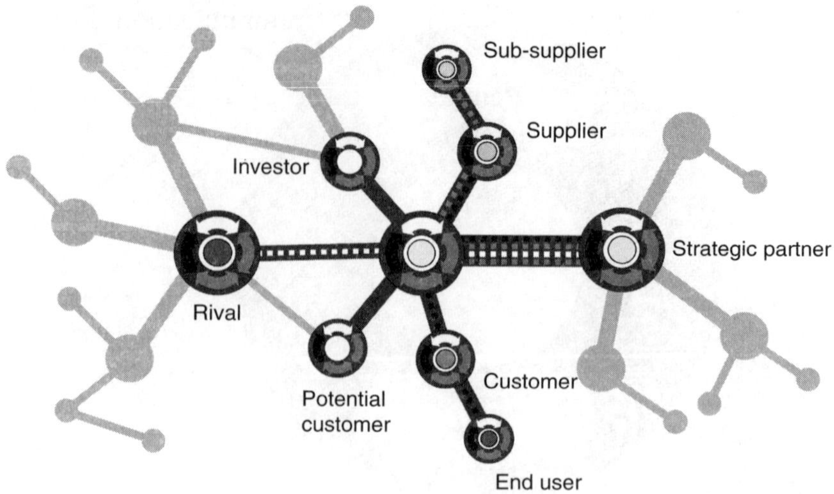

Figure 6.2 The developmental constellation

approaching a 'realistic' illustration would become exceedingly complex. The system of communication that constitutes the backbone of the innovation process as a systemic phenomenon involves every partner in the developmental constellation, but in a complex way: far from all partners are connected with every other partner by means of direct links.

Figure 6.2 is intended to visualize the emerging developmental constellation at one point in time. Over time, people, materials, rules, routines and communication itself change in non-linear ways. As the innovation process progresses, the developmental constellation may become broader, more differentiated and more complex. At other times in its life cycle, there might be shakeouts, restructuring and reorientation with respect to the partners involved, their positioning and their understanding of the development path that they are following.

Figure 6.3 shows one developmental constellation embedded in its institutional context. Innovators are influenced by, and themselves influence, a number of actors in their environments. Each actor plays multiple roles, for instance concurrent roles as customer and supplier, and as client of public agencies. Several societal sub-systems may have a significant impact on the efforts of innovators. Among these are clearly the legal system, the system of education and research, and the political system. Public institutions and government agencies may also have an impact on a large number of developmental constellations, whether this influence is intended or not.[4]

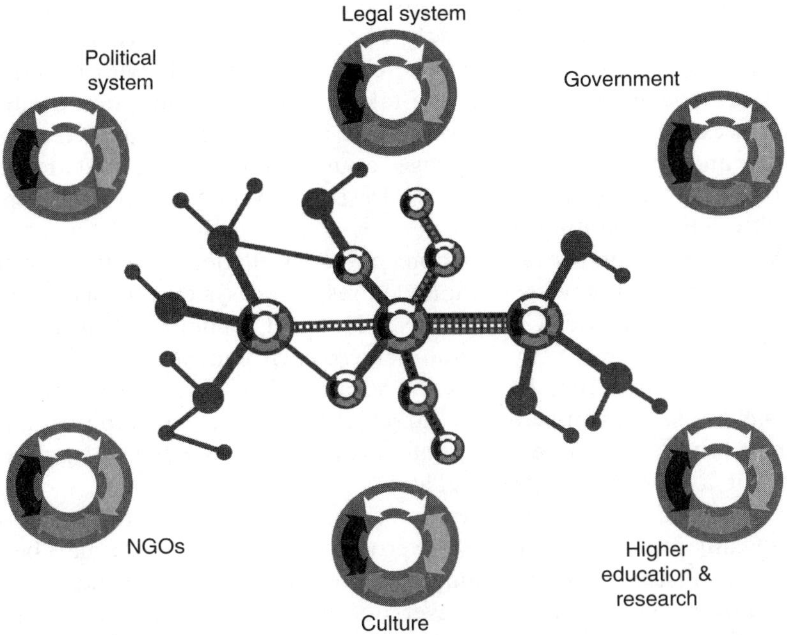

Figure 6.3 The developmental constellation in its institutional context

The developmental constellation can be seen as akin to an industrial cluster (Beccatini, 1978; Piore and Sabel, 1984; Porter, 1998a, 1998b), as well as to what Malerba defines as a sectoral innovation system (Breschi and Malerba, 1997; Malerba, 2002). Contrary to these, however, the developmental constellation is focused on just a single specific, functionally integrated set of innovation efforts. In addition, the institutions that are relevant in the environment of a specific developmental constellation at one point in time may be only a small sub-set of the institutions that exist in the sector where the developmental constellation is operating, and the actors in the constellations may be located across completely different sectors. Communication establishes the essential links between all the involved partners, and relevance with respect to the ongoing innovation effort is important, not that actors belong to the same cluster or the same sector.

MULTI-LEVEL HETEROGENEITY

The complexity and the dynamics of the overall communicative situation force repeated selection with respect to which interfaces should actually be developed in a developmental constellation. Only a limited number of partners can be dealt with effectively. The developmental constellation may be characterized analytically by the choices being made with respect to interface operations by the actors taking part in the innovation effort. Interfaces may be easier to develop and to maintain when partners are similar and 'speak the same language', as understanding and trust are more readily available in such situations. However, diversity is an obvious precondition for generating novelty.

A very large number of developmental constellations may be operative in a dynamic and growing economy. In respect of systems theory, each of these has its own specific environment. However, a number of institutions are important elements in several of these environments. Attention from public institutions may be a scarce resource that diverse developmental constellations are competing to attract. It is, however, an empirical question to what extent a specific public institution constitutes 'a significant other' in a developmental constellation.

The overall ability of developmental constellations to bring out significant results from a very heterogeneous set of actors may be an essential factor in determining the innovation performance of an economy. Heterogeneity is, however, not only important inside constellations. Such constellations will have diverse objectives, different kinds of involved actors, and diverse resource needs. While a number of developmental constellations are unrelated, several will be related, directly or indirectly. For example, there may be constellations that pursue quite similar goals, and that depend on some of the same resources. In other cases, one developmental constellation may produce effects that are beneficial for the efforts done by one or more other developmental constellations.

This shows that innovators may not only benefit from information about efforts of actors in the development constellation of which they are part. In order to manage innovation efforts effectively, an understanding of the strategic position where one is located with respect to other development efforts, will often be crucial. The ability to look ahead and to predict who will be potential allies and who will be rivals some time in the future may be as important in innovation as predicting the opponent's next moves in a game of chess.

CONCLUSION

Taking stock of innovation systems research, Lundvall and his associates made the following statements:

> a principal task for future research based on the concept of national systems of innovation is to adapt it in such a way that it . . . helps to stimulate policy learning. We will argue that a major step in this direction is to broaden and deepen the concept and make it more dynamic. A narrow focus on the role of science and science-based activities is not what is most needed. We need a concept that covers all aspects of competence building in socio-economic activities. We also need to deepen the concept by getting a better understanding of processes of interactive learning. Finally, we need to find ways to capture the formation and evolution of innovation systems from their birth to their death. (Lundvall et al., 2002, p. 216).

In this chapter, such issues have been approached by challenging established ways of thinking about innovation systems. It has been argued that theories of innovation systems must deal with systems and their environments, and in terms of self-organization, rather than continuing to define a systemic 'wholes' based on parts.

The combination of sociology, systems theory and innovation theory is a very powerful alliance and a most promising venue for strengthening the innovation systems concept, both theoretically and as a policy tool. Geels has been praised for understanding this, but his conceptual approach has been criticized.

Innovation processes are not controlled from one strategic apex: innovation is carried out in developmental constellations. The direction of development, the resultant force of innovation efforts, is an aggregate of interrelated efforts to produce viable change. While both government and academic science may play vital roles, none of these can be heralded as the 'mother of all innovation'. They do not enter into all developmental constellations, and they do not play the same role in all the constellations that they actually come into contact with. It is in the creative processes of interactive learning, inside developmental constellations and across them, that the future is being produced.

How should one proceed in order to achieve the theoretical shift in innovation systems theory in which the focus is removed from an erroneously assumed strategic apex of innovation, on to the multiple strategic loci of actual innovation processes? Here, innovation systems theory has been related to the social systems theory of Niklas Luhmann. Rather than subsuming innovation systems under the category of sociotechnical systems, or seeing them as heterogeneous networks under construction, the proposed solution is to define innovation systems as a specific type of social system. Such social systems are anchored in what has here been defined as 'developmental constellations'. In line with this, innovation systems could be defined as *communicative, self-organized systems anchored in developmental constellations of heterogeneous sets of actors*. How this should be done more specifically, however, can only be referred to at this point as a topic for further research.

NOTES

1. Important strands of the research have been summarized in editorials in *Research Policy*. See vol. 31 (2002), pp. 185–90, and vol. 33 (2004), pp. 1253–8 (Acha et al.).
2. It is a particular good, since knowledge does not disappear as it is consumed, and it might not lose its value for one person even if it is consumed by others. Such properties provide the basic economic arguments for public support of science.
3. Only primitive, static illustrations are supplied in this chapter. More adequate illustrations would involve dynamic renderings and simulations. Relevant work is currently being undertaken to create software that can visualize the complex processes we are interested in (Moody et al., 2005).

   It may be appropriate to point out that neither knowledge nor money are specified as basic elements. Being defined as the capacity of human beings to handle complexity, knowledge is in a quite concrete sense embodied in the people who innovate. Other parts of what is usually considered to be knowledge, such as structured information in books, 'knowledge' embedded in artefacts, in organizational rules, and so on, are not ignored. Rather, such elements are subsumed under the headings 'communication', 'routines' and 'rules'. As for money, this is considered but a medium of communication. Hence, money matters are considered to be an element in the system of communication.
4. The idea of a third generation, horizontal innovation policy focuses on the need of all areas in a sector-divided policy system to take into account policy effects with respect to innovation (Legrand et al., 2002).

REFERENCES

Acha, V., O. Marsili and R. Nelson (2004), 'Editorial, Special Issue: What do we know about innovation?', *Research Policy*, **33**(9), 1253–8.

Arrow, K. (1962a), 'The economic implications of learning by doing', *Review of Economic Studies*, **29**, 155–79.

Arrow, K. (1962b), 'Economic welfare and the allocation of resources for invention', in R. Nelson, *The Rate and Direction of Inventive Activity*, Princeton, NJ: Princeton University Press.

Beccatini, G. (1978), 'The development of light industry in Tuscany: an interpretation', *Economic Notes*, **2**(3), 107–23.

Bijker, W.E., T.P. Hughes and T.J. Pinch (eds) (1989), *The Social Construction of Technological Systems*, Cambridge, MA: MIT press.

Bijker, W. and J. Law (1992), *Shaping Technology/Building Society*, Cambridge, MA: MIT Press.

Breschi, S. and F. Malerba (1997), 'Sectoral innovation systems', in C. Edquist (ed.), *Systems of Innovation: Technologies, Institutions and Organisations*, London: Pinter Publishers.

Callon, M., J. Law and A. Rip (1986), *Mapping the Dynamics of Science and Technology: Sociology of Science in the Real World*, Basingstoke: Macmillan.

Carlsson, B. (ed.) (1995), *Technological Systems and Economic Performance. The Case of Factory Automation*, Doordrecht: Kluwer.

Carlsson, B. and R. Stankiewicz (1991), 'On the nature, function and composition of technological systems', *Journal of Evolutionary Economics*, **1**(1), 93–118.

Cooke, P. (2002), *Knowledge Economies: Clusters, Learning and Cooperative Advantage*, London: Routledge.

Cooke, P., M. Heidenreich and H.-J. Braczyk (eds) (2004), *Regional Innovation Systems. The Role of Governance in a Globalized World*, London and New York: Routledge.

Coriat, B. and O. Weinstein (2001), 'Organizations, firms and institutions in the generation of innovation', *Research Policy*, **31**(2), 273–90.

Dosi, G. (ed.) (1988), *Technical Change and Economic Theory*, London: Pinter Publishers.

Durkheim, E. (1964), *The Division of Labor in Society*, New York: The Free Press.

Edquist, C. (ed) (1997), *Systems of Innovation: Technologies, Institutions and Organisations*, London and Washington: Pinter Publishers.

Freeman, C. (1987), *Technology Policy and Economic Performance. Lessons from Japan*, London and New York: Pinter Publishers.

Freeman, C. (1995), 'The "National systems of innovation" in historical perspective', *Cambridge Journal of Economics*, **19**, 5–25.

Freeman, C., J. Clark and L. Soete (eds) (1982), *Unemployment and Technical Innovation*, London: Pinter Publishers.

Freeman, C. and K. Pavitt (2002), 'Editorial, Special Issue NELSON + WINTER + 20', *Research Policy*, **31**(8–9), 1221–6.

Geels, F.W. (2002), 'Technological transitions as evolutionary reconfiguration processes: a multi-level perspective and a case-study', *Research Policy*, **31**, 1257–74.

Geels, F.W. (2004), 'From sectoral systems of innovation to socio-technical systems. Insights about dynamics and change from sociology and institutional theory', *Research Policy*, **33**(6–7), 897–920.

Giddens, A. (1984), *The Constitution of Society. Outline of the Theory of Structuration*, Cambridge: Polity Press.

Godø, H. (2000), 'Innovation regimes, R&D and radical innovations in telecommunications', *Research Policy*, **29**, 1033–46.

Hagedoorn, J. (1996), 'Innovation and entrepreneurship: Schumpeter revisited', *Industrial and Corporate Change*, **5**, 883–96.

Hassard, J. and J. Law (1999), *Actor Network Theory and After*, Oxford: Blackwell.

Hughes, T.P. (1983), *Networks of Power: Electrification in Western Society, 1880–1930*, Baltimore, MD: Johns Hopkins University Press.

Hughes, T.P. (1986), 'The seamless web: technology, science, etcetera, etcetera', *Social Studies of Science*, **16**(May), 281–92.

Joas, H. (1989), *Praktische Intersubjektivität. Die Entwicklung des Werkes von G.H. Mead*, Frankfurt am Main: Suhrkamp.

Kidder, T. (1981), *The Soul of a New Machine*, London: Penguin Books.

Latour, B. (1987), *Science in Action: How to Follow Scientists and Engineers through Society*, Milton Keynes: Open University Press.

Latour, B. and S. Woolgar (1979), *Laboratory Life. The Construction of Scientific Facts*, Princeton, NJ: Princeton University Press.

Legrand, L. and Associates (2002), 'Innovation tomorrow. Innovation policy and regulatory framework: making innovation an integral part of the broader structural agenda', *Innovation papers*, Report for the DG Enterprise, Luxembourg: Office for Official Publications of the European Communities, p. 218.

Lesser, E.L., M.A. Fontaine and J.A. Slusher (eds) (2000), *Knowledge and Communities*, Boston, MA: Butterworth and Heinemann.

List, F. and S.S. Lloyd (2001), *The National System of Political Economy*, Kitchener, ON: Batoche.

Loasby, B.J. (2002), 'The evolution of knowledge: beyond the biological model', *Research Policy*, **31**(8–9), 1227–39.

Luhmann, N. (1988), *Soziale Systeme. Grundriss einer Allgemeinen Theorie*, Frankfurt am Main: Suhrkamp Verlag.

Luhmann, N. (1995), *Social Systems*, Stanford, CA: Stanford University Press.

Luhmann, N. and D. Baecker (eds) (2002), *Einführung in die Systemtheorie*, Heidelberg: Carl-Auer-Systeme Verlag.

Lundvall, B.-Å. (1988), 'Innovation as an interactive process: from user–producer interaction to national systems of innovation', in C. Dosi et al., *Technical Change and Economic Theory*, London: Pinter Publishers, pp. 349–69.

Lundvall, B.-Å. (ed.) (1992), *National Systems of Innovation*, London: Pinter Publishers.

Lundvall, B.-Å., B. Johnson, E.S. Andersen and B. Dalum (2002), 'National systems of production, innovation and competence building', *Research Policy*, **31**(2), 213–31.

Malerba, F. (2002), 'Sectoral systems of innovation and production', *Research Policy*, **31**(2), 247–64.

Marx, K. (1962), *Das Kapital. Kritik der Politischen Ökonomie*, Berlin: Karl Dietz Verlag.

Mead, G.H. (1934), *Mind, Self and Society from the Standpoint of a Social Behaviourist*, Chicago: The University of Chicago Press.

Miettinen, R. (2002), *National Innovation Systems: Scientific Concept or Political Rhetoric*, Helsinki: Edita.

Moody, J., D.A. McFarland and S. Bender-deMoll (2005), 'Dynamic network visualization', *American Journal of Sociology*, **110**(4), 1206–41.

Mytelka, L.K. and K. Smith (2002), 'Policy learning and innovation theory: an interactive and co-evolving process', *Research Policy*, **31**(8–9), 1467–79.

Nelson, R. (ed.) (1982), *Government and Technical Progress. A Cross-industry Analysis*, New York and Oxford: Pergamon Press.

Nelson, R. (ed.) (1993), *National Innovation Systems*, New York: Oxford University Press.

Nelson, R. and S. Winter (1982), *An Evolutionary Theory of Economic Change*, Cambridge, MA: The Belknap Press.

Noble, D.F. (1977), *America by Design*, Oxford: Oxford University Press.

Noble, D.F. (1986), *Forces of Production. A Social History of Industrial Automation*, Oxford: Oxford University Press.

Okimoto, D.I. (1989), *Between MITI and the Market. Japanese Industrial Policy for High Technology*, Stanford, CA: Stanford University Press.

Orstavik, F. (2004), 'Knowledge spillovers, innovation and cluster formation: the case of Norwegian aquaculture', in C. Karlsson et al. (eds), *Knowledge Spillovers and Knowledge Management*, Cheltenham, UK and Northampton, MA, USA: Edward Elgar.

Parsons, T. (1968), *The Structure of Social Action*, New York: The Free Press.

Parsons, T. (1979), *The Social System*, London: Routledge & Kegan Paul.

Piore, M.J. and C.F. Sabel (1984), *The Second Industrial Divide. Possibilities for Prosperity*, New York: Basic Books.

Polanyi, M. (1962), *Personal Knowledge. Towards a Post-Critical Philosophy*, Chicago: The University of Chicago Press.
Porter, M. (1980), *Competitive Strategy. Techniques for Analyzing Industries and Competitors*, New York: The Free Press.
Porter, M. (1998a), 'Clusters and the new economics of competition', *Harvard Business Review*, **76**(6), 77–90.
Porter, M. (1998b), *The Competitive Advantage of Nations: With a New Introduction*, Basingstoke: Macmillan Business.
Rosenberg, N. (1982), *Inside the Black Box: Technology and Economics*, Cambridge: Cambridge University Press.
Schmookler, J. (1966), *Invention and Economic Growth*, Cambridge, MA: Harvard University Press.
Schumpeter, J. (1976), *Capitalism, Socialism and Democracy*, London: George Allen & Unwin.
Schumpeter, J. (1982), *Business Cycles: A Theoretical, Historical, and Statistical Analysis of the Capitalist Process*, Philadelphia: Porcupine Press.
Schumpeter, J. (1983), *The Theory of Economic Development: An Inquiry into Profits, Capital, Credit, Interest, and the Business Cycle*, New Brunswick, NJ: Transaction Books.
Schumpeter, J. (2006), *Theorie der Wirtschaftlichen Entwicklung. Nachdruck der 1. Auflage von 1912*, Berlin: Duncker & Humblot.
Simmel, G. (1983), *Soziologie: Untersuchungen über die Formen der Vergesellschaftung*, Berlin: Duncker & Humblot.
Smith, A. (1776/1981), *An Inquiry into the Nature and Causes of the Wealth of Nations*, Indianapolis: Liberty Fund.
Tönnies, F. (1988), *Community and Society (Gemeinschaft und Gesellschaft)*, New Brunswick, NJ: Transaction Publishers.
Tunzelmann, G. (1995), *Technology and Industrial Progress. The Foundations of Economic Growth*, Aldershot, UK and Brookfield, USA: Edward Elgar.
Van de Ven, A., H. Angle and M.S. Poole (eds) (1989), *Research on the Management of Innovation: The Minnesota Studies*, New York: Harper & Row.
Van de Ven, A.H., D.E. Polley, R. Garud and S. Venkataraman (1999), *The Innovation Journey*, New York: Oxford University Press.
Vanderstraeten, R. (2002), 'Parsons, Luhmann and the theorem of double contingency', *Journal of Classical Sociology*, **2**(1), 77–92.
Weber, M. (1978), *Economy and Society*, Berkeley, CA: University of California Press.
Wenger, E. (1999), *Communities of Practice. Learning, Meaning and Identity*, Cambridge: Cambridge University Press.
Winner, L. (1977), *Autonomous Technology. Technics-out-of-Control as a Theme in Political Thought*, Cambridge, MA: MIT Press.
Winner, L. (1986), *The Whale and the Reactor. A Search for Limits in an Age of High Technology*, Chicago and London: The University of Chicago Press.
Wolfrum, B. (1991), *Strategisches Technologiemanagement*, Mainz: Gabler Verlag.

7. Entrepreneurship and heterogeneity

Olav R. Spilling

INTRODUCTION

The purpose of this chapter is to discuss heterogeneity and entrepreneurship, and how entrepreneurship works as a heterogeneity-increasing mechanism as well as a heterogeneity-decreasing mechanism. The discussion is inspired by evolutionary economics as well as recent developments in more 'standard' economics in which there has been a growing interest in the issue of heterogeneity. However, turning to many of the basic works within the tradition of evolutionary economics and population ecology, one finds very little on the concept of heterogeneity (see for instance, Aldrich, 1999; Hannan and Freeman, 1989; Magnusson and Ottosson, 1997; Metcalfe, 1998; Nelson and Winter, 1982), although the principle of heterogeneity may be regarded as an underlying precondition for many of the evolutionary approaches (Andersen, 1994).

While the key focus in analysing evolution is on mechanisms related to variation and selection, the principle of heterogeneity may be more implicit in the analyses; that is, mechanisms inducing variation contribute to increased heterogeneity, while selection mechanisms contribute to reduced heterogeneity.

If there is little explicit focus on heterogeneity in the tradition of evolutionary economics, there is even less within the tradition of entrepreneurship research, where a main tendency is to analyse entrepreneurship as individualized processes in which the systemic dimension often is absent. Of course, there are exceptions to this, such as in the analysis of entrepreneurship related to the environment, and analyses of entrepreneurship based on a population ecological approach (Aldrich, 1999), or entrepreneurship analysed in the context of technological regimes (Audretsch and Acs, 1990, 1991). However, the concept of heterogeneity – or diversity as is often discussed as synonymous to heterogeneity – is scarcely mentioned. To the extent that the concept is discussed, the focus is less explicit. Against this background, it is of interest to present a systematic discussion of entrepreneurship related to the principle of heterogeneity and to develop an understanding of how heterogeneity is related to processes of entrepreneurship. Heterogeneity may be partly regarded as an input variable affecting entrepreneurial activity and performance, and partly regarded as the outcome of entrepreneurial activity. In line with what seems to be the tendency in many evolutionary approaches where the focus is directly on heterogeneity, this chapter focuses on evolutionary processes in general, and discusses how entrepreneurship should be understood within this framework. Based on this, we then discuss heterogeneity as a factor influencing entrepreneurial processes as well as a factor influenced by processes of entrepreneurship.

Taking the Schumpeterian understanding of the role of the entrepreneur in economic change as our point of departure, we start by discussing the concept of entrepreneurship, and the phenomenon of entrepreneurship as an evolutionary mechanism. We then proceed to discuss the importance of analysing entrepreneurship in context, and how mechanisms influencing the entrepreneurial processes may be related to technological regimes. Within this framework, some important aspects of the processes of entrepreneurship are discussed as well as how the principle of heterogeneity relates to these processes. However, entrepreneurship also strongly depends on some kind of conformity and community formation around the evolutionary processes. Entrepreneurial processes occur in a field characterized by opposing tendencies: on the one hand they challenge the contexts and systems in which they occur and contribute to growing heterogeneity; on the other, they take advantage of growing conformity and the formation of communities, implying the reduction of heterogeneity.

ENTREPRENEURSHIP – THE SCHUMPETERIAN VIEW

Within the field of entrepreneurship many definitions of the basic concepts of entrepreneur and entrepreneurship exist, and there are many different approaches in the field (Landström, 2004). Although this might encourage a broad review of various definitions, in this chapter we simply base our approach on Schumpeter's definition of entrepreneurship as outlined in his 'Theory of economic development' (1934/1996). Here, Schumpeter introduced a new economic theory in which he focused on the entrepreneur as a key agent of change and as the main driving force in economic development.

There are many reasons for taking Schumpeter as the point of departure, the most obvious being his strong influence on evolutionary thought and evolutionary approaches. Although there are many other contributors to our understanding of entrepreneurship in an evolutionary perspective, no one has been more influential than Schumpeter within this tradition (Swedberg, 2000).

Schumpeter's basis for defining the entrepreneur was his perception of a production process as one combining various resources in specific ways. This process of combining 'materials and forces' as he phrased it, is the key to understanding his concept of development. According to Schumpeter, development is something that basically comes 'from within' the economic system, and is related to changes in the way production is organized; that is, the manner in which resources are combined in new ways. Based on this, Schumpeter introduced the concept of 'new combination' (1934/1996, pp. 65–66) and defined development as the introduction of a new combination. His most acclaimed definition of entrepreneurship is related to five different approaches to new combinations: (1) the introduction of a new good or quality of good; (2) the introduction of a new method of production; (3) opening up a new market; (4) discovering a new source of supply of raw materials or semi-manufactured goods; (5) carrying out a new organization of the industry.

To be an entrepreneur is to take on the role of organizing the new combination and developing the new business activity based on this combination. Generally, this will involve starting up a new firm, but as Schumpeter indicated, the development of new combinations may also be organized in an existing firm. However, when Schumpeter wrote this book in his early years, the typical way of implementing new combinations was to start up new firms. The process of entrepreneurship is related to the process of developing the new organization based on exploiting opportunities provided by the new combination.

An important aspect of Schumpeter's view on entrepreneurship is that the process of developing a new combination would often imply disturbing the current order of things by breaking up the existing circular flow. As a rule, 'the new combination must draw the necessary means of production from some old combinations' (1934/1996, p. 68). He also generalized by saying that the introduction of new combinations implied 'the different employment of the economic system's existing supplies of productive means' (p. 68). As a part of this perspective, he emphasized that in a competitive economy the new combination meant the competitive elimination of the old combination (p. 67).

By defining the role of the entrepreneur in this way, Schumpeter pointed to the entrepreneur as a key agent for change, placing him at the centre of the evolutionary process. In his later works, Schumpeter (1943/1996, 1947) described the capitalist system as a method of change that can never be stationary, but is continuously evolving. This means that each stage of change is irreversible, 'as economic life goes on in a social and natural environment which changes and by its change alter the data of economic action' (1943/1996, p. 82). Each step of the change contributes to a new situation that forms new conditions for further activity. Expanding on this, Schumpeter labelled the capitalist system as one that incessantly revolutionizes from within (1943/1996, p. 83). This means that evolution implies that old structures are destroyed while new structures are created. He characterized this process as 'creative destruction', possibly the most well-known concept inherited from his works.

The feature of creative destruction 'is the essential fact about capitalism' (Schumpeter, 1943/1996, p. 83), and it very well illustrates the central mechanism of evolution. To develop something new, existing resources have to be organized in new ways, which means that organizing new activities and ceasing old ones are aspects of the same process of evolution. The role of the entrepreneurs and processes of entrepreneurship must be analysed in this context.

TECHNOLOGICAL REGIMES

The concept of context is fundamental for our analysis of entrepreneurship and heterogeneity. Although Schumpeter was not very explicit when defining the context in which he analysed entrepreneurship, and did not, for instance, address various and specific environmental factors that might have an influence on the entrepreneur, he very clearly related the role of the entrepreneur to the economic system. When discussing entrepreneurship, he did this in the context of the capitalist system, and his major concern was how the intervention of entrepreneurs contributed to disrupting the existing system and the current 'circular flow'.

Schumpeter's theory of economic development was originally written during the early parts of the twentieth century.[1] In this work, the role of the entrepreneur was considered to be much more important for economic development than was the case in his later works, in particular his work *Capitalism, Socialism and Democracy* (Schumpeter, 1943/1996). This shift in Schumpeter's view on the role of the entrepreneur demonstrates the importance of the systemic context for understanding the phenomenon of entrepreneurship. This was to become an important basis in the development of the concept of the technological regime (TR), which has been applied in a number of contributions to evolutionary analyses.

Although strongly influenced by Schumpeter, the concept of TR was introduced some decades later, in the early 1980s, as part of the emerging discipline of evolutionary economics (Nelson and Winter, 1982; Winter,

1984). As mentioned above, one important element in the discussion of TR was the significant shift that occurred in Schumpeter's perspectives on economic development. In his early works, Schumpeter focused mainly on the entrepreneur as an agent of change. By starting new businesses based on new combinations, entrepreneurs disrupted the existing circular flow and contributed to irreversible processes of change in the economy. The entrepreneurs were outsiders in the economy – they were newcomers in the field in which they contributed (Winter, 1984). As stated by Schumpeter, the new combinations are 'embodied in new firms which generally do not arise out of the old ones but start producing beside them' (Schumpeter 1934/1996, p. 66).

These perspectives on the role of the entrepreneurs stand in strong contrast to Schumpeter's later works, among them *Capitalism, Socialism and Democracy* (1943/1996). Here, the focus has shifted towards the large enterprise as the main driver of economic progress, which has taken over the entrepreneurial function, even tending to make the entrepreneurs obsolete (Winter, 1984).

As stated by Winter (1984), and subsequently by Freeman (1988) among others, the shift in Schumpeter's perspective may be seen as a reflection of historical stages in economic development. The larger corporations developed much stronger positions in the economy, and with their R&D departments became very important players through their contribution to innovation.

The concept of technological regime is defined by Winter (1984) as 'the setting of parameters of the evolutionary model'; that is, of mechanisms and structures that characterize and explain economic development. In relation to Schumpeter's two perspectives, two modes of development are identified: the entrepreneurial mode and the routinized mode. The entrepreneurial mode is characterized by a large number of new firms led by individual entrepreneurs, which are the main source of innovation. In the routinized mode, innovations typically come from existing and large firms. However, a shift from one mode to the other does not mean that entrepreneurs are a threatened species: 'They survive in a number of niches, sometimes in competition and sometimes in symbiosis with research-intensive giants' (Winter, 1984, p. 295).

Against this background, Winter raised the question concerning the circumstances whereby new firms led by individual entrepreneurs are a major source of innovation. Furthermore, what differences were to be found in exogenous factors that might dispose an industry towards an entrepreneurial or a routinized mode of development; could such differences between the two modes be traced to differences in technological regimes (see Box 7.1)?

BOX 7.1 VARIABLES FOR DESCRIBING TECHNOLOGICAL REGIMES (BASED ON BRESCHI ET AL., 2000)

Technological opportunities:
Likelihood of innovating for any given amount of money invested in searching for opportunities. A high level of opportunities provide powerful incentives for undertaking innovative activities and denote an economic environment that is not functionally constrained by scarcity. Potential innovators may come up with frequent and important technological innovations.

Appropriability conditions:
Possibilities of protecting innovations from imitation and of reaping profits from innovative activities. High appropriability means the existence of ways to successfully protect innovation from imitation. Low appropriability conditions denote an economic environment characterized by the widespread existence of externalities.

Cumulativity conditions:
Related to the fact that today's knowledge and innovative activities form the base and the building blocks of tomorrow's innovations: an innovation generates a stream of subsequent innovations that are a gradual improvement on the original, or that create new knowledge that is used for other innovations in related areas. High levels of cumulativeness are typical of economic environments characterized by continuities in innovative activities and increasing returns.

Knowledge base for industry:
Nature of knowledge underpinning firms' innovative activities. Technological knowledge involves various degrees of specificity, tacitness, complexity and independence, and may greatly differ across technologies.

Later, a number of other scholars followed up the discussion of technological regimes, among them Breshi, Malerba and Orsenigo (Malerba and Orsenigo, 1993, 1995; Breschi and Malerba, 1997; Breschi et al., 2000; Malerba, 2004). These state that the notion of TR provides a description of the technological environment in which the firms operate (Malerba and Orsenigo, 1993). Further, a shared understanding of the concept of TR appears to have developed as defined by the following factors: (1) the opportunity conditions; (2) the appropriability conditions; (3) the cumulativeness of technological knowledge; and (4) the nature of the relevant knowledge base (Malerba and Orsenigo, 1993, 1995; Breschi and Malerba, 1997, Malerba, 2004) (see Box 7.1).

TECHNOLOGICAL REGIMES AND DYNAMIC PROCESSES

As indicated above, the point of departure for the analysis of TR was the recognition of Schumpeter's two perspectives of evolutionary mechanisms – often identified as Schumpeter Mark I and Mark II (Breschi and Malerba, 1997, Breschi et al., 2000, Malerba, 2004). Mark I refers to early Schumpeter and the entrepreneurial regime, while Mark II points to later Schumpeter and the routinized regime. The point of these distinctions is to characterize variables providing determinants of the dynamic mechanisms: the type of dynamics occurring in such systems is often referred to as 'Schumpeterian dynamics' (Breschi and Malerba, 1997) or 'Schumpeterian patterns of innovation' (Breschi et al., 2000). There are close relations between the main dynamic processes (i.e. entries, exits, survival and growth), and the way in which innovation activities are organized.

In the entrepreneurial regime, entrepreneurs and new firms play an important role in the evolutionary process. Actors who are new to the industry perceive opportunities for organizing new combinations, and based on their new ideas introduce innovation into the economy by starting new firms. In so doing, the circular flow represented by existing firms is disturbed, and the basis for established firms is challenged. Through the process of continuous entry of new firms, the current methods of production and organization are challenged. Thus, the entrepreneurial regime may also be associated with a large number of exits, as the basis for the existing firms may be eroded as new firms enter.

In contrast to this, the routinized regime is characterized by the dominant role of incumbent firms, which means innovation activities are mostly handled in the large organizations that have developed systematic routines for organizing innovation processes. Barriers to entry are high and it is difficult for entrepreneurs outside the industry to enter and start up new firms. As the main process of evolution is related to the larger firms and their internal organization of innovation activities, this mode of evolution is also called 'creative accumulation' (Malerba, 2004).

Table 7.1 Characteristics of technological regimes – Schumpeter Mark I and Mark II

	Mark I	Mark II
Type of regime	Entrepreneurial	Routinized
Technological opportunities	High, easy entry	Low, high barriers
Appropriability conditions	Low	High
Cumulativeness conditions	Low	High
Nature of knowledge base	Based on applied, experience-based knowledge, less based on R&D; knowledge is easily accessible and exposed to a large number of actors	Less accessible, exposed to a smaller number of actors; based on R&D and protected by IPR regimes and long lead times
Major actors	Entrepreneurs, new firms	Large existing firms
Type of evolution	Creative destruction	Creative accumulation
Heterogeneity – input factors	High level of heterogeneity in factors related to technological opportunities and knowledge-base	Low level
Heterogeneity – output factors	High	Low

The two regimes – which may be looked at as ideal types – may be discussed based on the variables used for describing TR. An overview is provided in Table 7.1.

In the entrepreneurial regime, there are many opportunities that may be easily identified by new actors. Here, barriers to entry are low in the sense that many actors have access to the knowledge and other resources required for starting a new venture. This also means that appropriability conditions are low; that is, the opportunities for protecting innovations from imitation are low such that other actors may easily take advantage of experiences and knowledge developed in existing firms. There is a high spill-over of knowledge. Furthermore, cumulativity conditions are also low, meaning that the relative advantages of continuous innovation activities are limited, and the knowledge base is so organized that knowledge of importance to the industry is exposed to many actors and is thus fairly easy to adopt.

In contrast to this, the routinized regime is characterized by a low level of opportunities and high barriers to entry, while appropriability conditions as well as cumulativeness conditions are high. This means that the incumbent firms have significant advantages compared to new entrants, as they can take advantage of their previous experience in the field. There are opportunities for protecting innovations through patenting or other means, and there is a significant cumulativeness of knowledge in the sense that new knowledge builds on previous knowledge. It will take time and resources for new entrants to catch up.

HETEROGENEITY AND TECHNOLOGICAL REGIMES

Based on the discussion of the concepts of entrepreneurship, technological regimes and Schumpeterian dynamics, we may now enter into a discussion of heterogeneity and how the principle of heterogeneity may be related to processes of entrepreneurship. The concept of heterogeneity may be defined as the extent to which variety or differences exist among actors or other factors in the system or environment. It is appropriate to apply a high–low scale for classifying the level of heterogeneity.

First, the concept of heterogeneity may be related to the variables indicated for describing the technological regimes (see Box 7.1). Among these, it seems relevant to apply heterogeneity as a characteristic of technological opportunities and the knowledge base. There will be a high level of opportunities if there is a high level of heterogeneity among the factors that are of importance for organizing new combinations of relevance to the actual sector. For instance, if a variety of resources within a specific field is available, the potential number of new combinations within this field will be high.

Technological opportunities are understood here as being related to opportunities for entering the relevant market with new business activities. The concept of technology should be understood in the broad sense, as opportunities depend on a number of technological as well as non-technological factors, including human and social capital. In principle, the level of heterogeneity will affect the number of potential combinations between the various elements of relevance. The higher the level of heterogeneity – up to a certain point – the higher the number of potential new combinations.

As far as the knowledge base is concerned, this may be regarded as part of the technology or the technological system (see Carlsson and

Stankiewicz, 1991). The same interpretation of heterogeneity will apply to this factor as to technological opportunities; that is to say there is a high level of heterogeneity in the knowledge base if it is accessible in many forms, and if it is applied in many different ways and the knowledge gained from these experiences is open, not monopolized in few organizations. A high level of heterogeneity in the relevant system will give a high level of potential entrepreneurial opportunities. However, the notion of 'potential' should be emphasized. Although a large number of entrepreneurial opportunities exist, this does not necessarily mean that there will be a high level of entrepreneurial activity. To be realized, other conditions also have to be met, and these are discussed in the following sections.

The other two factors – appropriability and cumulativity conditions – may be more difficult to relate to the concept of heterogeneity, as for instance an attempt to determine a level of heterogeneity for these two variables will not make sense. However, the function of these two variables will coincide with a change in heterogeneity conditions. In a situation where appropriability conditions are exploited, this will imply a general reduction of heterogeneity, as opportunities for outsiders to take advantage of this knowledge will be reduced. Similarly, in cases related to a high level of cumulativity, this also will mean a reduction of heterogeneity.

It clearly follows from this that the entrepreneurial regime is characterized by a high level of heterogeneity. Both in terms of technological opportunity as well as the knowledge base, the regime will be characterized by a high level of heterogeneity compared to the situation in the routinized regime. This heterogeneity will also be reflected through the industry structures developed as a result of the evolutionary processes. Thus, in the entrepreneurial regime, there will be a high level of heterogeneity among the actors involved in the dynamic processes – individuals, firms and institutions.

In an article from 1995, Malerba and Orsenigo introduced the concepts of widening and deepening as characteristics of the two systems. The concept of 'widening' is associated with the entrepreneurial regime, and relates to the fact that the innovative base is continuously expanding through the entry of new innovators and the erosion of the competitive and technological advantages of the established firms. This may also result in an increasing population of firms. The 'deepening' pattern relates to the routinized regime and the dominant role of the established firms, which are continuously innovating and accumulating technological and innovative capabilities, yielding advantages over the non-innovating firms and potential entrants.

As an analogy to this, the widening pattern will imply an increasing level of heterogeneity, while the deepening pattern is related to reducing the level of heterogeneity.

VARIATION AND SELECTION

At the core of evolutionary theory are the assumptions of some basic mechanisms of evolution on which the whole system rests. These basic mechanisms are related to the creation of variation (i.e. to bring into existence some kind of new entities) and to the selection mechanism (i.e. processes that select among the new entities). These processes may seemingly work as contradictory processes, as the first implies increased variety, while the second implies a reduction of variety (Carlsson and Stankiewicz, 1991).

A number of authors have discussed these principles and summarized the mechanisms of evolution. The seminal work of Nelson and Winter (1982) represents the starting point of the evolutionary approach. Their approach was based on the three building blocks of organizational routines, search behaviour and the selection environment (Van den Bergh, 2004). Many authors have followed up on this and summarized the principles in different ways (see for example, Aldrich, 1999; Carlsson and Stankiewicz, 1991; Edquist, 1997; McKelvey, 1997; Peneder, 2001; Saviotti, 1997). Among these Edquist (1997, p. 6) has stated that evolutionary theories often include the following elements:

1. The point of departure is the existence of reproduction of entities like genotypes in biology or a certain set-up of technologies and organizational forms in innovation studies.
2. There are mechanisms that introduce novelties in the system (i.e. mechanisms that create diversity). This includes significant random elements, but may also produce predictable novelties (e.g. purpose-oriented development work). In biology, the novelties are mutations and in our context they are innovations.
3. There are mechanisms that select among the entities present in the system. This increases the relative importance of some and diminishes that of others. The selection process reduces diversity and the mechanism's operation may be the 'natural selection' of biology or the 'market selection' of competition as regards technical change. Together the selection mechanisms constitute a filtering system that functions in several stages and leads to a new set-up of, for example technologies and organizational forms. There might also be feedback from the selection to the generation of new innovations.

The synopsis of this is that a population, system or technology develops through mechanisms of variation and selection. Diversity is a necessary condition for change; that is, evolutionary change depends on the existence of diversity in economic actions (Peneder, 2001). Not only does this mean that initial variety is required, the continuing creation of variety, for instance through new information, is also required, as the initial diversity will be 'consumed' during the process of selection (Peneder, 2001). This means that variation and selection processes are working continuously and in some kind of interaction; it is neither a stage model nor a sequential model, but rather 'a sort of harmonica movement' (Van den Bergh, 2004).

Diversity, or heterogeneity, is a necessary condition for evolution. The establishment of new firms may be regarded as the manifestation of diversity (Audretsch et al., 2004), and it is the variety in the system that governs the pace and direction of change (Metcalfe, 2000). However, this does not mean that it makes sense to focus on heterogeneity in isolation. Heterogeneity per se will not create evolution; it is how the 'system' reacts on heterogeneity that is important, where such reactions may be described through the type of selection processes that are working. This means that evolution is determined by efforts to exploit heterogeneity, and that these efforts will depend on available competence and the system's learning processes (Cohendet and Llerena, 1997). The continuous interplay between variety creating and selection mechanisms governs evolution.

THE ENTREPRENEURIAL PROCESS

So far, we have focused only indirectly on entrepreneurship by analysing technological regimes and how entrepreneurship may be affected by the two types of regimes. We have given a few hints on the role of heterogeneity in this context. Taking this as the backcloth, we will now look into the details of entrepreneurial processes and develop a framework for discussing how variation and selection regimes act on entrepreneurs and entrepreneurial processes. In line with Schumpeter's understanding of the capitalist system as a system for change, we focus on entrepreneurial processes as integrated parts of the regimes, partly formed by the regimes, partly contributing to the evolution of the regimes.

In order to analyse entrepreneurship, it may be helpful to apply a stage model for analysing entrepreneurial processes, which will naturally focus on idea generation, start-up, growth and maturity. There are many versions of the stage model. Storey (1994, p. 121), for instance, refers to the following stages: inception, survival, growth, expansion and maturity. In textbooks on entrepreneurship one may find a variety of stages in the entrepreneurial process described (for instance Deakins and Freel, 2003). Thus, Vesper (1990) has pointed at 12 different milestones in venture creation.

Another approach is to focus on different activities undertaken during the processes of start-up. This approach was taken by Reynolds and White (1997). Distinguishing between more than 20 different start-up activities, their study revealed a diverse pattern among people starting new businesses. The different activities were not of the same importance to all starters, and neither was there a specific order in which the different activities were organized, although a main pattern was revealed. This means that the process of entrepreneurship is a diverse one that may be organized in many different ways, depending on the specificities of the new venture idea as well as the characteristics and resources of the entrepreneur.

In our discussion, we will adopt a simple approach based on the exploration–exploitation concept as discussed by March (1991), but adding to this an introductory stage of opportunity recognition. Our approach will then be based on the following three elements:

1. opportunity recognition – i.e. the identification of a potentially new business activity;
2. opportunity exploration – i.e. the analyses and development of the business activity;
3. opportunity exploitation – i.e. economically, to take advantage of the business opportunity by running the business.

This approach is illustrated in Figure 7.1, where the focus is on the entrepreneurial process, but in order to understand this process it has to be analysed in context; here, that of the technological regime. For our further discussion, however, it seems feasible to consider the technological regime as comprising two parts – the technological system and the entrepreneurial community. The technological system refers to the technological parts of the system, including hardware, machinery, and so on, and the knowledge base organized around it. The entrepreneurial community includes all potential actors of relevance to entrepreneurial processes; that is, the whole 'system' of existing and would-be entrepreneurs, and potential team members and their networks, including potential investors and other groups of contributing parties. In order to emphasize that actors do not act in isolation but are embedded in social structures, we apply the concept of community to illustrate this aspect of the entrepreneurial processes.

The way in which the framework is developed is inspired by Carlsson and Eliasson's concept of an experimental organized economy (see Eliasson, 1991, 2003; Carlsson and Eliasson, 2003; Carlsson et al., 2002), in which they made a similar distinction between the technological system on the one hand, and what they refer to as the 'competence bloc' on the other. This

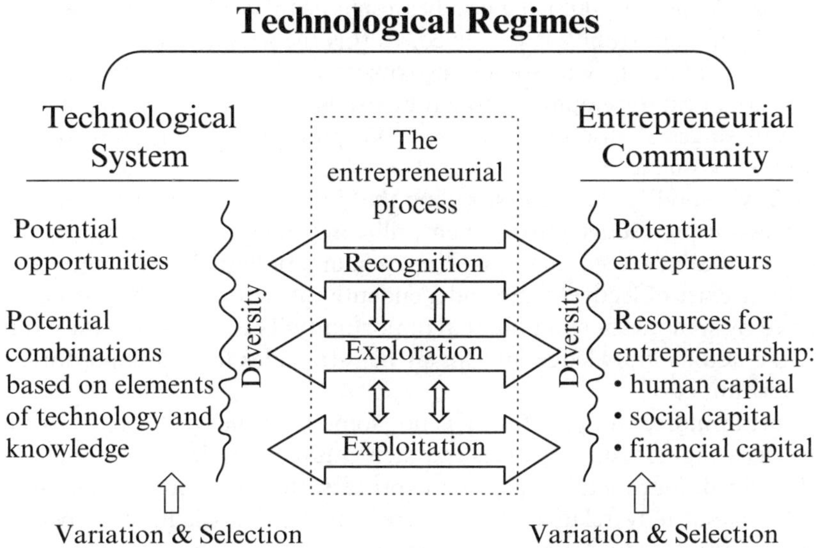

Figure 7.1 Analytical model for analysing the entrepreneurial process

is defined as the total infrastructure required for creating innovations and developing, organizing and exploiting new business ideas. The competence bloc is constituted by the human-embodied competence capital that determines the efficiency of organizing economic activities (Eliasson and Eliasson, 1996). Our concept of entrepreneurial community is not so very different from this, but we prefer to use the concept of 'entrepreneurial community' to indicate the parts of the evolutionary processes on which our main focus will be placed.

The most important aspect of this approach is that the entrepreneurial process is taking place at the interface between the entrepreneurial community and the technological system, and serves as a mediating function between the two spheres. This emphasizes that the understanding of entrepreneurship as an evolutionary process should not be treated as an individualistic process. The basic processes of entrepreneurship are the result of the complex interaction between the different types of actors and resources available in the actual environment.

A recent theoretical approach to the analysis of entrepreneurship has made the 'individual–opportunity nexus' as its key focal point (Shane, 2003; Shane and Venkataraman, 2000). Although this approach makes an interesting contribution to the understanding of entrepreneurship through its combination of an environment-oriented and an individual-oriented approach (Casson, 2005), it embodies a significant weakness in the way that the role of opportunity is perceived. In this theory, there is a tendency to focus on opportunity as something 'objectively' given,[2] and the idea that there is a linear development from the existence of opportunity through the various stages of discovery, decision to exploit, and so on, up to the performing business.

We will modify this by first saying that in an environment rich in components or productive means, there will – in principle – be a large number of potential new combinations. However, any potential new combination will not exist objectively and independently of actors, but has to be perceived by some actor as a potential opportunity. This perception is founded on the actor's knowledge and his or her strategies for searching for new opportunities.

Furthermore, an opportunity is not something that just 'exists'. It may start as a vague, visionary idea, and then gradually be developed and operationalized. During all stages of the process, the 'existence' of the opportunity is strongly related to the entrepreneur's knowledge, competence, networks, ability to organize, access to resources, and so forth. The opportunity will be developed through these processes. Often, there may be significant changes in the business idea, such that the outcome may be something quite different from the original idea (Klofsten, 2005). Although there are, of course, elements that exist independent of the entrepreneur, the capabilities of the acting entrepreneurs are crucial to the outcome of the process. It seems most adequate to say that opportunities do not generally exist, but are created during the entrepreneurial process and in the interaction between the entrepreneurs and their environments.

A second limitation of the 'individual–opportunity nexus' is its tendency to individualize the entrepreneurial process. Whatever the entrepreneurial process might be, it depends to a significant extent on the active collaboration of other actors, some of whom will also contribute actively to the process of assessing and exploring the business idea. Thus in various ways the entrepreneurial process may take the form of a collective process (Johannisson, 2004). For instance, this will be the case when the entrepreneurial process is organized by an entrepreneurial team or family. In many cases, other parties will be actively involved, including financial actors or potential customers, or a supplier who may invest significant resources into the entrepreneurial process. Consequently, the entrepreneurial process has an important dimension of collective activity, and the simplified picture of the 'lonely entrepreneur' has to be abandoned (Schoonhoven and Romanelli, 2001). Furthermore, as we have indicated in the model, communities will develop around different kinds of entrepreneurial activities, which will further underline entrepreneurial processes as embedded in sociocultural structures, and in this way are manifested as collective phenomena (Saxenian, 1994, 2000, 2001).

ENTREPRENEURSHIP AND HETEROGENEITY

The role of heterogeneity in relation to entrepreneurship may be discussed within the framework outlined above. First, it is obvious that the number of potential new combinations depends on the number of 'components', 'elements' or productive means that may be combined in the technological system. Mostly, we are thinking in terms of the relevant technology and knowledge system as providers of those elements that may be combined to create the new opportunity. In principle, the higher the number of potential opportunities, the higher the number of elements. That is, a high level of heterogeneity is a precondition for a high level of potential opportunities.

On the other hand, the initiation of entrepreneurial processes also requires actors with access to adequate resources. In order to initiate one specific entrepreneurial process, there should be at least one entrepreneur who has an adequate combination of resources and who recognizes the potential opportunity. Thus, the greater the degree of heterogeneity among the potential entrepreneurs, the higher the probability of initiating an entrepreneurial process based on one specific potential opportunity. Following on from this, the more heterogeneity there is in the technological system, and the more heterogeneity there is in the entrepreneurial community, the higher the potential for triggering new entrepreneurial processes.

However, the level of heterogeneity should not be as high as possible. Obviously, any field of technology or technological system needs to have some kind of structure. When we discuss heterogeneity, it is not about chaos, but some kind of organized, structured variation. If heterogeneity goes too far, the system will fragment and the number of potential new combinations will decrease. Similarly, the degree of heterogeneity in the entrepreneurial community should not be too high. If it is too high, the heterogeneity will contribute to the fragmentation of the entrepreneurial community, and to the reduction of the number of potential actors who may contribute to the development of new combinations.

As pointed out above, the entrepreneurial process takes place at the interface between two systems. Heterogeneity in both systems is a precondition for evolution. Evolution means the reduction of heterogeneity; that is, at each stage of the entrepreneurial process, a selection process occurs among alternatives on the technological as well as on the entrepreneurial side, and the new opportunity – the exploration of the new combination – is developed further.

In this sense, the evolution of the new combination will represent a manifestation of heterogeneity (Audretsch et al., 2004), and as the entrepreneurial process evolves, heterogeneity is reduced at the project level. On the other hand, the materialization of a new opportunity may provide a contribution to increased heterogeneity as it provides new elements in terms of technology as well as entrepreneurial practice. These elements may contribute to form the basis for exploring new opportunities. At the system level, this means there are ongoing processes of variety creation as well as selection processes – heterogeneity is continuously created and reduced.

One important aspect of the evolutionary process is the tendency towards harmonization of the entrepreneurial sphere and the technological sphere. There will be a tendency towards specialization by entrepreneurial actors in specific combinations, and within specific technological fields or types of business. There will be co-evolution of the technological system and the entrepreneurial community (Nelson, 1994). A specific entrepreneur will only have resources that are adequate for working with certain types of opportunity, and through experiences with these types of opportunities, the resources, including the experience-based knowledge derived from working on business opportunities, will develop in certain directions.

COMMUNITY FORMATION

An important part of the evolutionary process and the contextualization of entrepreneurship is the formation of the entrepreneurial community. The process of evolution implies the co-evolution of the technological system and the entrepreneurial community, and the processes of entrepreneurship may be regarded as an important mediator between the two systems.

The entrepreneurial community consists of all actors of relevance to the entrepreneurial process, including other business people such as managers, industrialists, technologists and investors, or actors with other relevant expertise in addition to the acting entrepreneurs themselves. Our understanding of the entrepreneurial community is closely related to the concept of 'competence blocs'; that is, the whole structure of the actors who have competence of relevance to developing the technological system (Carlsson and Eliasson, 2003), although we concentrate our focus on the part of the 'bloc' that is of particular interest to the processes of entrepreneurship and commercialization. Furthermore, to be part of a community, it is necessary to have some kind of active relationship, often socially embedded, with other members of the community.

The main point is that entrepreneurship is a socially embedded process, and the entrepreneurial performance will strongly depend on the extent to which potential entrepreneurs are included in relevant communities. As various technological systems turn out to be highly specialized, there will also be specialization of entrepreneurial communities relative to the technological systems they are exploring and exploiting. There is a long tradition for focusing on entrepreneurship as embedded in social systems in this way, for instance in relation to industrial districts (Brusco, 1986, 1990; Bull et al., 1991; Asheim, 1992), or the role of entrepreneurs in local communities (Johannisson and Nilsson, 1989). More recently, the community dimension of entrepreneurship has also been analysed in association with the evolution of high-technology clusters (Saxenian, 1994, 2001; Schoonhoven and Romanelli, 2001).

In her analysis of Silicon Valley, Saxenian (1994) introduced the concept of 'technical community' to characterize social systems that were formed around emerging new industries. Her point was that quite specific communities were developed around these industries, based on a collective identity among the involved actors. The communities were characterized by different ways of informal cooperation and of social and professional networks, and a variety of gatherings and meeting places were developed. An important aspect of this community formation was the strong homogeneity among its participants. In the actual case of Silicon Valley this comprised young men of similar backgrounds, and with a shared distrust of the established industries and their regimes (Saxenian, 1994).

This was followed up in a later study (Saxenian, 2001), where the role of ethnic-based communities in the Silicon Valley area was studied, identifying the formation of specific communities around different immigrant groups (from China and the West Indies). These were seen to have had strong influence on the development of specific industries. Interestingly, the specialization of the groups in different industries may be partly related to their skills in the English language.

These communities have played important roles in supporting specific types of entrepreneurial activity, and the communities may be regarded as a basic precondition for the high performance of entrepreneurial activity in the area. The function of the community is to support specific values and cultures; networks are developed, experiences are exchanged, different role models are available, and different arenas provide important learning environments. Thus these systems provide an arena for communities of practice (Wenger, 1998) – that is, communities of entrepreneurial practice, which include processes of collective learning of importance for developing the entrepreneurial performance among the members of the community.

From one perspective, the formation of a new community represents the creation of variety and the introduction of an alternative to the current regime. Also, within the community, its development will to some extent be based on heterogeneity. A certain level of variation in terms of its members and their networks has to exist. From another perspective, however, the formation of new communities implies the evolution of a homogeneous system of people with some kind of common identity and shared beliefs, and with a common interest related to specific areas of industrial development. In this sense, homogeneity rather than heterogeneity is a precondition for entrepreneurial activity.

There is a balance between heterogeneity and homogeneity. Community members should represent different roles and different experiences, and have access to different types of resources. Given a sufficiently high level of heterogeneity, the community may provide an important basis for new entrepreneurs to learn from experienced entrepreneurs, and to draw on their resources when forming their new businesses. As discussed by Aldrich (1999), heterogeneity in networks is very important to nascent entrepreneurs. Heterogeneity increases access to information and resources, and if the network is too homogeneous it will be of limited value for organizing new businesses.

At the same time, however, there should be a sufficient level of homogeneity, as this is an important precondition for trust and interaction between the community members. A central issue is that an adequate trade-off between heterogeneity and homogeneity is established (Audretsch and Thurik, 2001). If heterogeneity is developed beyond a certain point, the entrepreneurial community will disintegrate and the entrepreneurial performance will decline. And vice versa – if homogeneity develops too far, the potential for forming adequate constellations of resources to explore new opportunities will decline, and the entrepreneurial performance will suffer.

CONCLUSIONS – SUMMARY

A characteristic feature of entrepreneurship is that it will always be associated with some level of uncertainty; the risk of failure may be higher the less the level of knowledge existing in the actual field of business. Although an important function of an entrepreneurial community is to develop knowledge about the various aspects of entrepreneurial processes, there will never be a final recipe for entrepreneurial success. The economic system is continuously evolving; innovation and entrepreneurship are irreversible processes that are changing the system and creating new preconditions for evolution. Thus old knowledge relating to entrepreneurial processes will gradually become outdated, and have to be updated with new knowledge based on new experiences.

The main challenge to the entrepreneurial community is not to try to reduce the risk of each individual project to the lowest possible level, but to maintain a certain level of heterogeneity in terms of new business concepts. The point is not that every project should have success, but to promote a sufficient level of variation among the projects in order to secure the process of collective learning. In this perspective, a business 'failure' may be just as important as a 'success'.

This means that the entrepreneurial community should stimulate a high level of flexibility. As pointed out by Bahrami and Evans (1995), a high level of individual 'failures' may actually be an important precondition for the success of the system as a whole. The point is that a continuous flow of new business concepts will provide a high level of heterogeneity, and thus a high level of collective learning. In a system based on 'flexible recycling', people are open to making mistakes. When a concept turns out not to work, it is reformulated into something with a higher chance of success; 'entrepreneurs learn, just as scientists do, much from failed experiments' (Bahrami and Evans, 1995, p. 81).

The idea of regarding evolution as based on experimental activity is more systematically developed by Carlsson and Eliasson in their analyses of the experimentally organized economy (Carlsson and Eliasson, 2003; Eliasson, 1991). One key aspect of this framework is the virtually unlimited scope of opportunities, and thus a high level of ignorance about opportunities as well as about how to exploit them. It is commonplace to make mistakes, and progress is made through experimentation. Based on this, Carlsson and Eliasson indicate two types of error that can occur in the selection and retention processes in the system: one is to keep losers too long; the other is to lose winners. In line with this, the function of the entrepreneurial community may be interpreted as serving as a filter for identifying and sorting out different business opportunities. The challenge related to the process of evolution is to adjust – or calibrate – the selection and retention mechanisms in order to optimize the system.

As has been discussed throughout this chapter, entrepreneurship may be regarded as a process occurring at the interface between the technological system and the entrepreneurial community. While the original perspectives on entrepreneurship provided by Schumpeter (1934/1996) were not very explicit regarding the systemic relationships around processes of entrepreneurship, later contributions provided by evolutionary economics, like those concerned with technological regimes (Winter, 1984; Malerba and Orsenigo, 1993; Malerba, 2004) and experimental organized economies (Bahrami and Evans, 1995; Eliasson 1991; Carlsson and Eliasson, 2003), have provided significant contributions to understanding systemic mechanisms of importance to entrepreneurship.

Heterogeneity is a fundamental precondition for entrepreneurship. In systems characterized by a very low level of heterogeneity, there will be few opportunities for entrepreneurship. This applies to the technological system as well as the entrepreneurial community. Heterogeneity in the technological system provides potential for identifying and exploring more opportunities. The higher the level of heterogeneity – up to some optimum level – the higher the number of potential opportunities. However, none of the opportunities may be exploited without entrepreneurs. Although there are independently existing elements that are perceived as providing potential opportunities, their recognition depends fully on the entrepreneurs' perceptions, and these perceptions will strongly depend on previous experiences and access to various forms of resources.

Furthermore, the probability of an entrepreneur identifying a potential opportunity depends strongly on the heterogeneity of the entrepreneurial community. The greater the number of potential entrepreneurs and the more diverse the networks and access to various resources in the community, the greater the potential for establishing an adequate entrepreneurial organization for exploring and exploiting new business opportunities.

This is the input dimension of heterogeneity, and entrepreneurship may be perceived as a process in which heterogeneity is consumed (Peneder, 2001). Entrepreneurship represents a selection mechanism that contributes to the reduction of heterogeneity. However, there is also an output side. The outcome of the entrepreneurial processes will be new business concepts (either successes or failures) that will contribute to increased heterogeneity. In the technological system, heterogeneity will be increased as a result of the entrepreneurial action, which may lead to new technologies, new artefacts, and so on. In the entrepreneurial community, the new experiences derived from the entrepreneurial process will contribute to greater heterogeneity.

Thus the process of entrepreneurship may be regarded as a heterogeneity-reducing mechanism as well as a heterogeneity-increasing mechanism. This indicates that entrepreneurship operates in a field characterized by opposing forces. On the one hand, heterogeneity is a fundamental precondition for entrepreneurship; on the other, entrepreneurship also depends strongly on some kind of conformity and community formation around the evolutionary processes. From this point of view, entrepreneurial processes will take advantage of growing conformity and the formation of communities, implying the reduction of heterogeneity. Simultaneously, however, entrepreneurship will also challenge the contexts and systems in which it occurs, and contribute to growing heterogeneity.

NOTES

1. The original version, in German, was published in 1911; the English version was first published in 1934.
2. Shane and Venkataraman (2000) phrase it the following way: 'Although the recognition of entrepreneurial opportunities is a subjective process, the opportunities themselves are objective phenomena that are not known to all parties at all times' (p. 220), and Shane (2003): 'Opportunities have an objective component that does not exist solely in the mind of the entrepreneur' (p. 6).

REFERENCES

Aldrich, H. (1999), *Organizations Evolving*, London: Sage Publications.

Andersen, E.S. (1994), *Evolutionary Economics. Post-Schumpeterian contributions*, London and New York: Pinter.

Asheim, B.T. (1992), 'Flexible specialisation, industrial districts and small firms: a critical appraisal', in H. Ernste and V. Meier (eds), *Regional Development and Contemporary Industrial Response. Extending Flexible Specialisation*, London: Bellhaven Press, pp. 45–63.

Audretsch, D.B. and Z.J. Acs (1990), 'The entrepreneurial regime, learning, and industry turbulence', *Small Business Economics*, **2**, 119–28.

Audretsch, D.B. and Z.J. Acs (1991), 'Small firm turbulence under the entrepreneurial and routinized regimes', in L.G. Davies and A.A. Gibb (eds), *Recent Research in Entrepreneurship*, Aldershot: Avebury, pp. 11–28.

Audretsch, D.B., P. Houweling and A.R. Thurik (2004), 'Industry evolution. Diversity, selection and the role of learning', *International Small Business Journal*, **22**(4), 331–48.

Audretsch, D.B. and A.R. Thurik (2001), 'What's new about the new economy? Sources of growth in the managed and entrepreneurial economies', *Industrial and Corporate Change*, **10**(1), 267–315.

Bahrami, H. and S. Evans (1995), 'Flexible re-cycling and high-technology entrepreneurship', *California Management Review*, **37**(3), 62–89.

Breschi, S. and F. Malerba (1997), 'Sectoral innovation systems: technological regimes, Schumpeterian dynamics, and spatial boundaries', in C. Edquist (ed.), *Systems of Innovation. Technologies, Institutions and Organizations*, London and Washington: Pinter, pp. 130–56.

Breschi, S., F. Malerba and L. Orsenigo (2000), 'Technological regimes and Schumpeterian patterns of innovation', *The Economic Journal*, **110**(April), 388–410.

Brusco, S. (1986), 'Small firms and industrial districts: the experience of Italy', in D. Keeble and E. Wever (eds), *New Firms and Regional Development in Europe*, London: Croom Helm, pp. 184–202.

Brusco, S. (1990), 'The idea of the Industrial District: its genesis', in F. Pyke, G. Becattini and W. Sengenberger (eds), *Industrial Districts and Inter-Firms Co-operation in Italy*, Geneva: ILO, pp. 10–19.

Bull, A.C., M. Pitt and J. Szarka (1991), 'Small firms and industrial districts, structural explanations of small firm viability in three countries', *Entrepreneurship and Regional Development*, **3**, 83–99.

Carlsson, B. and G. Eliasson (2003), 'Industrial dynamics and endogenous growth', *Industry and Innovation*, **10**(4), 435–55.

Carlsson, B., M. Holmén, S. Jacobsson, A. Rickne and R. Stankiewicz (2002), 'The analytical approach and methodology', in B. Carlsson (ed.), *Technological Systems in the Bio Industries. An International Study*, Boston/Dordrecht/London: Kluwer Academic Publishers, pp. 35–42.

Carlsson, B. and R. Stankiewicz (1991), 'On the nature, function and composition of technological systems', *Journal of Evolutionary Economics*, **1991**(1), 93–118.

Casson, M. (2005), 'The individual – opportunity nexus: a review of Scott Shane: *A General Theory of Entrepreneurship*', *Small Business Economics*, **24**(5), 423–30.

Cohendet, P. and P. Llerena (1997), 'Learning, technical change, and public policy: how to create and exploit diversity', in C. Edquist (ed.), *Systems of Innovation. Technologies, Institutions and Organizations*, London and Washington: Pinter, pp. 223–65.

Deakins, D. and M. Freel (2003), *Entrepreneurship and Small Firms*, Maidenhead: McGraw-Hill.

Edquist, C. (1997), 'Systems of innovation approaches – their emergence and characteristics', in C. Edquist (ed.), *Systems of Innovation. Technologies, Institutions and Organizations*, London and Washington: Pinter, pp. 1–35.

Eliasson, G. (1991), 'Modeling the experimentally organized economy. Complex dynamics in an empirical micro–macro model of endogenous economic growth', *Journal of Economic Behavior and Organization*, **16**, 153–82.

Eliasson, G. (2003), 'Global economic interaction and regional attractors of competence', *Industry and Innovation*, **10**(1), 75–102.

Eliasson, G. and Å. Eliasson (1996), 'The biotechnological competence bloc', *Revue d'Economie Industrielle*, **78**(4 trimestre), 7–26.

Freeman, C. (1988), 'Introduction', in G. Dosi, C. Freeman, R. Nelson, et al. (eds), *Technical Change and Economic Theory*, London and New York: Pinter, pp. 1–8.

Hannan, M.T. and J. Freeman (1989), *Organizational Ecology*, Cambridge, MA: Harvard University Press.

Johannisson, B. (2004), 'Entrepreneurship in Scandinavia: bridging individualism and collectivism', in G. Corbetta, M. Huse and D. Ravasi (eds), *Crossroads of Entrepreneurship*, Boston/New York/Dordrecht: Kluwer Academic Publishers, pp. 225–41.

Johannisson, B. and A. Nilsson (1989), 'Community entrepreneurs: networking for local development', *Entrepreneurship and Regional Development*, **1**(1), 3–19.

Klofsten, M. (2005), 'New venture ideas: an analysis of their origin and early development', *Technology Analysis and Strategic Management*, **17**(1), 105–19.

Landström, H. (2004), *Pioneers in Entrepreneurship and Small Business Research*, New York: Springer.

McKelvey, M. (1997), 'Using evolutionary theory to define systems of innovation', in C. Edquist (ed.), *Systems of Innovation. Technologies, Institutions and Organizations*, London and Washington: Pinter, pp. 200–22.

Magnusson, L. and J. Ottosson (eds) (1997), *Evolutionary Economics and Path Dependence*, Cheltenham, UK and Lyme, USA: Edward Elgar.

Malerba, F. (2004), 'Sectoral systems of innovation: basic concepts', in F. Malerba (ed.), *Sectoral Systems of Innovation. Concepts, Issues and Analyses of Six Major Sectors in Europe*, Cambridge: Cambridge University Press, pp. 9–41.

Malerba, F. and L. Orsenigo (1993), 'Technological regimes and firm behavior', *Industrial and Corporate Change*, **2**(1), 45–71.

Malerba, F. and L. Orsenigo (1995), 'Schumpeterian patterns of innovation', *Cambridge Journal of Economics*, **19**(1), 47–66.

March, J.G. (1991), 'Exploration and exploitation in organization learning', *Organization Science*, **2**(1), 71–87.

Metcalfe, J.S. (1998), *Evolutionary Economics and Creative Destruction*, London and New York: Routledge.

Metcalfe, J.S. (2000), 'Restless capitalism, experimental economies', in W. During, R. Oakey and M. Kipling (eds), *New Technology-Based Firms at the Turn of the Century*, Amsterdam: Pergamon, pp. 4–16.

Metcalfe, J.S. (2004), 'The entrepreneur and the style of modern economics', in G. Corbetta, M. Huse and D. Ravasi (eds), *Crossroads of Entrepreneurship*, Boston/New York/Dordrecht: Kluwer Academic Publishers, pp. 33–51.

Nelson, R.R. (1994), 'The co-evolution of technology, industrial structure, and supporting institutions', *Industrial and Corporate Change*, **3**(1), 47–63.

Nelson, R.R. and S.G. Winter (1982), *An Evolutionary Theory of Economic Change*, Cambridge, MA, and London: Belknap Press.

Peneder, M. (2001), 'Dynamics of initial cluster formation: the case of multimedia and cultural content', in OECD, *Innovative Clusters. Drivers of National Innovation Systems*, Paris: OECD, pp. 303–13.

Reynolds, P.D. and S.B. White (1997), *The Entrepreneurial Process*, Westport, CT and London: Quorum Books.

Saviotti, P.P. (1997), 'Innovation systems and evolutionary theories', in C. Edquist (ed.), *Systems of Innovation. Technologies, Institutions and Organizations*, London and Washington: Pinter, pp. 180–99.

Saxenian, A. (1994), *Regional Advantage. Culture and Competition in Silicon Valley and Route 128*, Cambridge, MA and London: Harvard University Press.

Saxenian, A. (2000), 'The origins and dynamics of production networks in Silicon Valley', in R. Swedberg (ed.), *Entrepreneurship. The Social Science View*, Oxford: Oxford University Press, pp. 308–31.

Saxenian, A. (2001), 'The role of immigrant entrepreneurs in new venture creation', in C.B. Schoonhoven and E. Romanelli (eds), *The Entrepreneurship Dynamic. Origins of Entrepreneurship and Evolution of Industries*, Stanford, CA: Stanford University Press, pp. 68–108.

Schoonhoven, C.B. and E. Romanelli (2001), 'Emergent themes and the next wave of entrepreneurship research', in C.B. Schoonhoven and E. Romanelli (eds), *The Entrepreneurship Dynamic. Origins of Entrepreneurship and Evolution of Industries*, Stanford, CA: Stanford University Press, pp. 383–408.

Schumpeter, J.A. (1934/1996), *The Theory of Economic Development*, London: Transaction Books.

Schumpeter, J.A. (1943/1996), *Capitalism, Socialism and Democracy*, London and New York: Routledge.

Schumpeter, J.A. (1947), 'The creative response in economic history', *Journal of Economic History*, **VII**(2), 149–59.

Shane, S. (2003), *A General Theory of Entrepreneurship. The Individual–Opportunity Nexus*, Cheltenham, UK and Northampton, MA, USA: Edward Elgar.

Shane, S. and S. Venkataraman (2000), 'The promise of entrepreneurship as a field of research', *Academy of Management Review*, **25**(1), 217–26.

Storey, D.J. (1994), *Understanding the Small Business Sector*, London and New York: Routledge.

Swedberg, R. (2000), 'The social science view of entrepreneurship: introduction and practical applications', in R. Swedberg, *Entrepreneurship. The Social Science View*, Oxford: Oxford University Press, pp. 7–44.

Van den Bergh, J.C.J.M. (2004), 'Firm behaviour and organisation from an evolutionary perspective', In H.L.F. de Groot, P. Nijkamp and R.R. Stough (eds), *Entrepreneurship and Regional Economic Development. A Spatial Perspective*, Cheltenham, UK and Northampton, MA, USA: Edward Elgar, pp. 15–45.

Vesper, K.H. (1990), *New Venture Strategies*, Englewood Cliffs, NJ: Prentice Hall.

Wenger, E. (1998), *Communities of Practice: Learning, Meaning and Identity*, Cambridge, MA: Cambridge University Press.

Winter, S.G. (1984), 'Schumpeterian competition in alternative technological regimes', *Journal of Economic Behavior and Organization*, **5**, 287–320.

8. Heterogeneity interpreted and identified as changes in the populations of firms

Svein Olav Nås and Tore Sandven

INTRODUCTION

The term 'heterogeneity' refers to differences between units of observation, such as persons, organizations, routines or firms. The perspective concerns differences between firms, but the differences identified may originate in properties of components of the firms, including the composition of employees, industry, size, organization, routines or other capabilities. The aim of this chapter is to consider how one can identify, operationalize and describe the development over time of some aspects of heterogeneity of firm populations, and to consider if heterogeneity seems to increase or decrease over time. We argue that in a dynamic, evolutionary setting, the maintenance and reproduction of heterogeneity are necessary if incentives for change and growth are to be preserved.

The work refers to previous analyses of persistent differences between industries and firms over time, for instance Schmalensee (1989) on intra-industry differences in profitability, or Jensen and McGuckin (1997) who utilize micro data to study heterogeneity at firm and plant level, doing away with the commonly used representative firm. The results show persistent differences in average profitability over time between industries, but with even larger differences between firms within industries. Schmalensee's empirical observations challenged the dominant neo-classical general equilibrium theory, and contributed to paving the way for alternative approaches such as disequilibrium theories and evolutionary reasoning.

The current analysis attempts to apply an evolutionary perspective, and addresses heterogeneity in terms of changes over time in numbers of firms and employees, their industry affiliation and the occurrence of different *business demographic events* during their existence; type of entry, reorganizations and type of exit. In this respect it differs from previous studies of entry and exit in the Norwegian context, for instance in Salvanes and

Tveterås (2004) and Tveterås and Eide (2000). To do this, time series data on populations of firms are required. The data utilized here cover matched employer–employee register data for Norwegian firms and their employees over the period 1995–2001.

Firms are studied at two levels: the formally registered enterprises (equivalent to the firm level discussed by Jensen and McGuckin) and the establishments – or plants – they consist of. For each of the units there is information on industry affiliation and identification of all employees each year. On this basis it is possible to distinguish different events that occur in the firms and to follow their development over time. This approach allows the identification of events like births, spin-offs or spin-outs, reorganizations, or exits. Such events affect the composition of firms in given industries, and the composition of industries in the economy – which is how heterogeneity is interpreted in the present context. The intention is to address measures of dispersion rather than to describe central tendencies. Details of the methodology are presented below.

In the following, some reflections on the concept of heterogeneity and forces affecting the presence of heterogeneity are presented in the next section. The third section addresses possible operationalizations of heterogeneity within the present framework. The fourth section presents the data and methodology utilized, and the fifth section contains the analytical results. We conclude in the sixth section.

INTERPRETATIONS OF HETEROGENEITY

The concept of heterogeneity fits evolutionary reasoning about diversity and selection, as it highlights variation in properties and opportunities rather than representative averages or other measures of central tendency. To be meaningful, diversity and selection are only of interest in a dynamic setting where selection mechanisms are allowed to work over time, and the existence and generation of variety can feed the selection mechanisms. To be viable, such a system must not only demonstrate heterogeneity at the outset, but needs also to be able to regenerate heterogeneity. If regeneration of heterogeneity fails, the risk is to evolve into a system characterized by monopolies or a declining number of actors. The monopoly case is considered less dynamic as the incentive to further development posed by competition from other existing firms is absent. For this reason it is of interest to investigate whether the degree of heterogeneity increases or decreases over time, and to better understand the types of events that contribute to this process.

In practice it can be observed along several dimensions that heterogeneity exists and persists over time. The differences, for instance, in profitability have turned out to be at least as large within industries as between them (Schmalensee, 1989; Geroski, 1995). A series of different characteristics of firms and plants are reviewed and discussed by Jensen and McGuckin (1997), even though the concepts of learning and path dependency are not explicitly raised. In our view a reasonable interpretation is that there are mechanisms at work that limit learning and copying of best practice to an extent that convergence does not occur, as discussed by Nelson (1996).

We can think of several forces at work simultaneously, where the outcome in terms of heterogeneity is determined by the relative strengths of those forces. First, duplicating or copying from competitors, or diffusion of any new or better technology works in the direction of reducing heterogeneity, in the sense that it is rational for all actors to select the same, most effective technology – other things being equal. If copying is relatively easy, the incentive for generating new knowledge or business ideas is limited, and we would expect decreasing heterogeneity. If copying is difficult, it is in itself an incentive for developing new knowledge as the probability of appropriating the benefits is higher. One should note, however, that copying does not only require access to the best practice technology in itself, but also the ability to implement it in one's own organization. The latter depends on the existing routines, capabilities, capital structures and market conditions of each single firm. Allowing – and expecting – heterogeneity to be present, what functions as the most efficient technology will therefore be different for each firm. The ability to implement and profit from each technology – even if it is the most efficient (other things equal) – should therefore be expected to differ between firms. For a particular technology to become dominant the added efficiency must be large enough to offset such firm-specific effects by driving non-adopters out of business and allowing time to diffuse and replace the old technology.

A second force is technological advances made by the firms themselves, by researching new knowledge or adapting or combining existing knowledge to new uses. This works in the direction of creating additional heterogeneity, but interacts with the extent to which copying and appropriation of resulting knowledge are regarded as easy. The possibility of pursuing a strategy of technological advance depends on the prevailing technological opportunities in the given industry and market, and will even vary between firms as they are restricted in their choices by their historically acquired capabilities.

Third, the selection mechanisms at work discriminate the viable solutions. If competition is fierce and profitability low, there is little room for differences in solutions and costs. The expectation is that the most effective technology will be selected, possibly in combination with a development in the direction of segregation of markets, allowing differentiated products and firms. Even though strong competition is generally viewed as a driving force for economic growth and development, it can also be argued that it may lead to limited resources for creating new solutions or preserving differences – and one may end up in a situation where the winner takes all and monopolizes the market, with the consequence that incentives to change vanish. Prior to such a shakeout, however, fierce competition may signify strong incentives to innovate since the alternative most probably means going out of business. The question then is how the different forces balance each other out.

Regulations, like antitrust acts, belong to this equation. This introduces the last theme, as part of the selection environment; how the market structure affects the relative strength of the actors. An industry dominated by large incumbent firms is usually more difficult to enter than an industry with many small firms. Likewise, industries dominated by standardized products usually experience fiercer competition, lower profitability and more emphasis on economies of scale. The size distribution of firms depends on many things, where the existence of economies of scale, age of the industry, technology applied and technological opportunities are central elements. We consider high age of industries, presence of scale economies and limited technological opportunities all to work in the direction of less heterogeneity within the given industries.

A necessary element of the argument concerns the existence of uncertainty and the firm's ability to learn and appropriate or protect its own knowledge and the resulting innovations. When uncertainty is present – for example regarding about what may be considered the best technology/product/market in the future – the solutions that different firms come up with will differ and thereby contribute to increased heterogeneity. In fact, no firm is free to choose any particular development path, but must rely on utilizing – and developing – the skills it already possesses. Firms and economies are dependent on their recent development paths that constitute their current position and consequently what can be done in the future. Path dependency stems from the accumulation of routines, skills and knowledge, external and internal relations and physical investments, as argued by, for instance, Malerba and Orsenigo (1993) and Lundvall (1992). In this way we may say that the existence of continuity in terms of development paths of existing and new entrant firms is a necessary condition for sustaining heterogeneity. However, path dependency may also result in sustaining monopolistic competition – depending on the initial state – to the degree that it blocks entry of new firms.

Heterogeneity means that there are potential challenges to the incumbent actors or leaders, resulting in a struggle to keep or even improve on the actual position. Therefore, heterogeneity can contribute to increasing the level of competition and innovation, even though the actual effect depends on market structure and concentration rates. Another way of assessing heterogeneity can be as a kind of insurance, as a heterogeneous economy is less locked into particular solutions since it is possible to choose between several different and existing optional development paths. This is not necessarily true for the individual firms, as they are more limited when it comes to the number of activities and technologies. For society at large such an insurance argument is clearly relevant.

Not only for the society but for the existing firms the presence of heterogeneity may open up opportunities. Persistent heterogeneity – as is observed in terms of both within industry and between industry differences – means that there is room for different ways of operating. Therefore, it is possible to survive over time without exactly copying the leader in a given market by, for instance, differentiating the product, utilizing different production or distribution technologies, or utilizing small-scale advantages. This is mainly due to path dependency and the generally underestimated costs associated with the transfer of knowledge.

We often think of heterogeneity with reference to single countries. As a final point in this context, we should be aware that the internationalization and liberalization of trade are affecting the presence of heterogeneity. Even a single-firm industry in a given country may be part of a highly competitive and diversified environment on the world markets. Opening up an economy to the world markets may initially increase heterogeneity in the sense that the number of firms that (potentially) operate in the global market becomes larger. The resulting heterogeneity in individual countries will usually be different, however, as domestic producers may expand abroad, reorganize production, or lose out in the competition. This way, the domestic heterogeneity may increase or decrease as a result of trade liberalization or deliberalization, depending on initial conditions. Over time, selection will alter the composition of firms, of which some are survivors, others are newcomers and yet others are reorganized. Increased market size resulting from a shift from national to international markets may pull in the direction of economies of scale and larger size, thus possibly reducing the number of actors. On the other hand, the potential market even for niche producers may become sufficiently large to sustain a larger number of smaller firms.

In what follows internationalization and the other factors are not directly addressed. They have been presented in order to have an idea about the kinds of underlying dynamics that may explain the patterns of firm demographic changes we expect to unravel from our data. In the rest of the analysis the focus is on how heterogeneity can be operationalized with the available data, and on how different kinds of changes in the population of firms contribute to the heterogeneity we are able to operationalize and observe.

IDENTIFYING HETEROGENEITY BY FIRM DEMOGRAPHIC CHANGES

As pointed out above, heterogeneity can be defined by reference to a series of different aspects of firms or markets; types of knowledge or technology, profitability, which markets firms operate in, the number of firms in each market, the distribution of firms between industries, the size distribution of firms, and so on. The present approach addresses the outcomes of the different processes discussed above, by highlighting the *trends in the composition of firms of different kinds* in the economy. Choosing the firm as the main unit of analysis one can argue that a certain number of firms are necessary for heterogeneity to exist. These firms should be different in some sense. If there is a large number of firms where all are more or less similar, as the assumption of the representative firm in the mainstream economic theory stipulates, one can argue that a situation with a few large but different firms may constitute a higher degree of heterogeneity.

To assess the population of firms one has to clarify the definition of 'the firm'. It may be the formal entity – enterprises – that exists as a separate legal unit in registers. Enterprises may consist of one or several production facilities – usually referred to as establishments. Establishments can be said to constitute the routines and/or physical infrastructure that take care of actual production. Enterprises, on the other hand, often take care of strategy and innovation planning, in addition to representing the economic value that is publicly traded and liable for paying wages, taxes, and so on.

Enterprises represent different sets of routines from establishments. They are generally less stable over time; physical production facilities may exist without much change over time even if ownership and the structure at the enterprise level change. The current analysis focuses on establishments as the units of analysis. This way a more precise description of industry affiliation is obtained, and the underlying stability of production units during periods of turbulence in ownership or organization is captured. However, the enterprise level is also taken into account. The persistence, birth or closure of enterprises is used to classify the types of business demographic changes that may occur. This brings the more turbulent changes of enterprises – as compared to establishments – into the analysis.

A differentiating characteristic of firms is their recent development path. Above, it was pointed out that the abilities of a firm depend on acquired knowledge and experience, and that these are likely to be difficult to copy or transfer. Such differences are not directly comparable from a statistical point of view since they are difficult to operationalize and compare across firms. We can, however, observe other differences in the event histories that we expect to be related to their capabilities. They include the age of the firms (how long they have survived), how they were established, if they have experienced recent changes in structure such as takeovers, mergers, spin-offs (defined below), and so on. As part of the data allowing us to track developments in firms over time we are able to classify firms in terms of such changes. Thus we utilize this information to study the development and comparative performance of firms that have experienced different types of changes.

In most studies of firm demography only certain aspects of the dynamics are taken up, mainly because of limitations in the data at hand. The most common methods of data collection are survey questionnaires and administrative information on the outcomes of particular policy measures. As a consequence of the differences in approaches, the results are frequently incomparable; across countries or institutions. An advantage of such approaches is that in each case the responsible data collector is free to identify precisely those cases that are of interest by utilizing all available supplementary information (Callan, 2001). Against this background the present work pursues a different approach, by developing precisely defined characteristics for demographic changes on the basis of fully covering administrative registers of firms (Nås et al., 2001). The registers are matched employer–employee data containing information on enterprises, establishments and their employees over a number of years. This information is used to classify demographic events occurring in the firms and to track their development over time.

The methodology for classifying events is summarized in Box 8.1 in the next section. Changes in the population of establishments and enterprises are identified and characterized in two stages. First, changes in identification numbers of the Norwegian population of firms between two (consecutive) years are identified at the level of establishments and enterprises. On this basis, nine different categories of change are defined, ranging from 'no change' to 'totally new establishments', closures and different kinds of transfers or transformations of units. Among the new establishments the category where a new establishment is also a new enterprise is further subdivided by tracking the previous employment of the personnel involved. This is how spin-offs, Greenfield births,[1] and other new establishments are identified. The methodology not only allows the identification of new or changed establishments and enterprises, but also tracing their history to identify the firms where new or changed establishments originated.

The analysis is limited to the business enterprise sector on the results side, but it is accepted that the public sectors contain delivering establishments from which new establishments spin out, or personnel switch jobs.[2] Agriculture and forestry are also excluded, mainly for technical reasons.[3] Lastly, establishments are left out when they have only one employee in both years we use to identify changes. This limits the number of observations considerably, but is necessary in order to distinguish spin-offs from any kind of new establishment. It also contributes to reducing 'statistical noise' in the registers.

The same basic information that is used to classify events is also used to describe the outcomes of the demographic process, and in particular to develop operationalized indicators on heterogeneity. This is taken up in the next section.

MEASURING HETEROGENEITY OF FIRM POPULATIONS

In the previous section we argued for utilizing firm demographic events as characteristics of heterogeneity in firm populations, and as a device for tracking the evolution of these populations. In this section we suggest how heterogeneity can be operationalized within this framework. How can we define that a given (national or regional) population of firms is becoming more or less heterogeneous? It is not sufficient merely to describe existing differences in heterogeneity; it is necessary to study how the heterogeneity – or lack of heterogeneity – develops over time in interaction with the forces of selection. Survival is a core factor for the resulting structure. However, survival can be both in the form of continuation of an independent unit and in terms of the reorganization or transformation of a unit. The demographic approach allows the identification and comparison of these cases. An overview of the categories and their definitions are found in Box 8.1.

What we attempt to do is to describe how the composition of firms classified by firm demographic events develops within given industries, and how the composition of industries evolves in the economy as a whole. In so doing it becomes important how industries are grouped and classified. The available standard industrial classification system (SIC) has shortcomings, but constitutes the sole possibility to control for type of activity.[4] The classification is far more detailed when it comes to manufacturing than is the case for services. Changes between industries will therefore tend to be more frequent in manufacturing than in services, if we don't bear this problem in mind. Broadly defined industries (as for many service

BOX 8.1 CLASSIFYING BUSINESS DEMOGRAPHIC EVENTS BY MEANS OF REGISTER DATA

The initial step in identifying and classifying the different kinds of changes in the population of enterprises and establishments involves utilizing their formal identification numbers. Changes in identification numbers are used as the first filter to identify the different types of changes. On the basis of this information nine different categories of events are distinguished. Classification of changes is carried out by comparing two adjacent years.

The second step involves tracking the labour in one particular class of new establishments in order to further subdivide by their origins. This procedure is applied to the class of establishments that are new since last year, and at the same time belong to a new enterprise. The whole process results in the following categories of enterprises that are being used throughout the work:

1. No change
An existing establishment continues within the same existing enterprise.
2. Transformation
An existing establishment continues and becomes a new independent enterprise, and the old enterprise is closed down.
3. Takeover
An existing establishment continues within another existing enterprise, and the old enterprise is closed down.
4. Move
An existing establishment continues within another existing enterprise, and the old enterprise survives.
5. Spin-out
An existing establishment continues and becomes a new independent enterprise, and the old enterprise survives.
6. Entrepreneurial new
A new establishment comes into existence as a new independent enterprise.
7. New by expansion
A new establishment comes into existence within an existing enterprise.

8. Complete closure
An establishment is closed down and the enterprise it belongs to is also closed down.
9. Partial closure
An establishment is closed down but the enterprise it belongs to survives.

Subdividing category 6 'Entrepreneurial new' by labour tracking adds the following category:

6.1 Spin-offs
A minimum of 2 employees come from the same delivering establishment. They must make up at least 30% of the new establishment and less than 50% of the delivering establishment (year 0). The delivering establishment must have at least 1 employee after the spin-off (year 1).
6.2 Greenfield births
Not more than 1 employee can come from the same delivering establishment.
6.3 Other new
All cases that are not characterised as spin-offs, Greenfield births or corrections are assigned to this residual category.
6.4 Corrections
Even if a new ID-number is assigned to an establishment, it is considered to be a continuation (not new) if 80% of employees in the new establishment were found in the same closed-down establishment the year before, or if 80% of employees in a closed establishment since the previous year are found in the new establishment.

industries) also tend to have larger numbers of firms or employees included, appearing to be more heterogeneous compared with the more narrowly defined industries found in manufacturing. To cope with this problem we try to define reasonably homogeneous industries in terms of activity and size, and keep the classification stable over time. Instead of comparing industries, we focus on changes over time within industries, but also on changes in the relative sizes of industries.

The focus on heterogeneity also calls for measures that capture differences rather than central tendencies such as averages, as discussed, for instance, by Ijiri and Simon (1977). In what follows the differences between averages and measures of variation are addressed. Special cases or outliers should also be addressed, such as, for instance, fast growers (often referred to as gazelles). The concrete indicators include the following:

- Number and distribution of employees per industry.
- Number and distribution of enterprises and establishments per industry.
- Distribution of size (number of employees) of firms in industries.
- Relative contribution of exiting firms and reorganized firms to entering firms (*N* firms, *N* employees):
  - in same industries,
  - in new industries,
  - contributing to more or less concentration in industries,
  - fast growers,
  - resulting age structure of population of firms.

EMPIRICAL RESULTS

The composition of different demographic events is illustrated in Tables 8.1 and 8.2. The Norwegian results are compared to the situation in three other Nordic countries in 2000 in order to consider if the Norwegian case is different from what we find elsewhere. The majority of establishments are unchanged, but significant numbers have undergone different kinds of change. The numbers of closures and entries generally balance each other out, even if this balance varies over time. The basic pattern is essentially the same in all countries, which strengthens the confidence in the methodology.

A core factor for determining the resulting composition of firms is survival. To address the question of survival and whether the different kinds of startups differ, we first look at the development of entrant firms of different types in Norway each year from 1996 to 2002. These can be compared to establishments already existing in 1995; that is, ones that were created before 1996. A general finding is that entrants suffer from high infant mortality rates (Caves, 1998, p. 1954). Survival rates increase steadily after the first year. This is confirmed in Figure 8.1 where we see that, in general, the older the establishment, the higher the probability of survival.[5]

Introducing the categories of change of the entrant establishments helps explain the variations in survival rates. Here we look at the probability of being a survivor in 2002, using the change categories of spin-offs, Greenfield, other new establishments, and in addition new establishments within existing enterprises. Only establishments entering after 1995 are

Table 8.1 *Changes in establishments and enterprises 1999–2000, Sweden, Finland, Denmark, Norway (Number of establishments)*

Change	1	2	3	4	5	6	7	8	9		
Country	No Change	Trans-formation	Take-over	Move	Spin-out	Entrepreneurial new	New by expansion	Complete closure	Partial closure	Total Year 0	Total Year 1
Sweden	152 940	1 004	743	358	813	11 857	3 209	8 908	2 773	167 539	170 924
Finland	76 939	743	422	280	436	5 763	1 809	5 589	1 502	85 911	86 392
Denmark	89 308	2 326	308	279	964	5 316	1 554	2 757	1 577	97 519	100 055
Norway	92 194	2 235	975	224	376	6 501	1 548	5 591	1 706	103 301	104 053

Source: Nås et al. (2001).

Table 8.2 Number of entrepreneurial new establishments 2001 (2000) by type of demographic event

	Sweden		Finland (2000)		Denmark		Norway	
	High-tech	Others	High-tech	Others	High-tech	Others	High-tech	Others
Spin-offs	311	1 355	115	503	166	783	135	760
Greenfield	334	4 431	247	2 513	189	2 020	195	3 338
Others	247	3 044	179	2 147	197	1 961	91	1 105
Corrections	47	525	4	55			2	106
Sum	939	9 355	545	5 218	552	4 764	423	5 309

Source: Nås et al. (2001).

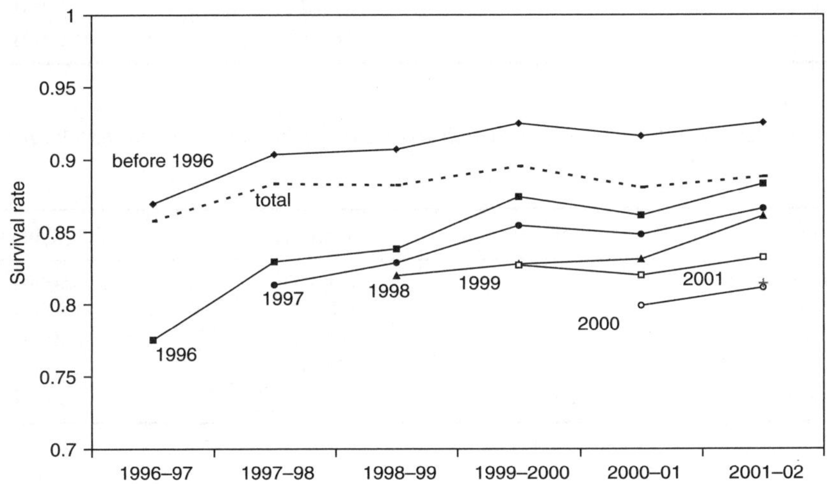

Figure 8.1 Survival rates for entrepreneurial new establishments, by cohort (Norway, 1996–2001)

included; that is, establishments for which we know the year of entry. Variables are establishment age, initial size (log of number of employees at start) and the type of entrant category.

The number of employees in the first year is used as control variable for size. Number of employees in subsequent years should be seen as part of the performance of the different categories of entrants.

First Table 8.3 presents the log likelihood of surviving as a function of age and initial size, without taking change categories into account. Both

Table 8.3 Likelihood of survival 2001–02 by initial size and age

Source	DF	Chi square	Pr>ChiSq
Intercept	1	1315.63	<.0001
Logeini	1	548.25	<.0001
Age	1	212.60	<.0001

Table 8.4 Likelihood of survival 2001–02 by initial size, age and change category

Source	DF	Chi square	Pr>ChiSq
Intercept	1	1092.87	<.0001
Logeini	1	192.90	<.0001
Age	1	241.43	<.0001
Newcat	3	160.13	<.0001

Table 8.5 Likelihood of survival 2001–02 by initial size, age, industry and change category

Source	DF	Chi square	Pr>ChiSq
Intercept	1	471.13	<.0001
Logeini	1	200.23	<.0001
Age	1	219.70	<.0001
Newcat	3	179.26	<.0001
Nace01	27	235.24	<.0001

age and initial size contribute positively to explaining the probability of survival. Introducing change categories in addition also contributes positively to explaining the survival rates as presented in Table 8.4. The contribution also holds, in fact is slightly strengthened, when controlling for industry as presented in Table 8.5.

The results are illustrated in Figure 8.2, where we show the probability of being a survivor from 2001 to 2002, by age and change category. Here we control for initial size and have chosen to present probabilities for an initial size of 10 employees. Probabilities increase with age and are significantly different between the change categories. As we control for size, this is not a pure size effect. However, the difference between the categories becomes smaller when controlling for size. The probability of survival is

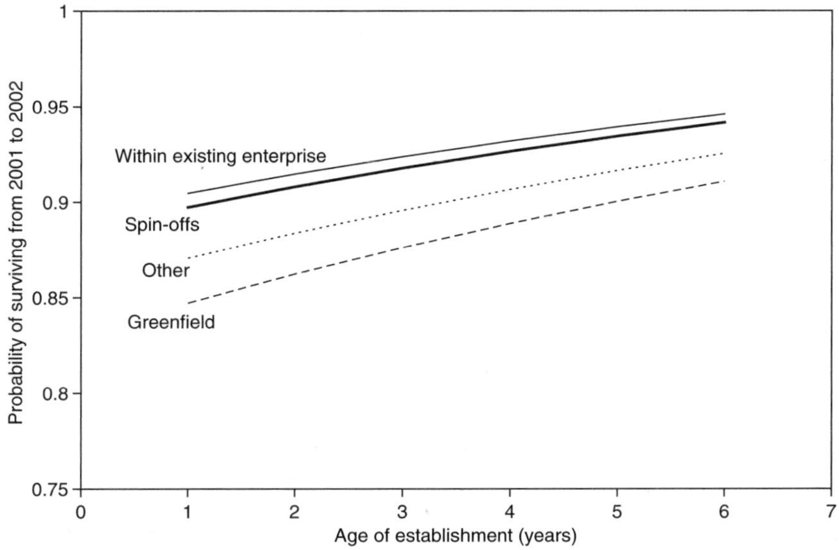

Note: * Controlling for initial size. Probabilities for initial size of 10 employees shown in figure.

*Figure 8.2 Probability of surviving from 2001 to 2002, by age and change category**

higher for new establishments created within existing enterprises, closely followed by spin-offs.

Greenfield births have the lowest probability of survival, and this is particularly true when size is not taken into account, as this group is dominated by small entrants. A possible explanation for this could be that the knowledge generated through previous experience helps survival. It can be expected to be present to a higher degree when a new unit is created within an existing enterprise, and in the spin-off cases where presumably some knowledge has been generated and put into practice. These cases clearly contrast the Greenfield cases where no common prior experience (that we can identify) is involved.

The differences in average sizes of new entrant establishments of different types are illustrated in Figure 8.3 for the 1996 cohort. The average size of survivors increases over time for all four types of entrants. This is partly the result of a selection process: the smaller the establishment, the lower the probability of survival. However, the main part of the increase reflects quite strong growth in employment among the survivors. After six years the average is about 6 employees for the Greenfield group and around

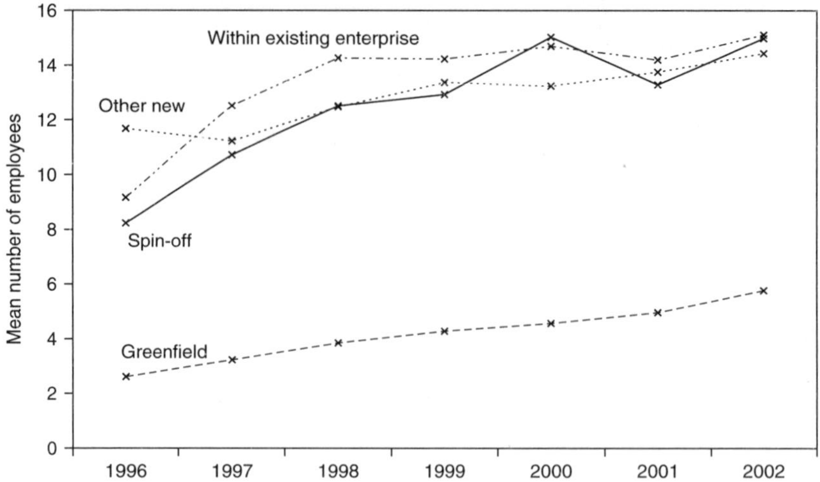

Figure 8.3 Development in mean number of employees in 1996 cohort, by type of entrant

15 employees for all the others. In this respect the Greenfield category is clearly different from the rest. As a result of their larger number they also dominate in terms of the number of surviving firms, both for each single year and cumulated over time. In other words, the higher number of new Greenfield births every year outweighs the higher mortality rate of these firms compared to the other change categories. This is illustrated in Figure 8.4, showing the accumulated proportion of entrant firms in Norway for the years 1996 to 2001. Accumulated over six years the Greenfields total almost 30 per cent of all establishments in 2001. By contrast, each of the three other categories make up some 3–7 per cent, with spin-offs at the bottom.

As a contribution to industrial renewal this clearly makes some difference, as close to 45 per cent of the establishments existing in 2001 had been established during the previous six years. However, compared to the incumbent firms most of the new entrants are small, such that the contribution to the composition of employment is smaller. This is illustrated in Figure 8.5 where the employment in each of the types of entrant firms is taken into account. Greenfield births still remain ranked first, contributing around 9 per cent of total employment in 2001 (again the accumulated effect of entrants over the previous six-year period). Spin-offs are found at the lower end, accounting for just under 4 per cent of employment in 2001.

In aggregate the four categories of entrant establishments discussed above account for around a quarter of total employment in 2001 when

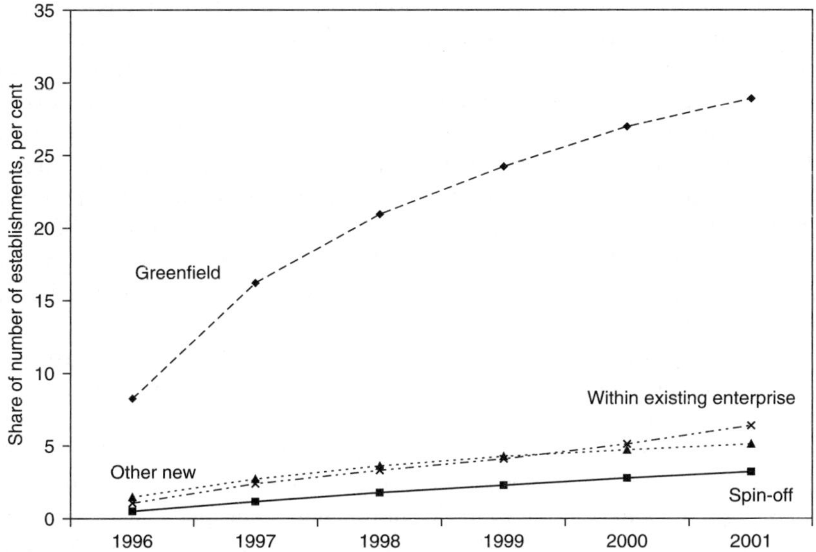

Figure 8.4 *Entrants' cumulative share of total number of establishments from 1996 to 2001, by category*

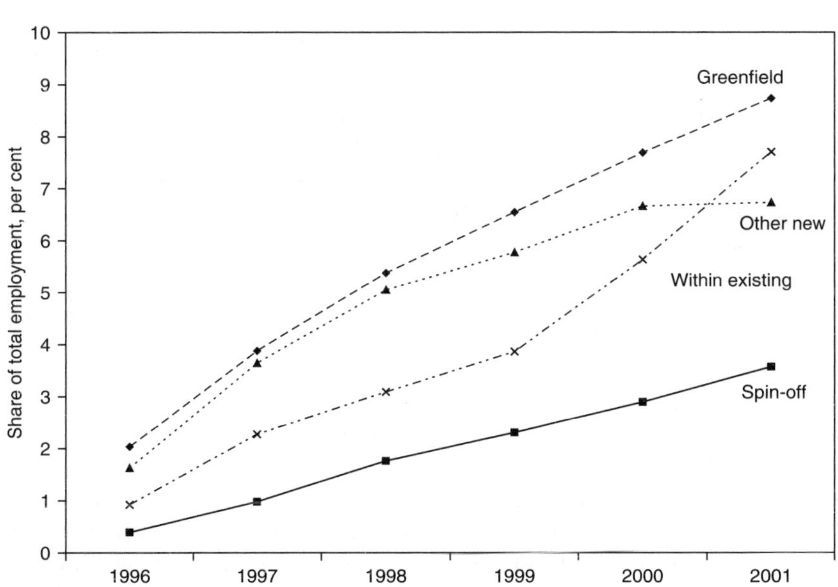

Figure 8.5 *Entrants' cumulative share of total number of employees from 1996 to 2001, by category*

changes over the whole six-year period are included. This is clearly a significant proportion and will consequently contribute to renewal – internally to the industry, and in terms of industrial structure. As was shown above there is far more dynamism within certain service industries than in manufacturing industry. However, part of this dynamism consists of exits, and to a large extent exits occur in the same industries that dominate in terms of higher proportions of entrants. Therefore, the resulting effect on the observed industrial structure is limited – with the level of aggregation of industries applied here. One should be aware, though, that the more disaggregated the industrial breakdown, the larger will be the changes in the observed structure.

In Figures 8.6 and 8.7 a breakdown into 19 different industries is applied, where a comparison is made of the share of employment in each industry affected by each major type of change over the six-year period for Norway. Just over 50 per cent of total employees in 1995 were not subsequently affected by any change. Some additional 25 per cent work in surviving establishments where some kind of change has taken place, whereas some 25 per cent experienced that the establishment was closed down by 2001. These percentages vary considerably between industries, from around 55 per cent exits in fish farming, down to some 10 per cent in larger processing industries (chemicals, pulp and paper) and even to 5 per cent in metals. Apart from financial services – with a high share in the unchanged category – the stability seems to be higher within manufacturing industries than in most service industries.

The resulting net changes in industrial structure are illustrated in Figure 8.8 and Figure 8.9 for the number of establishments and the number of employees, respectively. As can be seen, changes are moderate at this level of aggregation. The larger positive contribution is found for business services and computing, and for trade. A negative development is recorded for one service industry: transport and communications. For the manufacturing industries changes are smaller, but in most cases negative. Changes are generally larger when weighted by employment, apart for transport and telecommunications.

Besides the gross and net changes in industrial structures reported above we can also address the question of increasing or decreasing heterogeneity more directly by applying some measure of concentration, and study its development over time. The simple idea in this context is that increased concentration within industries means a reduction in heterogeneity, and decreased concentration indicates a rise in heterogeneity. This should of course be coupled with information on the competitive environment as pointed out in the introductory section. The Herfindahl index indicates the degree of concentration of employment for a limited number of

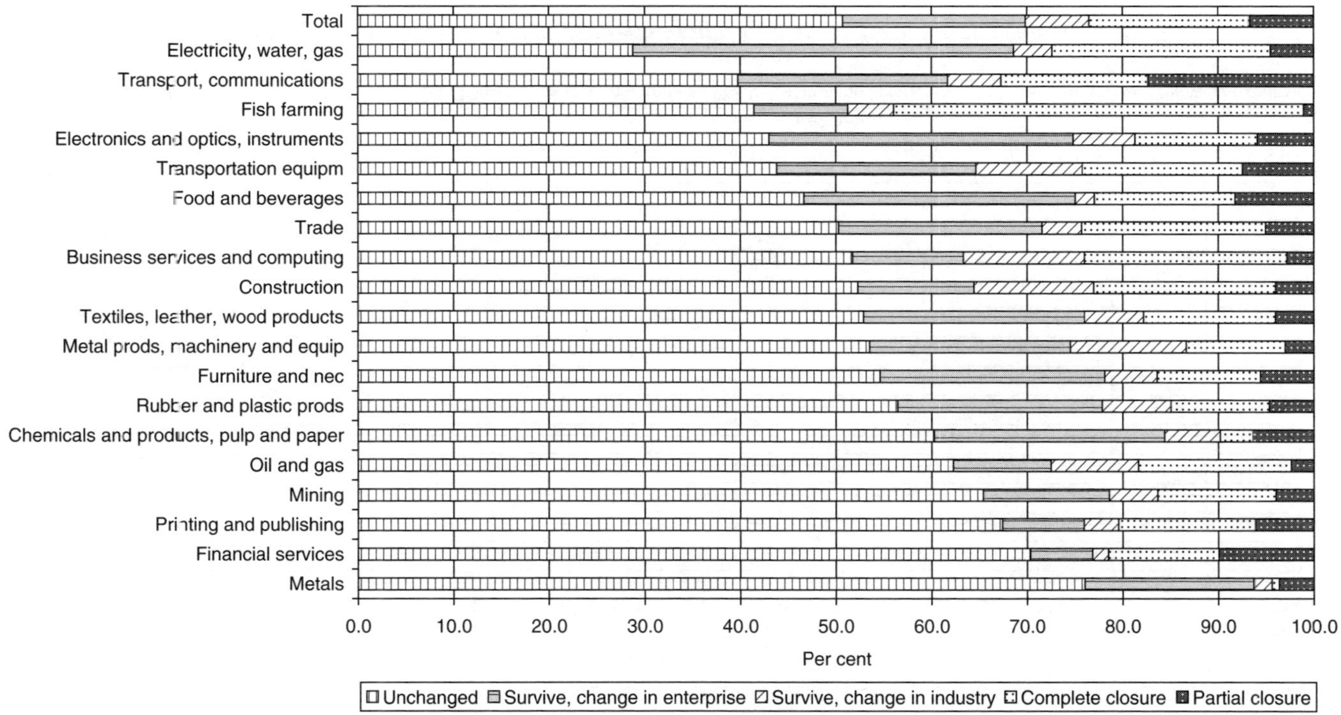

Figure 8.6 Share of employees in 1995 affected by different kinds of change, by industry (Norway 1995–2001)

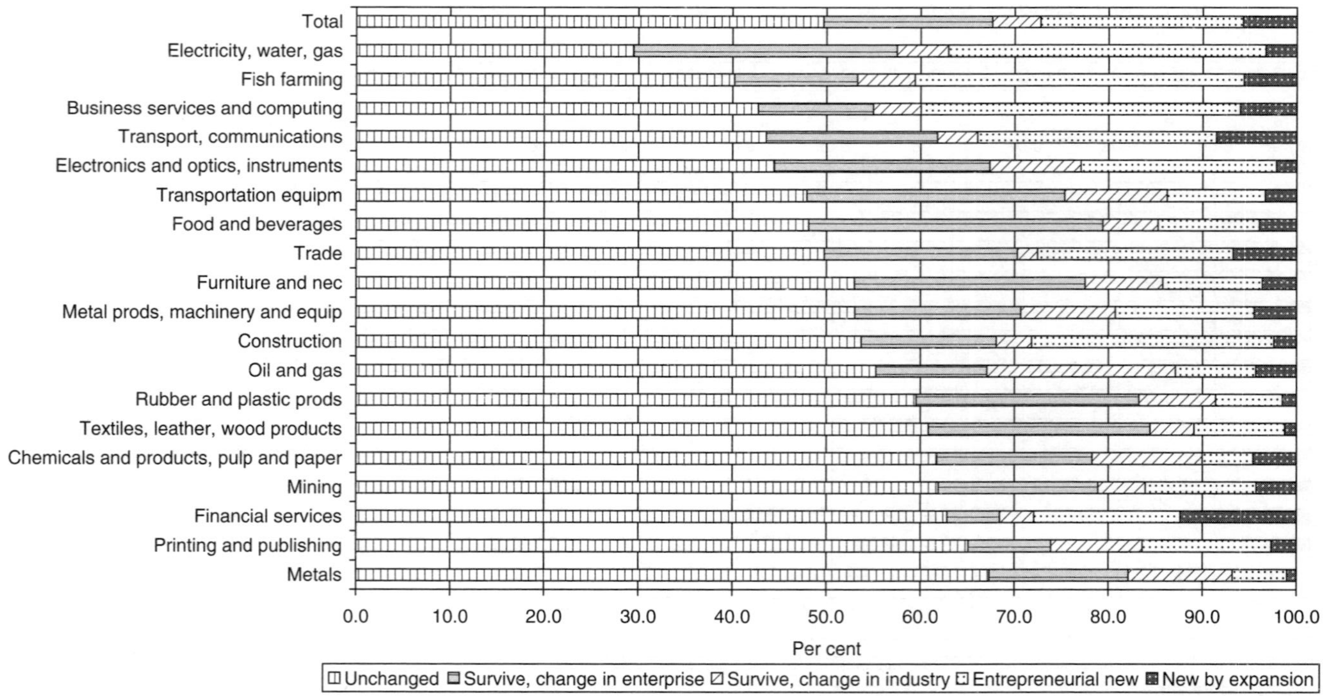

Figure 8.7 *Share of employees in 2001 affected by different kinds of change, by industry (Norway 1995–2001)*

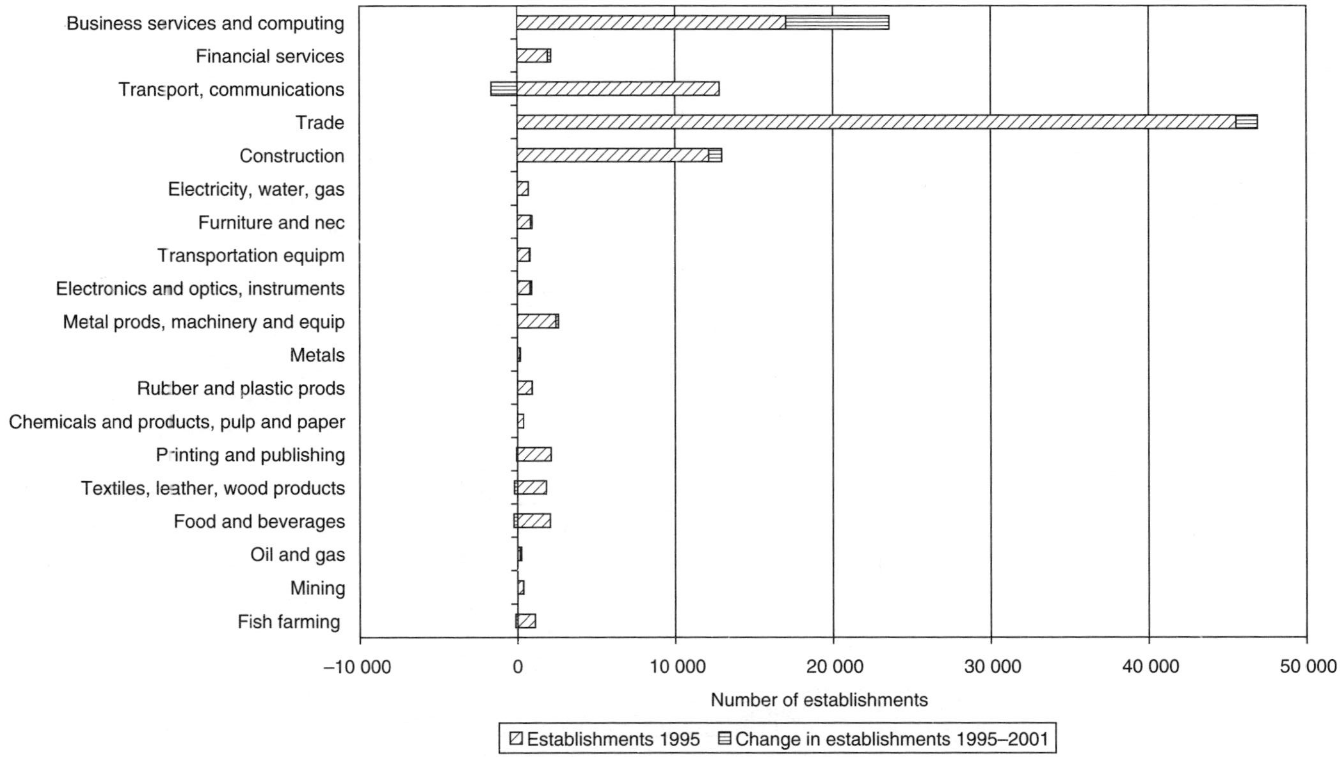

Figure 8.8 *Number of establishments 1995 and change 1995–2001, by industry (Norway)*

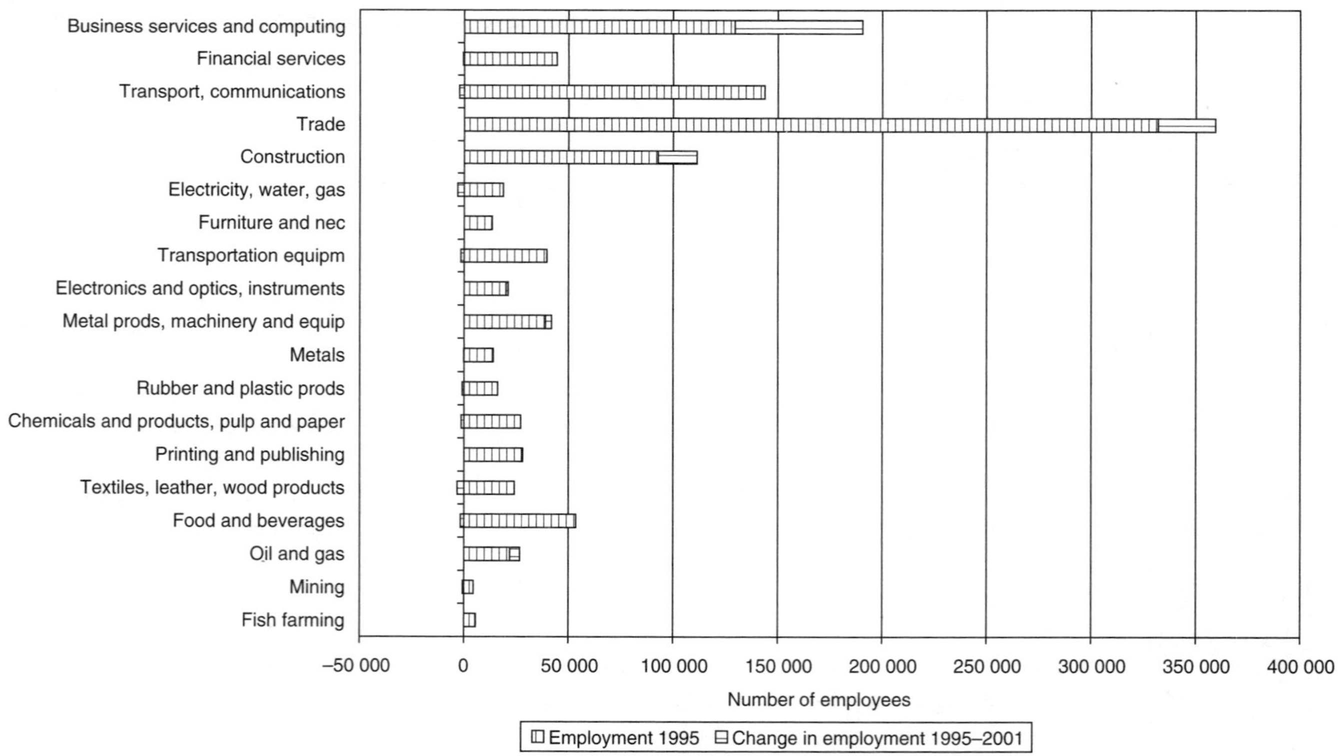

Figure 8.9 Employment 1995 and change 1995–2001, by industry (Norway)

establishments within industries. The Herfindahl index for the Norwegian case is characterized by very low concentration in all industries.[6] Over time there are only marginal changes in concentration rates, and the general trend is in the direction of less concentration – or more heterogeneity. Interpreting this result one should keep in mind that the unit of analysis is establishments – or 'plants' – which demonstrate a more stable survival pattern than do enterprises. At the level of enterprises it is still possible, and likely, that heterogeneity is reduced over time to the extent that each enterprise comprises an increasing number of establishments.

CONCLUSIONS

The analysis has shown that it is reasonable and possible to interpret and operationalize heterogeneity in terms of firm demographic evolution. The number of variables included so far can be supplemented from other sources to provide a more detailed picture. Even though a significant degree of entry, exit and reorganization of different kinds is documented, the composition of industries changes only moderately over time, indicating that the demographic events identified largely appear to reproduce existing structures rather than significantly altering them. Even if heterogeneity in terms of composition of industries in the economy only evolves slowly, heterogeneity in terms of composition of firms and employment within industries may change. However, the degree of concentration of employment in certain establishments within industry classes is generally low. The concentration ratios change only marginally over time, and in the direction of even less concentration. This is interpreted as an indication of preserved or even increased heterogeneity, with the reservation that competitive environments may affect how we consider the relationship between concentration and heterogeneity. One should also be aware that establishments are more stable units over time than enterprises, so that heterogeneity at the enterprise level may develop differently from what goes on at the level of establishments.

NOTES

1. Spin-offs are the formation of new firms by splitting out part of an existing firm. Greenfield births are the formation of totally new firms without origin in existing firms.
2. In principle the methodology can also be applied to public sector activities. These are usually insufficiently subdivided into 'enterprises' and 'establishments' to make all of the analyses meaningful. In this respect the practice will, of course, also vary between countries.

3. The main reasons are that the coverage of these sectors varies over time, and the institutional forms of businesses are not consistent over time. While some actors are registered as ordinary firms, others operate as self-employed or one-man firms. Including these sectors in the analysis would bias the analysis purely as a result of administrative practices.
4. SIC, Standard Industrial Classification, is an international standard. In the present work the Norwegian implementation SN92 is applied.
5. It is important to emphasize that there are three different types of effect behind such an aggregate result: period effects, age effects and cohort effects. This is not further explored in this context, however. In addition there is a size effect so that the probability of surviving increases with size at the time of establishment.
6. Postal services and telecom seems to be an exception in 1988 with a high concentration rate. Given the significantly lower rates in subsequent years we suspect this result to be an error in the underlying data.

BIBLIOGRAPHY

Callan, Benedicte (2001), 'Generating spin-offs: evidence from across the OECD', in OECD, *STI review No. 26*, *Special Issue on Fostering High-tech Spin-offs: A public Strategy for Innovation*, Paris: OECD.

Carroll, Glenn R. and Michael T. Hannan (2000), *The Demography of Corporations and Industries*, Princeton, NJ: Princeton University Press.

Caves, R.E. (1998), 'Industrial organization and new findings on the turnover and mobility of firms', *Journal of Economic Literature*, **XXXVI**, 1947–82.

Chamanski, Alexandre and Sigmund J. Waagø (2001), 'The organizational success of new, technology-based firms', Stavanger University College Report, September.

Cooper, A.C. (1973), 'Spin-offs and technical entrepreneurship', *IEEE Transactions on Engineering Management*, EM-18 (February) 2–6.

Dale-Olsen, Harald and Dag Rønningen (2000), 'The importance of definitions of data and observation frequencies for job and worker flows – Norwegian experiences 1996–1997', *Statistics Norway Discussion Papers No. 278*, July 2000.

Davidsson, Per and Frederic Delmar (2003), 'Hunting for new employment: the role of high-growth firms', in David A. Kirby and Anna Watson (eds), *Small Firms and Economic Development in Developed and Transition Economies: A Reader*, Aldershot: Ashgate Publishing, pp. 7–20.

Delmar, Frederic (1997), 'Measuring growth: methodological considerations and empirical results', in R. Donckels and A. Miettinen (eds), *Entrepreneurship and SME Research: On Its Way to the Next Millennium*, Aldershot: Ashgate Publishing.

Delmar, Frederic and Per Davidsson (1998), 'A taxonomy of high-growth firms', in P.D. Reynolds, W.D. Bygrave, N.M. Carter, S. Manigart, C.L.M. Mason, G. Meyer and K.G. Shaver (eds), *Frontiers of Entrepreneurship Research*, Wellesley, MA: Arthur M. Blank Center for Entrepreneurship, Babson College, p. 399–413.

Dietrich, G.B. and D.V. Gibson (1990), 'New business ventures: the spin-off process', in F. Williams and D.V. Gibson (eds), *Technology Transfer – A Communication Perspective*, London: Sage Publications

Eriksson, T. and J.M. Kuhn (2003), 'Firm spin-offs in Denmark 1981–2000 – Patterns of entry and exit', paper presented at the CAED conference in London, September.

European Commission (2001), 'Corporate and research-based spin-offs: drivers for knowledge-based innovation and entrepreneurship', proceedings of the expert workshop held in Brussels, 18 January.

Eurostat (2001), *Statistics on Science and Technology in Europe. Data 1985–1999*, Brussels: Eurostat.

Eurostat (2002), *Business Demography. Recommendations Manual*, Draft, December, Brussels: Eurostat.

Geenhuizen, Marina van and Peter Nigkamp (1995), 'A demographic approach to firm dynamics: formation of new firms and survival of old ones', *Research Memorandum 199525*, Vrije Universiteit, Amsterdam.

Geroski, P.A. (1995), 'What do we know about entry?', *International Journal of Industrial Organization*, **13**(4), 421–40.

Grorud, Ann-Kristin H. (2002), 'Bedrifts-og foretaksregisteret. Regler og rutiner for ajourhold', *Statistics Norway*, Notat 2002/57.

Ijiri, Yuri and Herbert A. Simon (1977), *Skew Distributions and the Sizes of Business Firms*, Amsterdam: North-Holland.

Jensen, J. Bradford and Robert H. McGuckin (1997), 'Firm performance and evolution: empirical regularities in the U.S. microdata', *Industrial and Corporate Change*, **6**(1), 25–47.

Lindholm, A. (1994), *The Economics of Technology-Related Ownership Changes*, Gothenburg: Department of industrial management and economics, Chalmers University.

Lundvall, B.Å. (1992), *National Systems of Innovation. Towards a Theory of Innovation and Interactive Learning*, London: Pinter Publishers.

Malerba, Franco and Orsenigo, L. (1993), 'Technological regimes and firm behavior', *Industrial and Corporate Change*, **2**, 45–71

Møen, Jarle (2002), 'Spin-offs and spillovers: tracing knowledge by following employees across firms', *NHH discussion paper 5/2002*.

Nås, Svein Olav, Anders Ekeland, Christian Svanfeldt and Mikael Åkerblom (2001), 'Knowledge transfer through labour mobility in the Nordic countries: structure and dynamics', in OECD, *Innovative People. Mobility of Skilled Personnel in National Innovation Systems*, Paris: OECD.

Nelson, R.R. (1996), 'Why do firms differ, and how does it matter?', in R.R. Nelson, *The Sources of Economic Growth*, Cambridge, MA: Harvard University Press.

OECD (1986), *OECD Science and Technology Indicators, No. 2: R&D, Invention and Competitiveness*, Paris: OECD, pp. 58–61.

OECD, (1996), *The Knowledge Based Economy*, Paris: OECD.

OECD (2001), *STI review No. 26, Special Issue on Fostering High-tech Spin-offs: A Public Strategy for Innovation*, Paris: OECD.

Parhankangas, Annaleena and Pia Arenius (2003), 'From a corporate venture to an independent company: a base for a taxonomy for corporate spin-off firms', *Research Policy*, **32**, 463–81.

Roberts, E.B. (1968), 'A basic study of innovators: how to keep and capitalize their talents', *Research management*, 11 (July), 249–66.

Roberts, E.B. and H.A. Wainer (1968), 'New enterprises on Route 128', *Science Journal*, **4** (December), 78–83.

Salvanes, Kjell Gunnar and Ragnar Tveterås (2004), 'Plant exit, vintage capital and the business cycle', *The Journal of Industrial Economics*, **LII** (2, June), 255–76.

Schmalensee, Richard (1989), 'Intra-industry profitability differences in US manufacturing 1953–1983', *Journal of Industrial Economics*, **37** (4, June), 337–57.
Spilling, Olav R. (2001), 'Vekstforetak i Norge. Om vekstforetak vekstforetaks dynamikk og politikk for vekstforetak', *Forskningsrapport 5/2001*, Handelshøyskolen BI.
Storey, D.J. (1994), *Understanding the Small Business Sector*, London: Routledge.
Svanfeldt, Christian and Jonny Ullstrøm (2001), 'Firm demography: mapping firm dynamics using human resource data', in OECD, *Innovative People. Mobility of Skilled Personnel in National Innovation Systems*, Paris: OECD.
Tveterås, Ragnar and G.E. Eide (2000), 'Survival of new plants in different industry environments in Norwegian manufacturing: a semiproportional Cox model approach', *Small Business Economics*, **14** (1), 65–82.
Utterback, J.M. (1974), 'Innovation in industry and the diffusion of technology', *Science*, **183** (February), 620–6.

9. IPRs and Norwegian enterprises: diversification of innovative efforts in Norwegian firms

Eric J. Iversen

INTRODUCTION

This chapter examines the generation of technological and commercial variety in the Norwegian economy using the complementary lenses of domestic trademark and patent data. Trademarks, which are increasingly used to understand economic activity, are useful in distinguishing products and services from rivals. They can be linked to the differentiation of commercial activity with an assumed innovative character. Patents on the other hand capture technologically innovative activity with an assumed commercial application. They can be linked to innovation, especially in R&D intensive fields such as pharmaceuticals.

Patent and trademark registration each reveal something about the ongoing differentiation of economic activity: the former emphasizes invention, the latter commercialization; the former tends to emphasize activity in manufacture, the latter activity in the service sector. There is therefore a significant degree of complementarity in these lenses. This chapter uses the combination to explore the heterogeneity that underlies longer term Norwegian industrial evolution. It first looks at the role of heterogeneity in terms of industrial change, in firm-level activities, and the role of the IPR systems. It then goes on to look at firm-level data illustrating how different Norwegian firms use the two systems.

HETEROGENEITY AND INNOVATION

In general terms the innovation process can be understood to involve the sustainable generation, distribution and utilization of new economically relevant knowledge that continuously accumulates and is recombined in the economy (David and Foray, 1995). The generation of variety is the engine of this evolutionary process and it is recognized to pay clear dividends in economic development. Saviotti (Saviotti and Metcalf, 1991) emphasizes the role of variety, suggesting that the increased net variety of goods and services may be more than a result of the evolution of the economic system; it may be considered a key aspect, closely complementing increased productivity efficiency. Trend growth in variety is associated with a build-up of complementary skills as well as new techniques; it is ultimately associated with changes in the competition landscape in the direction of greater choice and lower prices.

Heterogeneity and Industrial Evolution

The question of heterogeneity is central to the key question of how industries evolve. A persistent degree of heterogeneity of organizations is assumed to be desirable in terms of the knowledge bases, the productive behaviour, and the organization of firms. Together such factors help to promote industrial evolution. Indeed, a variety of social science subpopulations have increasingly studied the contribution of variety in organizations and the economy as a whole. A body of more sociologically rooted work has notably grown up to study different aspects of the population ecology of organizations (e.g. Hannan and Freeman, 1977; Carroll, 1985).

In economics, Marshall's early assertion that, 'the tendency to variation is a chief cause of progress' (Marshall, 1920/1962), has attracted renewed interest.[1] Such study has especially re-emerged in the Schumpterian tradition (see also Spilling, Chapter 7, this volume). Indeed, the emergence of 'new combinations', which are crucial to Schumpeter's story of economic development, may be directly linked to the tendency towards heterogeneity. And interest in this fundamental level of industrial dynamics has indeed evolved along a Schumpeterian vein of inquiry into a variety of areas. These include the sectoral composition of the economy, industrial demography and population ecology, stability of firm-size distributions, persistence differences between organizations, asymmetric firm performance, and so on.[2] During the past couple of decades this far-ranging inquiry has improved what is known about the heterogeneity of firms and its implications for different knowledge, competences and learning processes.

Such efforts to understand industrial evolution in this sense have at a fundamental level been inspired by evolutionary biology. One central link has been recognized (Melcalfe, 1994; Andersen, 2004) between the extent and effect of changing firms in industrial evolution and the Fisher theorem of natural selection, especially in Price's formalized approach. This link regarding the mechanism of change provides insight into the basic role of heterogeneity in industrial evolution and its link to innovation. According to Andersen, the Fisher theorem links the pace of evolutionary change directly to variance in the behaviour of a population, where Fisher, 'treats selection in terms of what has later been called replicator dynamics or distance-from-mean dynamics' (Andersen, 2004, p. 129). Price subsequently (e.g. 1972 until his death in 1975) developed a formalized approach based on measures to trace changes in such population characteristics. The Price equation uses the divergence from the mean in such metrics to partition populations and to trace changes. The equation can be linked to the simple models of Nelson and Winter and in the further efforts to isolate the 'selection effects' from 'innovation effects' as argued by Andersen (2004).

Industrial evolution involves transformation in underlying knowledge bases, in technologies, in actors and actor competences, in organization and relationships (e.g. with users), in products and processes, and in institutions. In this setting, the technological regime literature operates on the assumption that the potential for knowledge growth is conditioned by technological specificities (different technologies have different potentials) and by sector characteristics (the technological environment and the dimensions of demand constrain or otherwise shape development paths). Different technologies have different problem-sets associated with them, and solving these tends to define the 'routines' of the firms in the field (e.g. Orsenigo et al., 1997; Malerba and Orsenigo, 2002).

This view emphasizes the importance of interdependencies and complementarities in the economy. The premise is that industrial evolution can be more or less limited by the interaction between knowledge bases, the regime context (including appropriability conditions), and by demand. This dependence on complementarities requires a sufficient heterogeneity in the economy to perpetuate industrial evolution. The concern is with getting the creative destructive process to become a 'creative accumulation' process (Schumpeter Mode II, see Pavitt, 1984), and to avoid technological mono-cultures that threaten to reduce the learning dividends of a dynamic system.

Heterogeneity and the Firm[3]

But what are the mechanisms at the firm level that spur diversification of their activity within a country? Firm-level innovation processes are recognized to be shaped by an interaction between factors internal to the firm, such as strategy, physical resources and capabilities, and factors outside its boundaries. Diversity in the contextual aspects, for example, makes intercountry comparisons difficult (Smith, 2001). A major element of the external factors is the institutional conditions that both enable and restrict certain activities: such factors shape the way firms manage their resources.

In general, the process of differentiation stems from enterprises' attempt to distinguish what they sell from rivals in markets that are less than perfect. Schumpeter's emphasis on 'technological competition' (Schumpeter, 1942/1975) indicates that firms develop products and/or services that are new and stand out next to those of rivals; firms develop new processes and new techniques that improve the quality of what they sell; or firms explore new channels of distribution or new ways to influence consumer demand. Instead of focusing purely on efficiency, firms may take risks by investing in distinct and new products / processes / services, that, if successful, will allow them to charge more for what they sell without direct threat of competition.

Of course, many markets tend to blend the cost component, the technological component and the taste component. The importance ascribed to each will differ not only according to the main type of market (commodity, product or service) but also, to a certain degree, according to the maturity of the relevant product or service market. Mature markets characterized by little innovation will tend to behave more like commodity markets in which rivals compete principally on price (i.e. price-oriented competition), while the innovative aspect will be more important in emerging markets (technology-oriented competition).

Notwithstanding, three modes of competition can be distinguished. In competitive markets, firms can attract buyers by making what they sell:

- distinct in terms of price,
- distinct in terms of technological performance,
- distinct in the eyes of the consumer,
- distinct in a variety of ways that overlap these three categories.

In instrumental terms, competitive position depends on the firm's internal capabilities related to purchasing, finance, production, conducting R&D, marketing, distribution, sales and other functions important to its business. Organizational elements are also important, not least the firm's competitive strategy. Internal capabilities are necessary but not sufficient to establish and defend a firm's competitive position. Since competition is essentially a relational phenomenon, there will necessarily be factors outside the firm that will condition its position. Two general types of external factors can be distinguished. First, there are general aspects regarding the firm's competitive environment that are important and to which it has to be attuned and responsive. These are aspects that confront the firm but are difficult to observe. They include aspects of market structure both on the supply and the demand side, aspects such as the dynamics of demand, cost conditions, the existence of economies of scale, the size of capital requirements, and so on. A second set of external factors involves potential links the firm can develop with its surroundings. One important element is the role that access to public infrastructure and aspects of the regulatory framework plays in shaping the competitive environment. The availability and quality of the 'economic infrastructure', regulatory conditions and climate, and the nature and extent of institutional support are some of the external factors that are important (Guerrieri and Tylecote, 1994). Together internal and external factors contribute to the balancing of generation, selection, and accumulation processes of new knowledge.

Heterogeneity and Institutions

The role of institutions is important here, not least in relation to small firms whose limited resources might nip innovative activities in the bud (see Johnson, 1988). In general, it can be said that institutions hold three basic functions in relation to innovation. They can reduce uncertainty by providing information, they can help to manage conflicts, and they can provide incentives for example to promote R&D investment (Edquist and Johnson, 1997). In general, intellectual property rights (IPRs) have a role to play in this 'economic infrastructure' in organizing knowledge production, in promoting new R&D, and in promoting further utilization, as well as in coordinating use of new knowledge, while avoiding underutilization losses.

At the firm level, the assumption is that IPRs can augment the position of a firm by helping it to protect the distinctiveness of its products and/or services both in terms of the underlying technological originality – notably through patents and utility models – and in terms of their distinctiveness in the eyes of the customer – notably through trademarks and industrial designs. The use of relevant types of IP protection can potentially aid the competitive position of the firm by affording it the room to cultivate its distinct qualities without threat from direct competition from imitations. This suggests that IPRs may play important roles in managing IP in cases of technological competition. In terms of the economy as a whole, the way IPRs do this implies both costs and benefits for different actors. On the one hand, IPR protection brings with it social costs in the form of higher prices (monopoly pricing); on the other, IPRs provide the economy with an incentive to innovate (based exactly on the prospects for the innovative firm for monopoly pricing). The monopoly profits provided by IPRs may have an added advantage for the economy as a whole if they are ploughed back into higher levels of production and innovation.

EMPIRICAL EVIDENCE OF TRADEMARK AND PATENT USE

These traces of novelty provide two complementary lenses on the differentiation of economic activity. This section builds on Iversen (2003) to take stock of this effort in the Norwegian case as viewed through these lenses. It empirically explores the extent of heterogeneity of Norwegian innovation and commercialization processes as viewed through domestic patenting and trademark data.

First, some further aspects of the lenses should be noted. Patenting, commonly associated with the role of an appropriation mechanism, emphasizes the diversification of technological ideas that the resulting focus on R&D activities promotes. This focus on the role of patents, however, tends to overlook or downplay the contribution of important areas of economic activity. The emerging prominence of the service sector is one area whose importance, although not ignored, is reduced by the patent lens while other technological areas tend to be magnified.[4] The patent lens for example underplays such major shifts as the ascendancy of the service industries in the economy.

The widespread use of patents as an indicator of innovation has therefore rightly been accused of missing the sectoral change in the economy (see Saviotti and Metcalf, 1991). The trademark lens provides a substantially different look at firm-level differentiation, especially in terms of making products and services distinct in the eyes of the consumers. A trademark can be registered for 'signs' that differentiate products and services from rivals. The applicant can apply to have a proprietary name trademark for a defined product class. A registered trademark keeps competitors from mimicking the identity of the product on the market. In this way the mark becomes associated with the quality of a distinct product or a service that the consumer has come to recognize. It becomes a signal to consumers and provides a basis for marketing that cultivates the product's distinctiveness in the mind of the consumer (see Landes and Posner, 1987, which focuses on the impact of trademarks on 'search costs'). The trademark is particularly relevant to the increasingly important role played by the service sector (e.g. Greenhalgh and Rogers, 2006). Trademarks thus complement nicely the traditional use of the patent lens with its recognized sector bias.

THE APPROACH AND DATA

This section focuses on the domestic use of these instruments in the ten-year period from 1994 to 2003. It uses a unique dataset of domestic patents and trademarks (provided by the Norwegian Patent Office) that has been associated with full-count registry data for Norwegian enterprises. The approach was developed for the Wipo report on IPR use by small and medium-sized enterprises in Norway (Iversen, 2003), but is updated here. The enterprise-level information includes information about firm size, industrial activity, number of companies, annual turnover, and so on. It comes from a unique, publicly assembled registry covering all active Norwegian companies. The registry is put together by Statistics Norway on the basis of firm-level information from the Brønnøysund Register Centre's (http://www.brreg.no/english/) register of Norwegian enterprises and companies, and the National Insurance Service's (Rikstrygdeverket) registry of active employees and employers. This database gives us a picture of all enterprises (and subsidiary companies) that formally pay wages to at least one person (a registered workforce of about 2 million). Further information is found in Appendix 9.1.

In this presentation, counts of applicants are normalized for Norwegian applications.[5] The introduction of normalized counts presents a more accurate picture of Norwegian patenting and trademark activity, although it makes comparison with previous work more difficult.

Patenting

Three aspects of Norwegian patenting activity are of interest in terms of heterogeneity: the general tendency of Norwegian patenting (temporal and spatial), the participation of different size-classes, and the relative technical spread of Norwegian patenting in terms of foreign patenting in Norway.

Regional profile

The overall volume of 'Norwegian patent applications'[6] expanded by 5 per cent from 1994–98 to the next five-year period, 1999–2003. A total of 12 628 patents were applied for, involving 15 094 applicants (an average of 1.2 applicants per application). Based on the Norwegian Patent Office's internal number, 7290 separate entities were involved. The 17 Norwegian counties or 'fylker' are represented throughout the period. The overall ranking of counties remains largely the same across the periods, with the majority of applicants concentrated in a few of the more urban counties (see Figure 9.1).

The six most patent-intensive counties account for 70 per cent of Norwegian applicants. Oslo and the adjoining county of Akershus together account for more than a third (35 per cent) of the total volume of applications. This profile, however, corresponds to underlying features, particularly distributions of employment and relative R&D intensities. Oslo and Akershus make up the most R&D intensive area at the county level, with

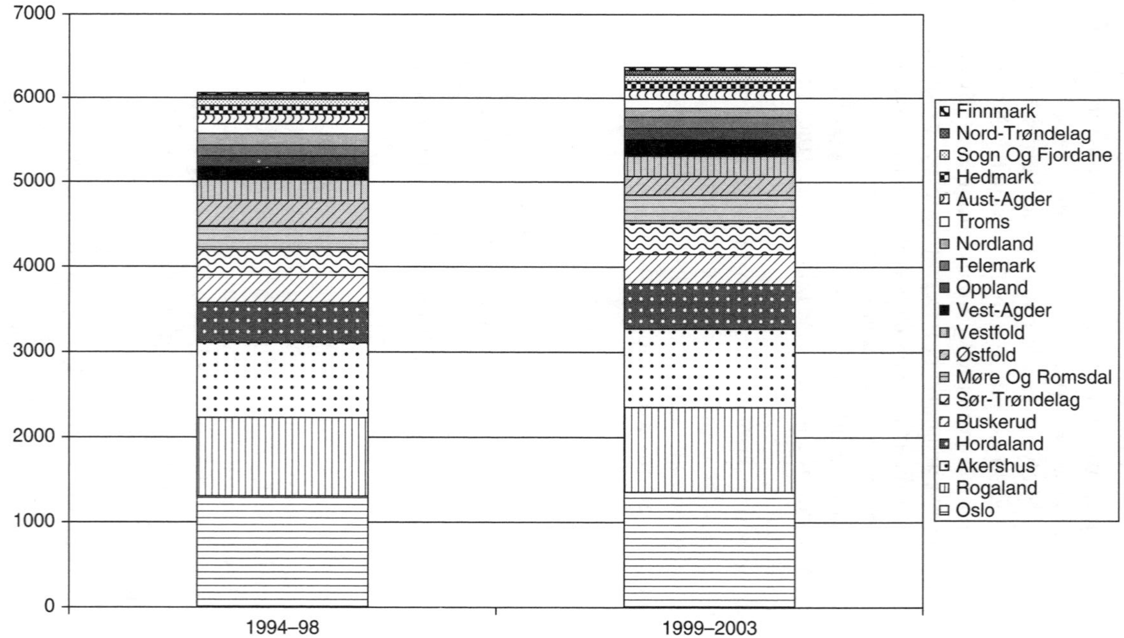

Note: * 12 628 is the normalized counts of applicants; 160 apparently foreign applicants collaborating in patents with Norwegian applicants are not presented. A further 47 Norwegian applicants could not be linked to county.

Source: NIFU STEP patent database built on Norwegian Patent Office data.

*Figure 9.1 Primary assignees (12 628) in Norwegian patent applications (normalized counts) by district of origin (= Fylker): 1994–2003**

a combined average R&D expenditure per capita that is over twice the national average of 5400 NOK in 2001, with Oslo at 13 300 NOK and Akershus at 9000 (The Research Council of Norway, 2003, p. 23).

The patenting share of Rogaland, which is the seat of the oil industry, increased 8 per cent across the two periods, while that of Sør Trøndelag, the home of the technical college, and of Sintef, grew by 21 per cent. The figure indicates that while the participation of applicants expanded in Oslo during the period, its overall share decreased from 22 per cent to 21 per cent of the total. The patent lens thus indicates that inventiveness is concentrated in several counties associated with high overall levels of R&D, but that it is also spread throughout the country. The overall expansion of patenting activity, which rose by a third through the period, is relatively evenly spread across the country. Applicants in urban areas however clearly dominate.

Technological specialization

The proportion of domestic patenting (i.e. at least one applicant with a Norwegian address) has remained stable for more than a decade at around 20 per cent of the total annual volume of applications registered by the Norwegian Patent Office. This minority situation is familiar for many small countries and is especially associated with the aggressive patent strategies of large pharmaceutical companies. As a group domestic applications differ technologically from those filed from abroad. Table 9.1 breaks down

Table 9.1 Patents applied for in Norway by technical area in NACE equivalents: 1994–2003

Sectors	Norwegian	Foreign	Total	NACE industries
Consumer goods	1 064	1 573	2 637	15, 16, 17, 18, 19, 36
Motor vehicles and equipment	1 507	1 343	2 850	34, 35
Material and process engineering	2 195	5 773	7 968	20, 21, 25, 26, 27, 28
Chemicals and petroleum products	852	8 553	9 405	23, 24 (not 24.4)
Machinery and equipment	3 713	8 112	11 825	29
ICT, electrical equipment, instruments	2 718	9 681	12 399	30, 31, 32, 33
Pharmaceuticals	375	12 774	13 149	24.4
Unknown	117	1	118	Missing IPC
TOTAL	12 541	47 810	60 351	

Source: NIFU STEP patent database built on Norwegian Patent Office data.

the 60 350 patent applications the Norwegian Patent Office received in the period 1994–2003 by technical area using a recently developed correspondence between patent IPC classes and industrial (NACE) classes (Schmoch et al., 2003).

The largest concentration of domestic filings is found in machinery and equipment, while relatively few patents are filed in the areas of chemicals and pharmaceuticals. In contrast, foreign filings were most active in pharmaceuticals followed by chemicals and petroleum products. This is particularly due to the fact that the chemical and pharmaceutical industries rely heavily on patenting to protect their products, which are expensive to develop but relatively inexpensive to imitate. The globalization of markets means that patent protection is sought globally. Patenting in this field is dominated by large multinational companies with large patent portfolios. This raises the concentration of foreign applications, which becomes especially pronounced among the small volumes of domestic patents in small countries.

The asymmetry between domestic and foreign filings in the country provides a basis on which to gauge domestic technological specialization. Technological specialization is typically expected to be more pronounced (see Andersson and Ejermo, 2006) but also stable in small countries (see Dosi, 1988). Table 9.2 uses the established index developed by Grupp to study Revealed Patent Advantage, based on a Revealed Patent Advantage approach (Archibugi and Pianta, 1992). This specialization index uses a band of +/− 100, which is symmetrical around 0 (corresponding to no specialization). In it, larger positives (such as 72 for Textiles, etc.) correspond to greater degrees of specialization of domestic filings, while larger negatives (such as −95 for Pharmaceuticals) correspond to higher specialization among foreign filings.

Dividing the period 1994–2003 into two parts, this approach illustrates the degree of specialization and tendencies towards change in what can be called the 'market for technology' in Norway. Two general dimensions emerge: there is some evidence of shifts in the balance across the two periods but otherwise a general persistence in Norwegian patent specialization, and there is a high degree of domestic specialization in two areas across the periods. A similar approach is used in Kaloudis (Chapter 12 in this volume), where EPO data are used instead of domestic patent data.

The index indicates relative stability in the spread of specialization. The negative and positive indexes generally retain the same sign across periods (with the exception of food and beverages). There is some fluctuation in individual indexes, including a drop in the specialization level for electrical equipment and a rise in office machinery. Paper and wood products mark a reduction in the dominance of foreign filings, as does inventive activity linked to television and related products.

Table 9.2 'Revealed Patent Advantage' in Norwegian domestic patenting: two periods 1994–2003

NACE 1.1	Sectors	1994–98	1999–2003
15, 16	Food, beverages & tobacco products	–2.8	2.2
17, 18,19, 36	Textiles, clothes, furniture	69.4	72.0
20, 21	Paper and wood products	−35.1	–18.4
23	Petroleum products & nuclear fuel	–42.6	–71.5
24 (not 24.4)	Chemical (excl. Pharmaceuticals)	–75.3	–66.9
24.4	Pharmaceuticals	–97.7	–95.6
25, 26	Rubber, plastics, and non-metallic products	20.8	23.3
27, 28	Metals and metal products	41.1	36.1
29	Machinery	39.6	33.6
30	Office machinery & computers	20.8	31.6
31	Electrical equipment	29.4	6.0
32	Television, radio and electronics	–49.0	–22.6
33	Instruments	15.0	9.5
34, 35	Vehicles, vessels and parts	73.1	70.2
NA	Unknown	90.6	91.0

Source: NIFU STEP patent database built on Norwegian Patent Office data. Patent IPC classes are translated to NACE 1.1 equivalents. This version of RTA is based on Hariolf Grupp's approach.

A high degree of specialization is found in two areas. The first, involving consumer goods such as textiles and furniture, means that the inventive activities of Norwegian firms tend to be more focused on the Norwegian home market for consumer goods. The second area of high specialization of inventive activities is concerning vehicles and vessels, primarily shipping related. The considerable patenting activity from major oil exploration and exploitation companies is however divided between several areas of this index, such as machinery (including energy machinery) and chemicals (especially under the subclass of rubber and plastic products). The index is not necessarily a good indicator of the industrial activity of the applicant.

Applicant type

The patenting activities of large firms form a popular focal point for innovation. The Yale surveys for example focus on the R&D units of relatively large firms, while the Community Innovation Survey in Europe, which includes questions about patenting, excludes firms of under ten employees and uses a stratified sample among the firms of up to 100 employees (in Norway). However, the majority of firms in many countries are small ones, which can form an important but often invisible or overlooked source of innovations in the industrial dynamics of a country (see also Nås and Sandven, Chapter 8, this volume). At the same time, many patent applications – especially at the domestic level – are sought by individuals without obvious commercial affiliations.

It is therefore worthwhile to link the volume of academic patent applications by the type of patent applicant in order to get an idea of the contribution of different segments of the economy to overall inventive activity. This type of population accounting provides an idea of where the inventive activity is coming from. What is the mix? Does one type of firm dominate; large firms for example, with dedicated R&D activities and resources to follow up in the bid to turn inventions into innovations? (See Figure 9.2.)

Figure 9.2 illustrates the breakdown of Norwegian patent applications by size-classes through the period. It shows that although unaffiliated individuals accounted for 43 per cent of Norwegian domestic patenting in the period, there is a clear downward trend in their participation. Large firm patenting is remarkably stable across the period, accounting for roughly 18 per cent of the total.

The patenting activity of SMEs, on the other hand, grew strongly during the period, and fluctuated more. It accounted for at least 31 per cent of the total. The size-classes of two other categories of enterprises could not be established for different reasons (see Appendix 9.1). The categories of size and firm unknown, which account for a further 7.6 per cent, are likely to be small firms without formal employment. If so, small firms are involved in nearly 40 per cent of Norwegian domestic patenting. SMEs were most significantly involved in mechanical engineering, instruments, rubber and plastics, and television and electronics.

Trademarks

Trademarks provide a substantially different view of heterogeneity in the Norwegian economy. Whereas the patent lens provides a look at differentiation in terms of technically oriented invention, the trademark lens provides a look at commercialization activities based on making what the applicant sells distinct in the eyes of the consumer. The use of trademarks involves a substantially different part of the population of Norwegian firms. There are many fewer manufacturing firms, and many more firms in the tertiary sector, not least in retail and wholesale. The overlap seems to be of the order of 5–10 per cent.

The basis and the approach are much the same as for the patent database.[7] There are however differences to be noted. First, trademarks do not

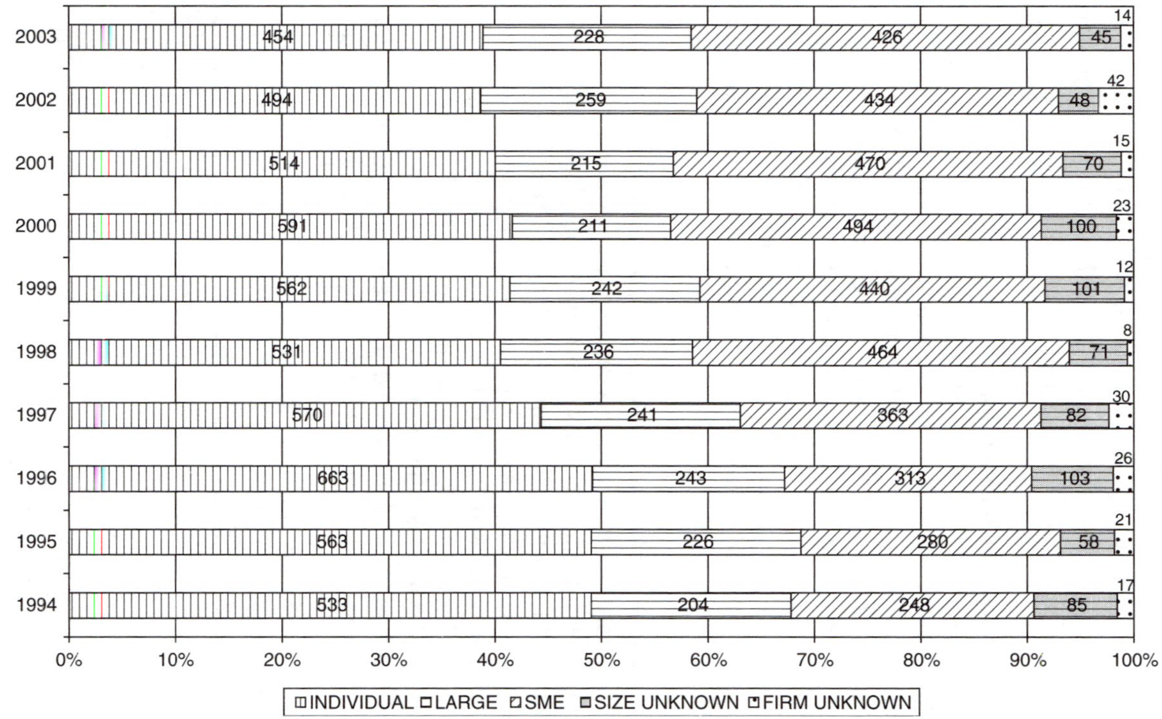

Source: NIFU STEP patent database built on Norwegian Patent Office data.

Figure 9.2 Norwegian patent applicants by type and size-class: 1994–2003, normalized counts

have technical fields that we can translate to technical areas. Instead it has 45 fields of application that are not immediately helpful in analysis. Second, the status is less fine-grained. These are ignored for present purposes. The underlying industrial activity relies solely on the link to the applicant's industry where available. We know only whether the trademark has been registered or has not (yet) been registered. Finally, the incidence of multiple applicants is much smaller than in the case of patenting.

Trademark applications and registrations

There were 2900 Norwegian trademark applications in 2003, as against 2550 in 1996. In the period 1994–2003 the annual number of Norwegian applications fluctuated from a low of 1828 in 1994 to a high of 3800 in 2000. Smoothing these fluctuations, Norwegian applications grew 32 per cent from 1994–98 to the subsequent five-year period 1999–2003. Foreign applications expanded by 54 per cent across the two periods, while the Norwegian share dropped from 35 per cent to 30 per cent. The corresponding share of trademarks registered for Norwegian applicants fell from 25 to 21 per cent (see Table 9.3).

Regional comparison of patent and trademark applications

The degree of commercialization activity represented by trademarks can be seen in relation to the inventive activity behind patenting. Trademark applications are an indicator of market competition. Trademark activity is therefore expected to move in conjunction with geographical markets and, to a high degree, in scale with them. Indeed, the major population centres – Oslo and environs, Trondheim, and Bergen – account for nearly three-quarters of the total number of Norwegian applications.

Figure 9.3 demonstrates the regional distribution of Norwegian trademark applicants. The dominance of the area around the capital, Oslo and

Table 9.3 Trademark applications and registrations per year for domestic and foreign applicants: 1994–98 and 1999–2003

Trademarks		1994–98	1999–2003
Applications	Norway	12 295	16 172
	Foreign	34 966	53 766
	Percent Nor	35	30
Registered	Norway	7 937	10 531
	Foreign	31 725	50 989
	Percent Nor	25	21

Source: NIFU STEP trademark database built on Norwegian Patent Office data.

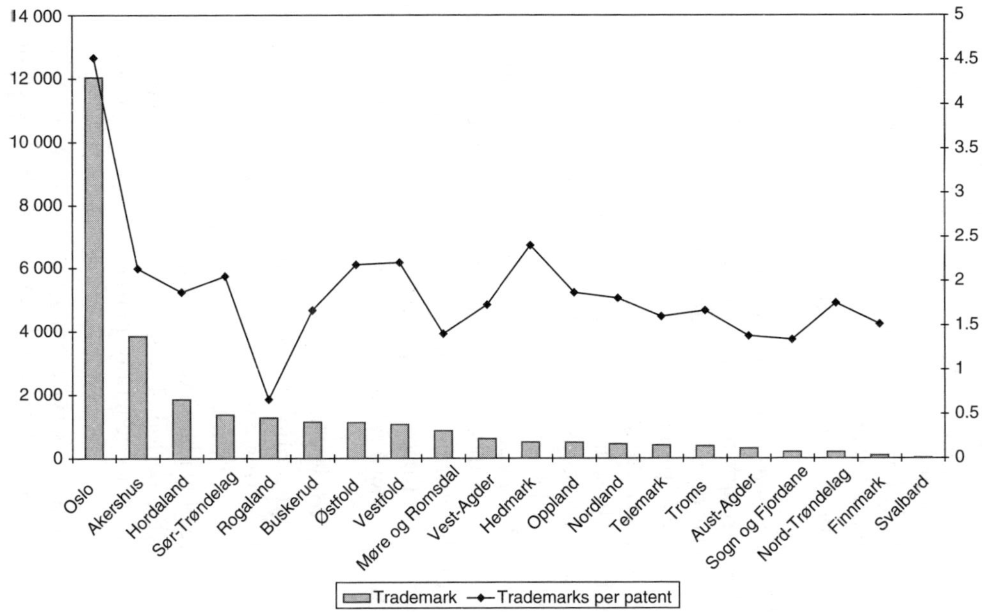

Note: The patent applications included 160 foreign applicants. These are excluded, together with a total of 90 applicants whose geographic location could not be established (42 for trademark and 47 for patent applicants)

Source: NIFU STEP trademark database built on Norwegian Patent Office data.

Figure 9.3 Patenting and trademark applications by district of origin: 1993–2004. Two-handed axes: the left shows fractional counts for trademark applications (normalized counts); the right shows the proportion of trademarks per patent applications

Akershus, is even more striking than with patent applications. Trademark applications generally outnumber patent applications two-to-one for the period. The relative levels, however, vary considerably at the regional level, indicating that these activities reflect something about economic activity beyond scale in a given region. The two-handed figure illustrates the regional variations between patent and trademark levels. The line indicates the proportion of trademarks to patent applications (the right-hand axis). Oslo sets itself out as the commercial centre, where trademarking leads patenting four-to-one. In Rogaland, a moderate-sized Norwegian county where the oil-industry is based, there are a disproportionate number of patent applications in relation to trademark applications.

Norwegian trademark applications by field of applicant

The business areas of trademark applicants overlap to a modest degree with patent applicants. The degree of overlap is of the order 5 to 10 per cent, indicating that the populations involved are quite distinct. Organizations involved in diversification in the eyes of the market are thus substantially different from those involved in technological differentiation involving patenting. Larger more diversified firms are more likely to apply both for patents and trademarks.

Trademark users are predominantly service sector firms. The number of trademark applicants who are individuals with no evident affiliation makes up around 14 per cent of the overall volume of trademark applications (in contrast to over 40 per cent in the case of patent applications). Figure 9.4 presents the breakdown of enterprise applicants by industry for the two five-year periods.

The use of trademarks in Norway increased by about a third (32 per cent by normalized Norwegian applications) from the mid-1990s to the first part of the new century. Manufacturing enterprise accounted for a substantial 3300 applications or 11 per cent of the trademark applicants identified here. This group, which is more readily identified with patenting activity, grew by 10 per cent across the ten-year period. The largest applicant group is that of retailers and wholesalers who generally market the wares of foreign producers.

The more knowledge intensive service enterprises (see Wiig Aslesen, Chapter 10, this volume) are actually more intensive users. The combination of business, computer/telcom, and research and teaching services (which includes research institutes, educational institutes, as well as some public sector services) outnumber retail and wholesale industries, accounting for over 27 per cent of all domestic trademark applications. In addition, trademark use in these sectors grew much faster than the average for the period. Both computer and telecom services, and research and teaching

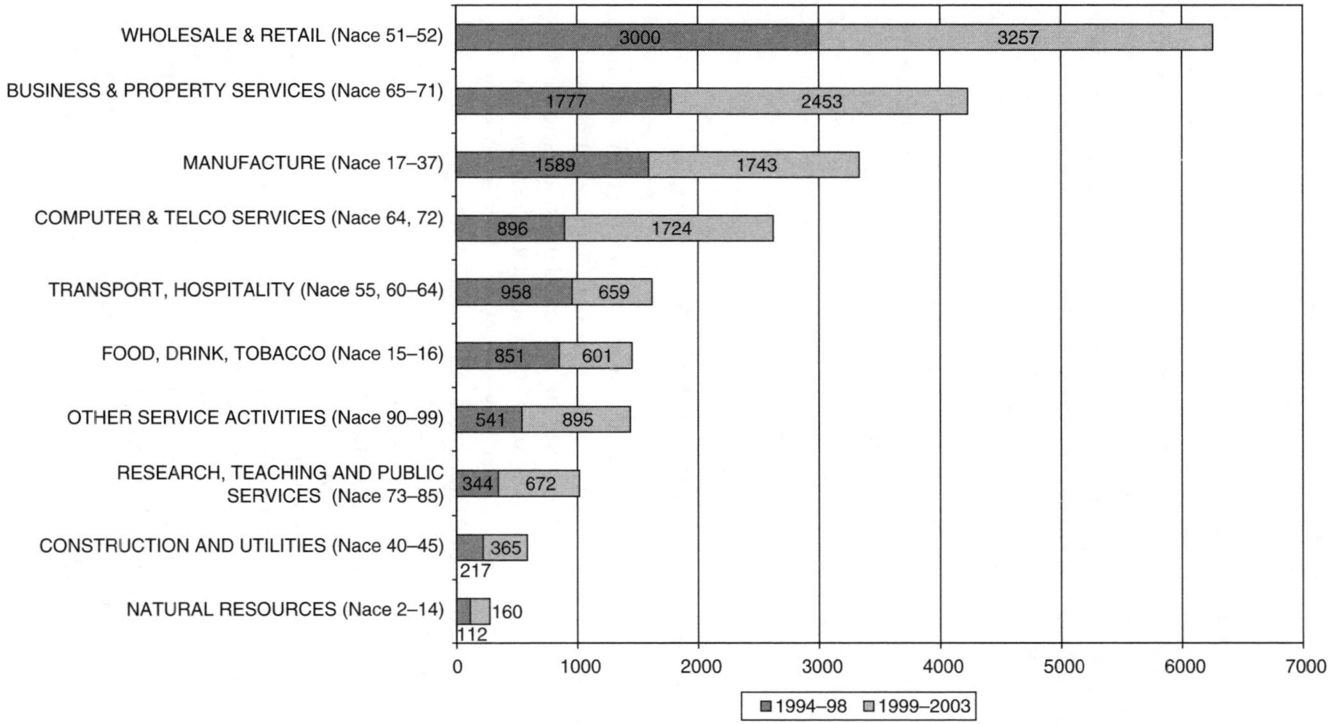

Source: NIFU STEP trademark database built on Norwegian Patent Office data.

Figure 9.4 Norwegian trademark applicants by industry (N = 28 482): 1994–2003

services more than doubled during the period, the latter in part a testimony to effects of the Norwegian dot-com era.

Norwegian trademark applications by size of applicant

Small firms dominate domestic trademark applications, accounting for over 55 per cent of domestic trademark applications in the ten-year period. As above, the actual proportion is likely to be higher, given that enterprises whose size could not be identified in this exercise are likely to be small firms. Application levels among small firms fluctuated the most, especially during the economic boom from 1998 to 2001. At the height of the boom (2000), small firm applications numbered 2150, or 140 per cent above their level four years before and four years after (see Figure 9.5).

Trademark applications among large firms on the other hand remain relatively stable through the period, at about 18 per cent of the total. A major difference with patenting is the much lower proportion of individuals involved in trademark activities. Their proportion was however more stable here than with patenting, strengthening suspicions that a substantial number of these individuals are actually single-person enterprises.

CONCLUSIONS AND DISCUSSIONS

The importance of the role of diversification processes in the economy has been recognized at least since Marshall classed it as 'a chief cause of progress' in the 1920s. (Marshall, 1961, p. 355, cited in Cohen and Malerba, 2001, p. 587). The relationship between 'the tendency to variation' in innovative activities and economic progress is however harder to show. This chapter has explored the diversification of innovative activities through the complementary lenses of patent and trademark activity. These lenses allowed us to focus on how different firm-types (size-groups, geographical locations, and industries) contribute differently to the differentiation of inventive activity (the patent lens) and the differentiation of commercialization activity (the trademark lens).

This exploration forms a potentially useful way to uncover patterns both in these aspects of the important diversification process as well as in the contribution of different firm types to it. We note the general escalation in the overall levels of inventive and commercialization activity. Comparing the late 1990s with the early 2000s, the chapter illustrates the regional diversification of inventive and commercialization activity. Five areas where patenting intensity is high were identified, but we do not find signs that this activity is being centralized; Oslo's share of domestic patenting fell slightly, for example. The pattern is largely the same for firms involved in trademark activity.

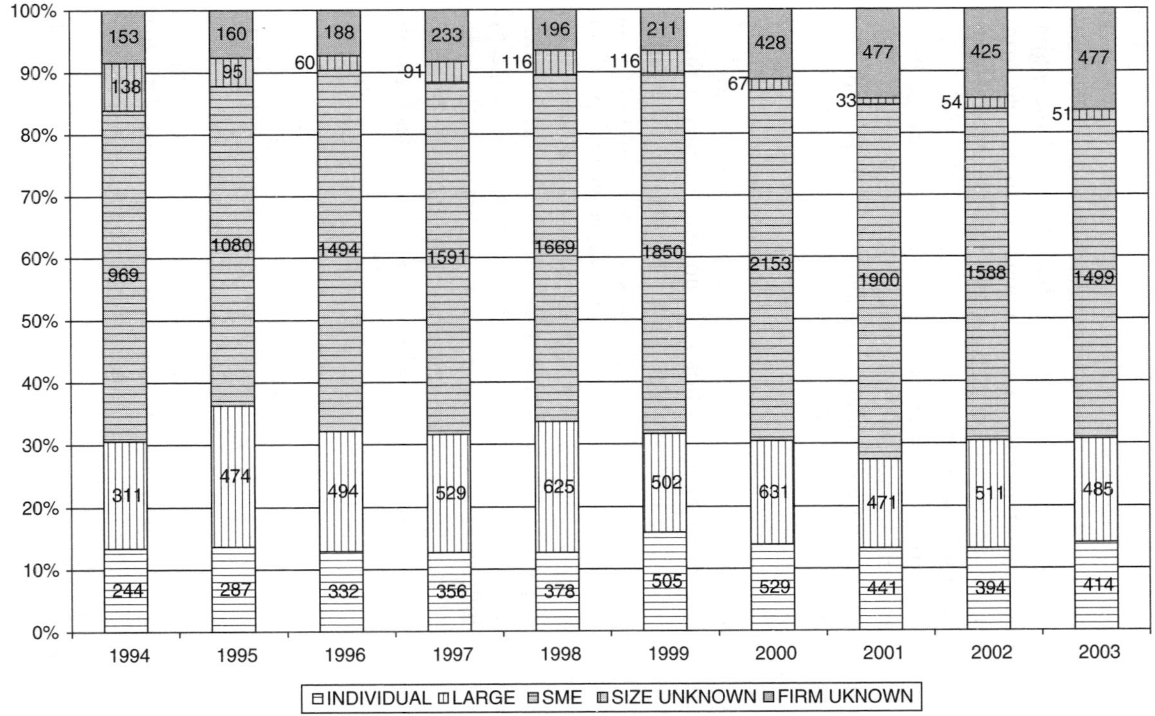

Source: NIFU STEP trademark database built on Norwegian Patent Office data.

Figure 9.5 *Trademark applicants by size-class (N = 28 475), normalized counts*

The chapter took particular pains to look at changes in technological specialization of Norwegian patent applicants. The pattern here was relatively stable. We noted the relative emphasis among domestic applicants to patent in the field of machinery and equipment and consumer goods, while the level of patenting in pharmaceuticals was relatively low. We also looked at the industrial activities of different size-classes that are involved in inventive activity. This revealed the relatively broad spread of the majority of small firm patenting in areas such as instruments, ships and electronics. A disproportionate percentage of small patentees however were found in the field of machines and machinery

Trademark activity provides a different look at the diversification of economically important activity across firm-classes. The major difference is that it affords a look at the tendency of service-based firms to diversify themselves in the eyes of the consumer.

Trademark activity, like domestic patenting, increased through the period. Small firms are more prominent in the profile of Norwegian firms that apply for trademarks than in the profile for patenting firms in Norway. The chapter notes the large contingent of firms in wholesale and retail industries that use the trademark system. The level among other services, not least among 'knowledge intensive firms' in the area of financial services and consultancies, bears witness to a large degree of commercial diversification of small and large firms across the country. But the activity is not isolated to the service sector, just as patenting extended beyond the manufacturing firms.

This overall exploration thus indicates the extent of these two aspects of diversification for different sets of Norwegian firms. This should complement indications of heterogeneity from the other chapters in this book, and help to provide a well-rounded view of what types of activities it involves and what types of firms are involved in it.

APPENDIX 9.1 THE COMPILATION OF THE DATABASES AND THEIR ANALYSIS

The database analysis conducted in this study is based on coupling the identity of Norwegian applicants for trademark and patents with firm-level information available for a full count of Norwegian enterprises. We linked the Norwegian Patent Office databases covering patents and trademarks with publicly compiled registry data covering all Norwegian enterprises. This was clearly the best possible way to approach the question of who uses the patent and trademark systems in Norway. It was pursued because it could provide a totally unique and detailed picture of Norwegian applicants for these two types of rights.

Some problems were inevitably confronted, which required more work than anticipated. The major reason for this was the fact that the link between databases had to be done on the basis of names (and zip codes) of the Norwegian applicant; there was no reliable identifier in the applications that would allow a link with the public registry data (see recommendations). Since errors or variations occur in the names columns of the databases involved, this required different approaches to make the links, combined with a large degree of manual checking of the links. Below we provide more information about the NPO databases and the Registry data, how the databases were linked, and assumptions made in their interpretation.

The NPO Data

A. The selection of the trademark and patent data was done along the following lines:
The time-span covered all applications that were active from 1994 to 2003. By this we mean all applications that were received from 1 January 1994 to 31 December 2003, or any application that was granted during that period, regardless of application date.

The information included information about who had made the application (the names of all applicants and an unreliable identity number), where they were (applicant address and zip code), when (application date and, if applicable, grant/registry date), what the application was for (application titles and the primary IPC class), and how it stood (the status of the patent application, for example whether granted, whether withdrawn/rejected and under what conditions, or whether still under examination).

B. This data was then cleaned, and the following links made:
The zip codes were associated to county and district levels via the Norwegian Post's database. The primary IPC classes of the patent applications were associated with Technological Areas by a widely used Correspondence Key: the INPI/OST/ISI Key, Version 3. Or with NACE via Schmoch et al. (2003).

The Registry Data

This data was then associated with full-count registry data of Norwegian enterprises. The enterprise-level information used here includes information about firm size, industrial activity, number of companies (in later years), annual turnover, and so on. It comes from a unique, publicly assembled registry covering all active Norwegian companies. This type of registry is only found in a limited number of countries, especially the Nordic countries.

The registry is put together by Statistics Norway on the basis of firm-level information from the Brønnøysund Register Centre's (http://www.brreg.no/english/) register of Norwegian enterprises and companies, and the National Insurance Service's (Rikstrygdeverket) registry of active employees and employers. This database gives us a picture of all enterprises (and subsidiary companies) who formally pay wages to at least one person (a registered workforce of about 2 million).

A. The selection of the registry data was conducted along the following lines:
For enterprises (foretak) versus establishments (bedrift), the enterprise level was used and all values (number of employees and turnover) were aggregated up to this level. For industrial activity, each enterprise's industry is defined via the EU's NACE classification (Nomenclature générale des Activités économices dans les Communautés Européennes). The activities of enterprises previous to 1994 when the NACE was introduced in Norway have been linked to the previous classification system used by Statistics Norway (= ISIC). Industrial activity is based on each enterprise's main product. In aggregating up from establishment to enterprise, the dominant NACE class has been used (see NACE 74150, Holding company as special case). The most up-to-data classification was used if this had changed over time.

Zip codes were associated with county and district levels via the Norwegian Post's database, thus allowing us an additional criterion on which to check the identity of the applicants.

B. This data was then cleaned, and defined in the following way:
Defining 'Large' enterprises:
Large enterprises are basically those with a total of at least 100 employees. Three additional criteria are used to define what is considered 'large' here as well. The first supplement involves enterprises which include at least 19 'establishments'; these include Norwegian parts of large franchises. Enterprises defined under NACE 74150 (Holding corporations) that employ more than 30 are also considered large. These include diversified corporations whose management is defined as a separate enterprise. In order to pick up all large-scale operations, enterprises with a combined turnover of 99 MKR in at least one of the years for which we have turnover data (1997–1999) are also considered large.

There is a disruption at around 1995 in the data, both regarding NACE code and number of employees. One source of these difficulties is the transition to NACE from ISIC Rev 2 classification system. Another is the way the firm-level information was compiled. A third is the fact that several major Norwegian companies were undergoing restructuring at that time (for example the telecoms operator, Telenor). These potential sources of errors have been screened, and any remaining inaccuracies are not expected to affect the results.

Defining 'Individual'

The classification 'individual' is based on applications with no apparent affiliation with an enterprise or other organization. These are applications in which the assignee is listed on the basis of a first and last name, and which do not connect with the significant number of individually run enterprises when the county is also checked. This population potentially includes inventions made at universities, since Norwegian law currently allows academic researchers to own their inventions. The addresses were hand-checked to help prevent incorrectly classifying them.

Defining 'Firm unknown'

A number of IPR applicants whose names seemed to be that of an enterprise or institution did not match the AA register or Enhetsregister. In other words, these entities are not registered in the registry material. These may be companies that failed before they could be registered, or that are in the process of registering. Or there might be a mistake that makes it impossible to link the name in the application to the name in the registry database. This population was manually checked to try to isolate any apparent mistake. We assume that entities in this population are most probably not large companies.

Defining 'Size uknown'

Another population has been called 'size uknown'. These include entities that have an identifiable enterprise number but which cannot be connected with substantive information in the registry database we have. This indicates that there are no employees.

NOTES

1. Marshall (1920/1962) *Principles of Economics*, cited in Cohen and Malerba (2001).
2. See Malerba (2006) for an overview of this legacy and the field.
3. This section relies on Iversen (2003).
4. The sector specificity of the patent system is widely noted. See, for example, Scherer (1983).
5. Normalized counts are used for applications. This is particularly important for patent applications where multiple applicants are common. Trademarks tend to involve single applicants to a much larger degree.
6. Those involving at least one applicant with a Norwegian address.

7. It is based on the updated data from the Norwegian Patent Office database covering all domestic trademarks applied for and/or registered since 1990, which NIFU STEP has linked against the national registry of Norwegian enterprises.

BIBLIOGRAPHY

Andersen, E.S. (2004), 'Population thinking, price's equation and the analysis of economic evolution', *Evolutionary and Institutional Economics Review*, **1**, 127–48.

Andersson, Martin and Olaf Ejermo (2006), *Technology and Trade – An Analysis of Technology Specialization and Export Flows*, Working Paper Series in Economics and Institutions of Innovation No. 65, Royal Institute of Technology, CESIS – Centre of Excellence for Science and Innovation Studies.

Archibugi, D. and M. Pianta (1992), 'Specialization and size of technological activities in industrial countries: the analysis of patent data', *Research Policy*, **21**(1), 79–93.

Cantwell, J. and G. Vertova (2004), 'Historical evolution of technological diversification', *Research Policy*, **33**(3), 511–29.

Carroll, Glenn R. (1985), 'Concentration and specialization: dynamics of niche width in populations of organizations', *American Journal of Sociology*, **90**(6), 1262–83.

Cohen, W. (1995), 'Empirical studies of innovative activity', in P. Stoneman (ed.), *Handbook of the Economics of Innovation and Technological Change*, Oxford: Blackwell.

Cohen, W. and S. Klepper (1992), 'The tradeoff between firm size and diversity in the pursuit of technological progress', *Small Business Economics*, **4**, 1–14.

Cohen, W.M. and Franco Malerba (2001), 'Is the tendency to variation a chief cause of progress?', *Industrial and Corporate Change*, **10** (3), 587–608.

Coombs, R., P. Saviotti and V. Walsh (1987), *Economics and Technological Change*, Basingstoke: Macmillan.

David, P. and D. Foray (1995), 'Accessing and expanding the science and technology knowledge base', *STI Review*, **16**, 13–68.

Dosi, G. (1988), 'The nature of the innovation process', in G. Dosi, C. Freeman, R. Nelson, G. Silverberg and L. Soete (eds), *Technical Change and Economic Theory*, London: Pinter, pp. 221–39.

Edquist, Charles and Björn Johnson (1997), 'Institutions and organizations in systems of innovation', in Charles Edquist (ed.), *Systems of Innovation – Technologies, Institutions and Organizations*, London: Pinter Publishers/Cassell Academic.

Ernst, H. (2001), 'Patent applications and subsequent changes of performance: evidence from time-series cross-section analyses on the firm level', *Research Policy*, **30**, 143–57.

Greenhalgh, C. and M. Rogers (2006), 'Trade marks and productivity in UK firms', *Oxford Intellectual Property Research Centre Discussion Paper*.

Guerrieri, P. and A. Tylecote (1994), 'National competitive advantage and microeconomic behavior', *Economics of Innovation and New Technology*, **3** (1), 49–76.

Hannan, Michael T. and John Freeman (1977), 'The population ecology of organizations', *American Journal of Sociology*, **82**(5), 929–64.

Iversen, E. (2003), *Norwegian SMEs and the IPR System: Exploration and Analysis. The STEP-Group. A Study carried out for the World Intellectual Property Organization (WIPO)*, Switzerland: WIPO.

Iversen, E.J. and A. Kaloudis (2006), 'IP-valuation as a tool to sustain innovation', in Bosworth, D. and Webster E. (eds), *The Management of Intellectual Property*, Cheltenham, UK and Northampton, MA, USA: Edward Elgar.

Johnson, Björn (1988), 'An institutional approach to the small country problem', in C. Freeman and B.-Å. Lundvall (eds), *Small Countries Facing the Technological Revolution*, London: Pinter.

Kamien, K. and N. Schwartz (1982), *Market Structure and Innovation*, Cambridge: Cambridge University Press.

Landes, William M. and Richard A. Posner (1987), 'Trademark law: an economic perspective', *Journal of Law and Economics*, **30**(2), 265–309.

Laursen, K. (1998), 'Revealed comparative advantage and the alternatives as measures of international specialization', DRUID Working Paper No. 98–30.

Levin, R., A. Klevorick, R. Nelson and S. Winter (1987), 'Appropriating the return from industrial research and development', *Brookings Papers on Economic Activity No. 3*, 783–820.

Malerba, F. (2006), 'Innovation and the evolution of industries', in F. Malerba and U. Cantner (eds), Special Issue of JEE on Innovation, Industrial Dynamics and Structural Transformation: Schumpeterian Legacies, *Journal of Evolutionary Economics*, **16** (1), 3–25.

Malerba, F. and L. Orsenigo (1996), 'The dynamics and evolution of industries', *Industrial and Corporate Change*, **5** (1), 51–88.

Malerba, F. and L. Orsenigo (2002), 'Innovation and market structure in the dynamics of the pharmaceutical industry and biotechnology: towards a history friendly model', *Industrial and Corporate Change*, **11**, 667–703.

Marshall, A. (1920/1962), *Principles of Economics: An Introductory Volume*, 8th edn, London: Macmillan & Co.

Metcalfe, J.S. (1994), 'Competition, Fisher's principle and increasing returns in the selection process', *Journal of Evolutionary Economics*, **4**, 327–46.

Orsenigo, L., F. Pammolli, M. Riccaboni, A. Bonaccorsi and G. Turchetti (1997), 'The evolution of knowledge and the dynamics of an industry network', *Journal of Management and Governance*, **1**(2), 147–75.

Pavitt, K. (1984), 'Sectoral patterns of technical change: towards a theory and a taxonomy', *Research Policy*, **13**, 343–73.

The Research Council of Norway (2003), *Det norske forsknings-og innovasjonssystemet – Statistikk og indikatorer (The Norwegian Innovation System – Statistics and Indicators)*, The research Council of Norway.

Saviotti, Paolo (1996), *Technological Evolution, Variety and the Economy*, Cheltenham, UK and Brookfield, USA: Edward Elgar.

Saviotti, Paolo and Stanley Metcalf (eds) (1991), *Evolutionary Theories of Economic and Technological Change: Present State and Future Prospects*, London: Harvard University Press.

Scherer, F.M. (1983), 'The propensity to patent', *International Journal of Industrial Organization*, **1**(1), 107–28.

Schmoch, U., F. Laville, P. Patel and R. Frietsch (2003), 'Linking technology areas to industrial sectors', Final Report to the European Commission, DG Research, available at http://www.obs-ost.fr/pub/TechInd2003.pdf.

Schumpeter, J. (1912/1961), *The Theory of Economic Development: An Inquiry Into Profits, Capital, Credit, Interest, and the Business Cycle*, New York: Oxford University Press.

Schumpeter, J. (1942/1975), *Capitalism, Socialism, and Democracy*, New York: Harper Torchbooks.

Smith, K. (2001), 'Cross-country comparisons: comparing economic performance in the presence of diversity', *Science and Public Policy*, **28**(4), 267–76.

Sutton, J. (1998), *Technology and Market Structure: Theory and History*, Cambridge, MA: MIT Press.

10. Heterogeneity and knowledge-intensive business services in the city

Heidi Wiig Aslesen

INTRODUCTION

Western economies are often described as knowledge-driven, based on the production, distribution and use of knowledge and information (OECD, 2001). However, knowledge matures, so learning, creative forgetting, efficient sharing and transfer of new knowledge are vital for economic development and growth in modern economies. The definition of innovation as interactive learning (Lundvall, 1992) links modern innovation theory with the work of economic geographers, making evident how geographical and institutional organizations create marked differences in regions' cultures, institutions and regulations, all factors that affect learning and innovation. Geography matters for innovation and for competitiveness (OECD, 2001).

One important aspect of the knowledge-driven learning economy is that economic activity and growth have become more spatially concentrated and increasingly city based (Eurostat, 2002; Cooke, 2002). Large cities are often characterized as creative centres of particular significance for innovation and entrepreneurship, and thereby for national economic performance and growth (Acs, 2002; Fischer et al., 2001; Simmie, 2001). With the growing complexity of information and knowledge, and the greater uncertainty of the economic environment, the city is seen as an important source of competitive advantage for organizations operating in a globalized economy (Storper, 1997). The complexity of information and knowledge and the greater uncertainty require an economic environment that has the ability to deal with and promote all knowledge-intensive assets, namely the city. 'Knowledge Intensive Business Services' (KIBS) is a fast-growing industrial sector that has a concentrated location pattern. The growth of knowledge-intensive services has been especially confined to city areas, suggesting that this is where the demand–supply interaction is best developed (Daniels, 1991; Howells, 1988; Marshall, 1988). The city seems to play an important role for learning and innovation in the knowledge economy, suggesting a new centrality in the global economy.

The acceleration and complexity of technological change, and the growth in demand for advanced knowledge and skills, make the ability to make use of information, to create and to manage knowledge, a competitive advantage for firms. Knowledge-intensive service activities, held internally or bought on the market, are crucial for firms' ability to adapt to the changes in modern economies. While firms employ different types of knowledge from various sources in their innovation processes, these processes are strongly shaped by the knowledge and capabilities that are held internally and are a result of a process of cumulative learning, which is inherently imperfect, complex and path-dependent (Dosi, 1997). This delivers persistent heterogeneity between firms, possibly affecting regions' knowledge-based capacity to respond to change. A region's capacity to respond to change will be dependent on the ability to assemble, adapt and apply new arrays of expertise within and between agencies (Wood, 2005a). It is in relation to this that economic actors such as KIBS play a role in the learning economy, since they can operate as agents helping firms adapt to changes in the knowledge economy. In the predominantly service-based economy, the potential for change is dependent on the effective application of service skills. Technological skills comprise only one component promoting change. KIBS play an important role in the city, both as an economic activity in their own right, and in their role as service provider to other firms, with both roles being important aspects of the propensity for learning and innovation in the city.

This chapter examines the sources of agglomeration economies (the benefits of being located in the city) that are of importance to KIBS: What is the importance of heterogeneous sources of agglomeration economies (urban size and density), and what is the importance of homogeneous sources of agglomeration economies (specialization and co-location of similar firms)?

Further, the demand for KIBS is looked into in order to understand what makes knowledge-intensive services so important to city based firms.

This chapter addresses the following questions:

- What does KIBS emphasize as important agglomeration forces in city regions in Norway (especially Oslo)?
- In what way does KIBS help city actors adapt to the complexity and heterogeneity of the knowledge-driven learning economy?

The empirical material presented is based on research financially supported by the Research Council of Norway in the programme 'City Development – Driving Forces and Planning Challenges' ('Byutvikling – drivkrefter og planleggingsutfordringer'). The references given are based on collaborative projects between the author and other researchers receiving funding from the same programme, making available empirical material from many different sources.

The discussion commences with a presentation of the theoretical contributions providing input to an understanding of learning and innovation in today's knowledge-driven learning economy in particular. An introduction to the literature is presented, focusing on the interplay between KIBS and the innovation system of cities, and thereafter KIBS and the different roles these services have in respect of their clients. Finally the chapter presents empirical material from city regions in Norway emphasizing the forces and outcomes of KIBS in the city.

HETEROGENEITY AND DYNAMIC INNOVATION FACTORS

Cities are often viewed as melting pots where complex interactions between a varied set of actors take place. As such, large cities are seen as important for the value creation in society, and for society's economic development and ability to uphold global competition related to investments and production (Vatne, 2005). But what are the most important factors used to understand the innovation dynamics of a city region in today's knowledge-driven learning economy?

The contribution of Alfred Marshall (often referred to as the founder of neo-classical economics) in the late 1800s can be said to be the first attempt to research the system of industrial activity, and represents the historical and conceptual antecedent of cluster studies (Giuliani, 2005). Marshall's writing on industrial districts as a concentration of large numbers of small businesses of similar kind in the same locality (1920, p. 227) focuses on the importance of proximity in order to achieve external economies. External economies were generated by three factors: the local availability of inputs; the presence of a skilled labour force; and knowledge spillover. Marshall (1930) then also brings the specific territorial aspect of a geographical agglomeration of industrial production, sociocultural factors such as mutual knowledge and trust, and the industrial atmosphere, into consideration. Marshall maintained that these factors would have a positive effect on innovation in small firms within industrial districts, while also being aware that agglomeration economies could not be seen as a guarantee for innovation taking place.

Homogeneous and Heterogeneous Agglomeration Forces

Marshall's agglomeration concept has been developed; differentiating between the sources of spatial agglomeration economies (spillover) and has, as such, contributed to concepts more relevant to the city region. Hoover (1948) and Isard (1956) made a distinction between localization economies and urbanization economies, where localization economies were defined as external economies available to all local firms within the same sector, usually taking the form of Marshallian (technical) externalities, and arising from labour market pooling, the creation of specialist suppliers and the emergence of technological knowledge spillover (Henderson, 2003; Feser, 2002). The agglomeration economies are then a result of agglomeration forces that fit one sector in particular, and can be referred to as homogeneous agglomeration forces.

Urbanization economies were defined as external economies available to all local firms irrespective of sectors and arising from urban size and density. This was linked to the presence of social, political and cultural organizations that support the production and absorption of know-how, stimulating innovation behaviour, and contributing to differential rates of interregional growth (Harrison et al., 1997).

Further, the so-called Jacobs's externalities are a variant of urbanization economies. Jacobs's (1969) argument is that heterogeneity is an important mechanism for economic growth, and where external economies are available to all local firms stemming from a variety of sectors. The variety itself might be a source of knowledge spillover and innovation, in that a diverse industry base in an urbanized locality improves the opportunity to interact, modify and recombine ideas, practices and technologies across industries (Frenken et al., 2002). Jacobs's externalities emphasize that the diversity found in cities gives rise to new ideas. She refers to 'the ballet of the streets' – focusing on how the agglomeration of different people with different backgrounds and skills increases the possibility for personal interaction, creating a scene for interaction that increases the probability of developing new ideas, products or processes. Urbanization economies and Jacobs's externalities therefore represent spatial agglomerations based more on heterogeneity, and will therefore be referred to as 'heterogeneous agglomeration forces'.

Localization economies and Jacobs's externalities represent different agglomeration forces and therefore different end results in the economy. Frenken et al. (2002) summarize that localization economies are expected to spur incremental innovation and process innovation, and thus productivity increase, as knowledge spillover derives from similar firms. Jacobs's externalities are expected to facilitate radical innovations and product innovation since knowledge from firms in different industries is coalesced. Jacobs (1969) argues that the heterogeneous environments existing in cities provide a better breeding ground for new ideas because of the possibility of cross-fertilization. Both localization economies and Jacobs's externalities are expected to contribute to regional economic development, but in different ways.

Similar to the thoughts that underpin the term 'urbanization economies', evolutionary economics also emphasizes the important role of variety to create variety (Frenken et al., 2002) as an economic theory interested in out-of-equilibrium dynamics and structural change. It is further characterized by its emphasis on economic growth as a process of qualitative change in the economy, involving the introduction of entities previously not present in an economy, and the disappearance of others, also emphasizing the importance of the institutional framework often co-evolving with the structural composition of an economy (Nelson, 1995). The main sources of qualitative change are innovation in processes, products and services, changing the variety of the composition of an economy (Saviotti, 1996). In evolutionary economics the new varieties leading to new sectors are considered a necessary condition for long-term economic growth in a country, a 'function' spatially confined to urban regions (Frenken et al., 2002).

Immaterial Dynamic Factors and Extra-regional Ties

More recent contributions on agglomeration overlapping with the formulation of Marshall's industrial district, are characterized by the formation of dense networks of formal exchange and untraded interdependencies (Grabher, 1993; Storper, 1995, 1997), which refer to labour markets, public institutions and locally or nationally delivered rules of action, customs, understanding and values. The agglomeration forces are information networks and relations that go beyond economic transactions. The term 'untraded interdependencies' relates to the kind of interaction between actors that is not scheduled or formalized but that occurs as a result of proximity. These kinds of relations and dependencies generate region-specific material and non-material production assets. Storper's interdependencies arise through input–output linkages and through conventions, rules, practices and institutions that combine to produce worlds of production (Storper and Salais, 1997). The benefits of agglomeration were perceived as static efficiency gains. Agglomerated firms can also benefit from more dynamic advantages, linked to a higher propensity for technological learning and innovation in an agglomeration of firms. Knowledge spillovers link geographical space and innovation, and are conceived of as 'technological externalities', leaks of knowledge that diffuse into the economic system and are 'in the air', available as a public good (Giuliani, 2005, p. 26). This suggests that forces that affect economic growth and innovation have shifted from material to immaterial inputs, and in particular have focused on the positive externalities arising from knowledge spillover (Jaffe, 1986; Griliches, 1992). Storper and Venables (2002, p. 4) put it like this: 'Cities used to be centres of agglomeration of material production; now the motor force of agglomeration is the production and communication of ideas, knowledge and information.' Knowledge spillover does seem to be geographically concentrated at the regional level (Audretsch and Feldman, 1996; Jaffe, 1989; Jaffe et al., 1993). The locational advantages that cities provide to businesses lie in their institutional and organizational infrastructure, which makes networking among actors low in terms of transaction costs, and high in terms of knowledge spillover (Cooke and De Laurentis, 2002).

Knowledge spillover is linked to geographical space in that localized knowledge cannot easily be transported over distance. A central distinction will then be between knowledge that can easily be communicated between individuals (explicit or codified knowledge), and knowledge that is complex and hard to transfer between persons or groups of people (tacit knowledge). Polanyi introduced the concept of tacit knowledge (1967), suggesting that 'we can know more than we can tell'. Since tacit knowledge is generated locally, and is therefore place-specific, the spatial element becomes decisive in the communication of tacit knowledge (Hudson, 1999). Trust and personal ties lower transaction costs and facilitate knowledge exchange; repeated interaction over time is more likely to occur when actors are co-located and share the same history and culture (Gertler, 2004, pp. 75–6). Geographically bounded knowledge spillover can occur in both vertical and horizontal relations, both in user–producer relationships, between rivals and from universities and higher education institutions. Knowledge spillover within the urban system is often related to 'scientific knowledge spillover'; that is, to externalities from which firms can profit, such as being co-located with an agglomeration of universities and research centres (Acs et al., 1992; Jaffe, 1989).

The theoretical contributions above link potentials for learning and innovation to geographically bounded endogenous processes. However, other contributions focus more on districts as nodes in global networks (Amin and Thrift, 2002), emphasizing the importance of extra-cluster networking and knowledge to avoid negative lock-in (Grabher, 1993). Amin and Thrift (2002) are critical of the tendency in recent studies to frame urban economies in terms of lines and boundaries. In such studies, cities are seen as central points in knowledge-based agglomerations, with high intensities of face-to-face contact and strong internal flows of knowledge.

However, according to Amin and Thrift, cities cannot be understood as systems of boundaries, or as bounded economic entities. Instead, they treat an urban system as a 'relay station in a world of flows'. Cities are structured around flows of people, images, information, and money. Amin and Thrift (2002) emphasize that cities are always both local and global, and that proximity does not always matter since cities are assemblages of more or less distantiated economic relations. Even when economic activities seem to be spatially clustered within a city, they actually rely on a multiplicity of institutions and connections that stretch beyond these clusters. The flow of knowledge into a cluster can be driven both by actors by outside the cluster as well as by local actors who try to tap into outside knowledge (Cantwell and Iammarino, 2003, cited in Giuliani, 2005), such as multinational corporations.

Amin and Cohendet (1999) emphasize that local business networks are not the only source of tacit knowledge. Firms and individuals have a rich collection of tight external linkages. Bathelt et al. (2004) emphasize the need for clusters to develop pipelines to bodies of knowledge outside the cluster boundaries in order to overcome shortcomings in firms' knowledge and in the local knowledge base. The authors, however, value locally embedded knowledge combined with codified and accessible external knowledge.

Allen (2000) maintains that it is possible to develop and maintain close personal ties over distance, and that one can create what he calls 'communities of practice'. By this he means that even tacit knowledge can flow between people and communities that are located at a distance, bound together by shared experience and expertise; they may also belong to different organizations. These communities consist of groups of workers who have developed a shared understanding of how to carry out specific tasks (Brown and Duguid, 1998), and are increasingly seen as the key sources of knowledge formation, exchange and learning (Asheim and Coenen, 2005). The knowledge may flow across organizational, regional and national boundaries between members of communities of practice (Asheim and Gertler, 2005), extending Marshall's definition of 'industrial atmosphere' and of knowledge being 'in the air' to actors outside the cluster or region.

To sum up, this section has emphasized that in order to understand innovation dynamics in today's city region, additional factors must be included in Marshall's innovation system. First, cities are characterized by heterogeneity, and innovation studies have taught us that not only production factors (material factors) are important for innovation, immaterial factors are equally important. Further, the importance of dynamic efficiency gains, leading to learning and innovation, is today at the core of studies of agglomeration (as opposed to the more static efficiency gains). Extra-regional ties are especially important when focusing on endogenous learning processes in a city.

KIBS AND THEIR ROLE IN THE ECONOMY

The potential for multiple, heterogeneous knowledge flows and exchanges makes an important contribution to learning and growth in city regions. The economies of cities are made up of heterogeneous industrial concentrations comprising diverse and complementary industries, suggesting that, to be able to be competitive, cities must promote all the knowledge-intensive assets that are likely to guide their future national and global roles (Wood, 2005a). Cities' prospects need to be built on their inherited strengths and weaknesses in knowledge-based adaptability and innovation, which today are predominantly non-technological and service-based (Wood, 2005b). The increased demand for services in all industries is a key aspect of the growing importance of cities (Sassen, 2001). Cities are the preferred production sites for such services, whether at the global, national or regional level. Geographers, urban economists and planners have recognized the contribution of advanced services to regional and urban competitiveness and growth (Gottman, 1961, cited in Wood, 2005b).

This section therefore focuses on KIBS as an actor in the city. First we define KIBS and their role, and thereafter present contributions on how KIBS are seen as bridges in the innovation system of the city, spanning a variety of economic actors and geographical levels. Closely connected to this, but more at the firm level, we focus especially on what KIBS contribute to their clients; what actually is transferred from KIBS to their clients.

Knowledge Intensive Business Services

KIBS are services concerned with the supply and management of knowledge and intangible assets (so-called 'knowledge-about-knowledge'). KIBS are private sector firms providing expert knowledge to firms, and are thus seen as a vital source of information, advice and specialized knowledge for other industries (Toivonen, 2004). It is argued that KIBS provide a point of fusion between (1) general scientific and technical information, (2) internal experience and competence acquired by KIBS firms in interaction with clients, and (3) the more tacit knowledge located within the daily practice of client firms and sectors (Hertog, 2000). Knowledge-intensive business services are said to have a strategic role in stimulating innovation processes (Miles, 2003).

KIBS are often used as an indicator of the more general structural changes projected by the learning economy. KIBS are often seen as a driving force behind the spread of new knowledge in the innovation system, through information and communication technologies (ICT) and new ways of organizing industrial activity (post-Fordist principles). The adaptation of new technologies to many applications and markets involves, at each stage, a range of 'business service' expertise (Wood, 2005b, p. 433). This may be provided by firms themselves or by other organizations in the production chain, including specialist service organizations. KIBS are thus supposed to make a distinctive contribution to enhance competitiveness and innovation in other industries. The competitive value of services to regions therefore lies in their relationship to the success of other functions within the regional and wider division of labour (Wood, 2005b, p. 431). As noted by Wood (p. 435): 'Consultancies significantly promote non-technological adaptability of clients, for example in management procedures, response to IT, human resource development, and marketing and logistics' (Wood, 1996; Wood, 2002a, pp. 72–89).

KIBS in the Innovation System of Cities

KIBS are particularly seen as a facilitator of innovation and growth in large cities, and the sector is seen as an important component of metropolitan innovation systems (Daniels and Bryson, 2002; Edquist, 2005, p. 191; Fischer et al., 2001). The software industry and organizational consultancies in particular, have experienced large growth since 1980 (Keeble and Nachum, 2002, p. 88; Wood, 2002a, p. 35). At the system level, KIBS have been studied as a bridging institution that combines heterogeneous knowledge sources and knowledge users, and has been seen to form a node in a system of clients, cooperation partners, public institutions and R&D establishments (Werner, 2001). KIBS have been thought to act as orchestrators of innovation, and even orchestrators of whole innovation networks (Miles, 2001). In a network-type of knowledge system, KIBS have a crucial role in enabling the flow of knowledge between various nodes within the network (Howells and Roberts, 2000). KIBS are intimately connected to the agglomeration of corporate head offices, the latter being important clients for the former. Sassen (2000) argues for a mutual dependency between head office location and the location of KIBS. They are often thought of as forming a joint head office–corporate service complex characterized by interdependency, knowledge spillovers and networking, and located in major cities and capital regions.

So why, then, are KIBS important in innovation systems: what is their role? First, KIBS are perceived as bridges between business expertise and localized knowledge and capabilities, thus becoming problem-solving actors specialized in the provision of complementary knowledge inputs allowing the generation of innovations (Hauknes, 1998, p. 5). However, the possible impact that KIBS might have on the innovation performance and productivity of an economic system will depend on the relative share of knowledge-intensive industries in a system, and the degree and type of linkages they are able to create between heterogeneous actors in that system.

Second, KIBS can be viewed as bridges between the global, national and regional levels. Global KIBS facilitate the flow of knowledge down the hierarchy from global to regional knowledge; however, knowledge may also flow up the hierarchy (Howells and Roberts, 2000). A diversified portfolio of clients located locally, nationally and internationally, provides possibilities for knowledge diffusion, making consultants important satellites and disseminators of information between clients at different spatial levels, thus emphasizing their integration function in innovation systems (Wood, 2002a). Sassen (2001, p. 188) refers to a transnational managerial elite as 'the new transnational professionals' because they are 'members of a cross-border culture . . . embedded in a global network of . . . international financial centres'. The above-mentioned role of KIBS in an economic system can also be linked to theories of firms or actors operating as gatekeepers in economic systems, channelling extra-cluster knowledge into local, intra-cluster knowledge systems (Giuliani, 2002; Gambardella, 1993). As such, KIBS can also be viewed as non-technological gatekeepers of knowledge to the city.

KIBS' Role towards their Clients

For KIBS to be valuable at the regional level, it is of importance to diffuse heterogeneous knowledge to regional firms in such a way that it goes in as part of the clients' own knowledge base and consequently their internal knowledge-generating processes. We cannot expect all knowledge-intensive services to play an important role with relation to innovation in the client firm. This section focuses especially on the relation between KIBS and their clients, possibly spurring innovation.

From a client firm's point of view, when do they seek external assistance? When firms seek external knowledge, several studies have emphasized that the propensity of firms to establish knowledge linkages is associated with the degree of similarity/dissimilarity in their knowledge bases (Lane and Lubatkin, 1998; Rogers, 1983). Too distant knowledge bases might be incapable of transferring each other's knowledge. If firms' knowledge bases affect their ability to absorb and use external knowledge, it is questionable whether one can expect that knowledge can be diffused evenly as if 'it were in the air' (Giuliani, 2005 p. 52; Marshall, 1920). So, we might expect that firms seeking KIBS are looking for sources of advice and knowledge that they do not wholly possess internally. One could also expect that firms would seek advice that they are able to absorb, and as such KIBS can potentially be seen as contributing to innovation at the firm level for their clients. KIBS provide advanced services that are reflected in the very high proportion of experts with higher education found in KIBS firms (Toivonen, 2004). Clients employing these expert services need to have a certain competence level of their own. High internal competence is necessary in order to find out what kind of external knowledge-intensive services a firm may need (Aslesen and Isaksen, 2005). However, KIBS can also deliver standardized services to their clients, such as a standard software program, suggesting that their services are 'heavily standardised to the extent that much is driven by simple recipes' (Bryson et al., 2004, p. 100), suggesting that KIBS interaction may not always lead to knowledge-generating processes.

In general we know that globalization increases a firm's exposure to competition and puts increased pressure on its ability to have an efficient organization with the ability to launch new high-quality products. With a growing demand for quick adjustment, access to relevant knowledge becomes crucial, and firms' learning activities become important in a knowledge-based, or learning economy (Lundvall and Johnson, 1994). Firms therefore have an increased need for knowledge-intensive services (Daniels and Bryson, 2002); for research and product development, organizational changes, introduction of new technology, marketing, and so on. Firms in high-cost locations in particular need to employ knowledge, as such firms rarely compete on price alone. They have to develop a unique competitive advantage by learning and innovation. Quick adjustments are necessary, and access to relevant knowledge becomes vital. Through the use of new means (ICT) KIBS can provide their clients with access to information disseminated throughout society, and enhance connectivity and receptivity to the economic system (Toivonen, 2006). The importance of skills linked with locating and selecting the relevant information and using it in efficient ways has increased markedly (Toivonen, 2006), meaning that there is an increasing demand for highly qualified professionals who are able to provide comprehensive and customized interpretation of random data (Lundvall and Johnson, 1994). These are factors increasing firms' need for knowledge-intensive services.

KIBS' contribution can be linked to the fact that firms are rarely able to integrate all the knowledge needed for innovation themselves. A larger division of labour and a larger degree of specialization of firms' core activities will offset a need to outsource expert activities (Toivonen, 2006). At the same time as the progress of the division of labour increases the general need for expert services, external service providers – KIBS companies – have certain competitive advantages compared to the in-house production of knowledge-intensive services (Gallouj, 2002; Kox, 2002; Strambach, 2001).

Firms might need KIBS to guide and steer them in the heterogeneity of information and knowledge. KIBS might act as a facilitator for their clients in many ways (Bessant and Rush, 1995; Miles, 1999, cited in Toivonen, 2006). These include:

- Direct transfer of expert knowledge, that is, the traditional model of consulting practice.
- Experience sharing, carrying experiences and ideas from one context into another.
- Benchmarking, where the process of identifying and focusing on 'good practice' can be established through an intermediary.
- Brokering, putting different sources and users in contact.
- Diagnosis and problem-clarification, helping users articulate and define their particular needs for innovation in such a way that external resources and opportunities can be effectively used.
- Change agency, where organizational development can be undertaken with help from a neutral outside perspective.

The flows of knowledge resources induced and triggered by KIBS are not confined to those that are discrete and tangible, contractual, explicit, and codified and embodied in artefacts or texts (rather than in human beings and their practices). The functioning and role of KIBS can only be understood if we include process-oriented and intangible, non-contractual, implicit, and tacit- and 'human-embodied' forms of knowledge (Hertog et al., 1998).

KIBS IN THE CITY: THE CASE OF OSLO

What role do homogeneous and heterogeneous agglomeration forces play with regard to KIBS in the city? Further, what roles do more dynamic factors and factors related to knowledge spillover and extra-regional ties play with regard to learning and innovation among agglomerated KIBS?

Heterogeneous Agglomeration Forces and External Ties

KIBS are viewed as being an 'evident' and well-rooted component of the urban system; this is also the case in the Norwegian economy. In 2001, Oslo

– the capital region of Norway – had 40 per cent of the workplaces in the KIBS sector in Norway, while having 22 per cent of all jobs in Norway (Aslesen and Isaksen, 2004), suggesting a concentration of the KIBS sector in the largest city region in Norway, a tendency found in general in western economies.

What are the agglomeration forces driving KIBS to locate in the largest cities in the economy? In our study we have focused on three potential agglomeration forces for KIBS, and how KIBS view the actual outcome:

a. The density and variety in the client base.
b. A sufficient and diverse supply of employees.
c. The presence of a varied set of organizations that support production of knowledge and innovation.

The following presents empirical material focusing particularly on these different agglomeration forces and their outcome for KIBS in the Oslo region.

The Density and Variety in Client Base

Our empirical material shows that KIBS' most important clients are located in city areas, and this is seen to be one of the most important agglomeration forces for KIBS' concentration in Oslo. The density of clients was emphasized by almost all the KIBS firms interviewed as an important agglomeration force. Oslo, the capital of Norway, is populated by a large number of firms and actors from a variety of sectors and public service institutions, and therefore represents a large and diversified market for KIBS. A survey carried out among KIBS in Sweden found that the typical KIBS had clients in two or three sectors in the economy representing both manufacturing and service firms, indicating that they are well rooted in the innovation system (Nählinder, 2005, p. 119), and that they have a varied set of customers. We expect the same pattern to be true for the Oslo region.

In the telephone survey to consultants, we found that the most important external factor for developing competitiveness in KIBS is related to their interaction with and proximity to clients. The clients can therefore act as an important source for agglomeration economies for KIBS, as the client obviously has a positive effect on KIBS' own knowledge-generating processes. Through interviews, it was indicated that many KIBS–client interactions were often long lasting; either the project extended into another project, or the project itself lasted for several months. Many projects were of a character that needed much day-to-day exchange of information and knowledge-sharing between the actors, and proximity was perceived to ease complex projects. Projects involving collaboration and interactivity are also seen as factors giving the opportunity to modify and recombine practices and ideas between the actors involved in a project, possibly leading to innovation processes for both KIBS and their clients. The knowledge contributed by KIBS and acquired through projects with firms from different sectors and geographical locations is typically not contained within KIBS, but included as part of the continuous process of cumulative learning, resulting in developing a knowledge base of the firm that is used in future client projects, possibly creating value by the heterogeneity of the actors' knowledge bases.

The variety in client base, as a potential agglomeration force, and the actual outcomes described above are linked to local and endogenous factors. However, the literature also focuses on the importance of extra-regional ties and cities' roles as nodes in the global system of innovation. Agglomeration in cities could therefore also be explained by the possibility of linking up with global actors or networks. Looking at the portfolio of KIBS clients it was obvious that the largest proportion of clients was local, supporting the idea of proximate client relations as being important. At the same time, international markets were important; average sales to international markets were 17 per cent (Aslesen, 2003). The competition in these markets was perceived to be quite important for triggering innovation in KIBS firms in cities in Norway. This suggests that 'Communities of practice' between local and extra-regional KIBS enable possible tacit knowledge flows even when there is no proximity between actors, and extra-regional ties are important for knowledge formation, exchange and learning (Allen, 2000; Asheim and Coenen, 2005).

In general, the existence of a large and varied set of clients, as well as the level and range of a variety of demanding clients, can be seen as important agglomeration forces as well as yielding potential outcomes that can spur learning and innovation for the actors involved.

A Sufficient and Diverse Supply of Employees

Following up on agglomeration forces linked to heterogeneity, it is relevant to look at the characteristics of the workforce entering the KIBS sector. Continuous supplies of employees are of utmost importance for KIBS since the employment flows to and from the sectors are much higher than in the general economy (Aslesen et al., 2004). Stambøl (2005) has looked at gross entries to the KIBS sector in 1998–99, and found that slightly under half the recruitment to the KIBS sector came from other sectors in the economy (46 per cent). Sectors 'delivering' the most people to the KIBS sector in Oslo in 1998–99 were 'Financial services', 'ICT wholesale', 'Retail, recreation, culture and sport'. There was also some recruitment from 'Printing and publishing' and from 'Public administration', illustrating the variety of other sectors flowing into the KIBS sector, and potentially the role of KIBS as bridges between a heterogeneous set of knowledge bases in the urban economy. The transition directly from the education system is also definitely of importance to the KIBS sector. The data also demonstrate that the gross flows of workers out of the KIBS sector to other sectors were almost as high as the inflows to KIBS for several years in the 1990s, suggesting a supply of knowledge workers carrying with them 'process-oriented and intangible, non-contractual, implicit and tacit and "human embodied" forms of knowledge' (Hertog et al., 1998).

Besides recruiting people from different sectors of the economy, KIBS also recruit people from different geographical regions within Norway. Close to a quarter of staff recruited to the KIBS sector in 1998–99 was represented by people from other regions in Norway. The tendency of centralization of KIBS activity is clear (Stambøl, 2005); less knowledge is seen to flow from KIBS in Oslo to the other large cities in Norway (Aslesen et al., 2004).

The empirical material suggests that the supply of highly educated people either from other sectors in the economy or directly from the education sector is of importance to KIBS, indicating that these are important factors behind KIBS agglomeration in cities, as well as giving agglomeration economies for KIBS themselves with regard to learning and innovation. The material also sheds light on the role of KIBS as bridges in the innovation system of the city, both with regard to the client base and to the mobility rates in and out of KIBS from a range of sectors and geographical levels.

Organizations that Support Production of Knowledge and Innovation

In a telephone survey of 570 managers of software firms and organizational consultants, firms were asked which factors are important for the development and maintenance of the competitiveness of KIBS. Linkages with research institutes, universities and other higher education institutions were perceived to have little impact on KIBS competitiveness and were not seen as an important force for city location. This might be because of the characteristics of the innovation process in KIBS, often characterized as being an incremental, step-by-step process, where new elements and combinations of services are a part of overall, continuous processes of development guided by the strategy of the firm (Scarborough and Lannon, 1989; Sundbo, 1997), making formal R&D projects between KIBS and the knowledge infrastructure less relevant.

Homogeneous Agglomeration Forces and External Ties

In the following we examine agglomeration forces and potential agglomeration outcomes that could be linked to the existence of 'similar firms', either being partners and/or competitors.

From interviews with managers in KIBS we found that an explanation for locating their activity in Oslo was that it eases the implementation of projects, since projects often are carried out as joint efforts involving several KIBS. From interviews we found that regional management consultants in the Oslo region has formed strategic alliances and networks with other KIBS in the region (also more global consultancy firms). This enabled them to take on projects that needed more diverse knowledge input, and projects that were relatively large, and the alliances compensated for lacks in their 'service portfolio' (Aslesen and Langeland, 2003). Collaborative projects are easier to initiate quickly and implement in regions where a large number of potential project partners are available (Aslesen and Isaksen, 2005).

In the telephone survey to consultants, they were asked which factors were important for the development and maintenance of KIBS competitiveness. Here, factors linked to a concentration of similar firms were emphasized, such as the professional environment in the region, informal contact with other representatives of the sector, and participation in trade meetings and similar events (Aslesen and Isaksen, 2005). Agglomeration outcomes (development and competitiveness) were ascribed to such factors, suggesting agglomeration effects also by firms in the same sector.

The concentration of KIBS in the city also means outcomes such as a tougher competitive climate. In our survey we found that 44 per cent of the KIBS firms in the Oslo region considered the local competition to be strong (Jakobsen and Aslesen, 2004). The agglomeration of firms in the same sector seems to spur both collaboration and competition, potentially contributing to innovations in the KIBS sector.

Other channels of global knowledge through KIBS can be linked to the fact that important KIBS actors in the Oslo region are subsidiaries of multinational corporations. Through interviews it was obvious that branches and subsidiaries of global firms still have the largest turnover, number of assignments and number of employees (Aslesen and Langeland, 2003). The above examples show that the flow of knowledge into the city region can be driven by KIBS tapping into outside knowledge by various means, either by having international clients, or through collaborative projects with international partners, or with subsidiaries of multinational firms in Oslo. KIBS could be viewed as non-technological gatekeepers of global knowledge to city regions, disseminating their acquired knowledge to their clients.

This suggests that learning and innovation in KIBS not only can be ascribed to outcomes of agglomeration of firms locally. Extra-regional ties can also trigger learning and innovation in agglomerated KIBS. The complementarities between the 'local' and the 'global' are especially important in urban regions, through the knowledge-intensive exchanges on which they currently most depend (Amin and Thrift, 1992; Swyngedouw, 1996; Scott, 1998; Yeung, 1998; cited in Wood, 2005a).

HOW DO KIBS CONTRIBUTE TO LEARNING AND INNOVATION AMONG THEIR CLIENTS?

The previous section focused on agglomeration forces and agglomeration outcomes. One of the most important heterogeneous agglomeration forces, also giving valuable input to KIBS own knowledge-generating processes, is clients and client projects. This section looks closely into the KIBS–client relationship to see what types of linkage KIBS develop with their clients and who these clients actually are. This will assist in answering the question: 'What role do KIBS have in adapting actors to the complexity and heterogeneity of the knowledge-based learning economy of the city?'

Using the Community Innovation Survey we found that 45 per cent of innovative firms have employed consultants as a source of information in innovation, and 44 per cent of those making use of such consultants argue that consultants' relative importance is of 'some relevance' as a source of information for innovation, lending support to the assumption that KIBS affect the knowledge base of the client firms.

Survey results show that KIBS clients do seek 'specialised knowledge they themselves do not possess' (Aslesen and Isaksen, 2004), supporting the idea that when searching for external knowledge firms search for actors with 'more knowledge' than themselves. The KIBS–client relationship involves the transfer of both codified knowledge in the form of methods, tools and documents, and articulated experience from previous projects. Tacit knowledge is conveyed by the individual consultant, who uses his or her personal experience when adapting the general methods and tools of the consultancy firm to the needs of a specific client (Jakobsen and Aslesen, 2004).

We asked the consultants what kinds of service they provide to their clients and found that the largest proportion of KIBS took part in the 'development of new or change of existing products/services for the client', contributing to the 'introduction of new solutions for the clients' daily routines (administration, executive work, etc.)' and in 'competence building among clients', suggesting knowledge input to the recipient firm. Besides this, we found that consultants perceived themselves as having different roles in client innovation. This was especially confined to (1) the supply of complementary knowledge to facilitate innovation, (2) managing the innovation process, and (3) especially confined to giving advice on the direction and type of innovation to carry out. The main responsibility of a consultant is often to enable the client firm to use its expertise in the most efficient and profitable way. Clients often have expressed needs and ideas on change; the consultants often have the necessary competence to bring these ideas to fruition. Further, the consultant can be used strategically to administrate and organize innovation processes rather than to generate ideas for innovation in the client firm. The consultants can also provide enterprise-level strategic advice affecting the direction of a client's innovative activities through giving advice on the type of products or processes needed in order to improve the competitive standing of the client's business and to improve the decision-making platform on which to base future strategies (Aslesen and Isaksen, 2004).

The empirical findings show that KIBS are important as providers of a heterogeneous set of knowledge tuned to the needs of their clients. The effect of this knowledge contribution on the clients' internal knowledge-generating processes is hard to quantify, but the empirical material presented above suggests that it is of importance.

In general, what kinds of client firms find KIBS of value as a source of information in innovation? Wood (2002b) argues that some groups of companies in specific industries that are small and medium-sized, and located in more peripheral geographic areas, will experience increased difficulties because of limited access to external knowledge-intensive services, since the KIBS sector is concentrated in a limited number of areas. In our empirical material we found that the typical KIBS client often is relatively large in terms of employment and turnover, and most often competes in an international market and with a need to change in order to adapt to changing external conditions. Often, the clients have a solid capital base and operate in branches where pressure to change is strong. Consequently KIBS were often seen as important facilitators for changes imposed on the different sectors or industries. An example of this was the demand from semi-public organizations in the process of transition to private market actors – a process that has gained momentum in recent years – and then was especially confined to areas where public administration is mostly located.

Using the Community Innovation Survey (CIS) carried out by Statistics Norway (SSB) in 2001, Aslesen and Isaksen (2001) found a clear correlation between company size and the use of consultants as a source of information in innovation; only 17 per cent of companies with 10–19 employees use consultants occasionally in innovation, as opposed to 32 per cent of companies with over 250 employees. These differences suggest that the potential knowledge contribution by KIBS is not evenly distributed. The lower user-rate among smaller companies in Norway may be a result of the general difficulties with which small businesses obtain information about external actors, and absorb and exploit knowledge from these. However, some of these problems might be diminished by proximity to larger constellations of specialized KIBS.

There are geographical differences with regard to the importance attached to the use of consultants by innovative firms. A much larger proportion of companies within the Oslo area than in other parts of the country view consultants as an important source of information in innovation (Aslesen and Isaksen, 2004). This indicates that firms located in city regions have a higher propensity to use KIBS, suggesting the agglomeration of KIBS leads to more use of KIBS services. The telephone survey showed that two-thirds of consultancy clients hold the opinion that geographic adjacency to consultants stimulates increased use, indicating a demand for knowledge and expertise of a national and international standard, with local (on-site) attendance. The repeated use of consultants and personalization are common characteristics of the KIBS–client relationship (Jakobsen and Aslesen, 2004), and it is common to assume that this personalization and the distribution of tacit knowledge often take place in local business networks (Cooke and De Laurentis, 2002; Lambooy, 2002).

However, even if face-to-face contact and sustained interaction between clients and consultants is essential for successful consultancy, this does not mean that it is only co-location in a KIBS–client relationship that can facilitate organizational innovations among clients (Jakobsen and Aslesen, 2004). Focusing especially on the head office–corporate service complex, the head office was interviewed. From these interviews we found a differentiation in head offices' search for consultancy services. The search radius for services of strategic importance did extend beyond the region, to other national or international KIBS. More standardized KIBS input was often found locally (Jakobsen and Aslesen, 2004, p. 28), in line with Torre and Rallet's (2005) critique of the overemphasis laid on co-location in the research on proximity in inter-firm cooperation.

Our findings illustrate that clients can establish close links to externally located KIBS as well. Torre and Rallet (2005) emphasize organized proximity as imperative to all kinds of cooperation, explained by the logic of belonging (rules and routines of behaviour that facilitate cooperation between actors) and similarity (implying a shared system of representation), suggesting that organized proximity is more important than geographical proximity when searching for advanced services. However, the probability for head offices to have routines of behaviour and knowledge bases that match KIBS is likely to be higher than for 'average' firms. The possibility of having a greater search radius for external knowledge will also be a characteristic feature of head offices, suggesting that proximity to KIBS might ease the use for smaller firms and for firms searching for knowledge-intensive services that are not considered strategic. Other studies have also found that strategic interests increase the willingness to overcome obstacles possibly resulting from geographical distance of each partner (Hyypiä and Kautonen, 2005); the higher the level of strategic interest for the client and for KIBS, the less emphasis is put on the importance of geographical proximity in line with the way it is understood in the literature on 'communities of practice'.

KIBS AND HETEROGENEITY

This chapter has focused on the role of the city in the knowledge-driven learning economy. We started by introducing theoretical contributions explaining territorial agglomerations influenced by cost efficiency, transportation costs, scale advantages and market access, followed by contributors also adding more sociocultural factors to the explanation such as industrial atmosphere, maintaining that these factors could have indirect or direct implications for agglomeration outcomes such as innovation. Thereafter, different agglomeration forces were introduced, suggesting that outcomes of agglomeration can arise from labour market pooling, specialist suppliers and knowledge spillover to firms in the same sector, and/or from urban size, variety and density, giving externalities available to all firms.

When studying learning and innovation in cities, dynamic factors as well as the importance of extra-regional ties have been emphasized in the literature. The emphasis on knowledge spillover, learning and innovation, focuses on the importance of leaks of knowledge that diffuse into the economy as if 'in the air'. The city region is viewed as the centre of such spillovers, giving a more dynamic approach to the possible outcomes of agglomerations. An important question is whether localized tacit knowledge can be transported over distance. The emphasis on knowledge diffusion and on cities as 'relay stations in a world of flow' brings both dynamic factors and extra-regional ties into our understanding of the role of the city, suggesting that one cannot delimit the learning and innovation processes to geographically bounded endogenous processes.

Agglomeration Forces and Outcome

In trying to understand the agglomeration forces of importance to KIBS and the potential outcomes of these, our empirical material finds support in forces and outcomes of agglomerations that can be related to both factors emphasizing the density of similar firms in the region, and the potential for knowledge spillover between firms in the same sector, as well as the existence of a variety of sectors and urban size. This suggests the potential force of localization, urbanization and Jacobs's externalities for KIBS in the city. Heterogeneous agglomeration forces that are found to be of great importance for KIBS and their own knowledge development are the vast supply and variety of clients, also the most demanding such as corporate head offices. Further, the possibility to access the largest pool of highly educated employees is of importance for an industry employing mainly people with high levels of formal education, and for an industry having one of the highest mobility rates in the economy. Both these factors are seen as essential in KIBS' own knowledge-generating processes. Proximity to knowledge and innovation-producing institutions does not seem to be an important agglomeration force. However, the supply of newly trained people from these institutions is important. Other kinds of actors and explanations linked to more 'in the air' and 'milieu' explanations are emphasized. Overall, urbanization economies and Jacobs's externalities do play a role for KIBS in the city, giving a continuous influx of possibilities to generate new ideas, and to be highly innovative.

Being agglomerated in a region with firms in the same sector provides the possibility for monitoring trends and benchmarking own activity against other KIBS. Competition can spur innovation and change, but also push less fit firms out of the market, possibly making KIBS in the city among the 'fittest'. Collaboration opportunities are also a positive side of being co-located with similar firms. The 'industrial atmosphere' and the possibility of meeting informally are of great importance, and the knowledge spillover through the buzz of the city was especially emphasized by the firms. Since 'the core competence of professional service firms is the expertise, experience and reputation of their staff, the asset base is knowledge and the competitive advantage is reputation' (Bryson et al., 2004, p. 87), it is important to be where the buzz is.

However, agglomeration forces linking KIBS to city regions are not always linked to the possibility of being close to local actors. The city also has the role as a satellite of global knowledge and ideas through various extra-regional links and networks. These extra-regional ties also trigger learning and innovation in agglomerated KIBS, suggesting that as projects can be carried out between different geographical levels, proximity is not always necessary. The externalities from this are the spillover of a heterogeneous set of global knowledge into the innovation system of the city.

In sum, it seems that explanations for KIBS agglomerations into the city, and potential outcomes such as learning and innovation, must be found in the interrelations between the forces of heterogeneity and the forces of heterogeneity-reducing mechanisms such as the agglomeration of specific activities that favour specialization (including KIBS). Explanatory factors are found both in factors putting weight on proximity as well as on factors putting weight on extra-regional ties. Immaterial factors and 'world of flows' are also important factors linked to localization in cities. In trying to understand factors behind growth and innovation in cities, a complex matrix appears to be relevant, especially for KIBS.

KIBS, the Heterogeneity Enabler

In general, clients appreciate KIBS' various knowledge inputs, and KIBS can be viewed as an important agent in adapting clients to the heterogeneity of modern economies as well as being an actor combining heterogeneous knowledge in internal innovation processes. The heterogeneity of the city enables KIBS to recombine and bridge ideas, knowledge and practices from a varied set of nodes in the system. This knowledge goes as valuable input into client firms, however also feeding into KIBS' own innovation processes. KIBS' bridging role between heterogeneous nodes enables learning and innovation, and a possibility of cross-fertilization in the innovation system.

However, the importance attached to KIBS for innovation is higher for larger firms, suggesting that there are factors which facilitate project learning in these relations, and make KIBS important as an external source for learning and innovation. There are certain firm-internal capacities that must be put in place before KIBS can be said to have a strategic role in the firm's innovation processes in general (Aslesen and Isaksen, 2004). The importance attached to consultants is also higher for city-based firms, acknowledging KIBS as being 'a source of innovation and growth for urban business firms in particular' (Daniels and Bryson, 2002).

However, we also found that the importance of agglomeration and proximity must be differentiated. Proximity between KIBS and clients was most important when searching for standardized services. Future research should focus on whether non-innovation related use of KIBS can be ascribed to firms with low absorptive capacity and to firms with knowledge bases of certain types (i.e. lack of analytical knowledge).

Clients searching for more strategic consultancy services, giving possible input to innovation and growth, had a search radius beyond the region, suggesting an ability to overcome hindrances from geographical distance and an ability to take part in knowledge networks with knowledge-intensive actors. This indicates that KIBS' clients that have the greatest chance to obtain innovation-inducing input from the KIBS–client project will probably be firms that are quite 'similar' in knowledge base, behaviour and system of representation. This implies that even if clients come from a variety of sectors and geographical levels, those searching for the most knowledge-intensive input are, to a certain degree, represented by a homogeneous set of actors with common beliefs, opinions and languages as KIBS.

These actors seem to have certain characteristics that 'match' the language and knowledge bases of KIBS, questioning the degree to which KIBS actually bridge 'heterogeneous' actors, or if KIBS' role as innovation enabler can only be ascribed to 'homogeneous' actors with organized proximity, sharing communities of practice.

This role as heterogeneity-enabler for a homogeneous set of actors is, of course, also important since these actors then have the ability to act as nodes in the innovation system of the city, making national and global knowledge accessible at the regional level. However, if this is so, KIBS and their most advanced customers also act as non-technological gatekeepers of both global and national knowledge, setting the knowledge scene for other city actors. The question is: do these nodes attract the knowledge-intensive assets of relevance to guide the future role of cities?

REFERENCES

Acs, Z.J. (2002), *Innovation and the Growth of Cities*, Cheltenham, UK and Northampton, MA, USA: Edward Elgar.

Acs, Z.J., D.B. Audretsch and M.P. Feldman (1992), 'Real effects of academic research: a comment', *American Economic Review*, **82**, 363–7.

Allen, J. (2000), 'Power/economic knowledge: symbolic and spatial formations', in J.R. Bryson, P.W. Daniels, N. Henry and J. Pollard (eds), *Knowledge, Space, Economy*, London: Routledge.

Amin, A. and P. Cohendet (1999), 'Learning and adaptation in decentralized business networks', *Environment and Planning D: Society and Space*, **17**, 87–104.

Amin, A. and N. Thrift (1992), 'Neo-Marshallian nodes in global networks', *International Journal of Urban and Regional Research*, **16**, 571–87.

Amin, A. and N. Thrift (2002), *Cities. Reimagining the Urban*, Cambridge: Polity.

Asheim, B.T. and L. Coenen (2005), 'Constructing regional advantage at the northern edge', paper presented at the 5th Triple Helix conference, 18–21 May, Turin, Paper ID: A 189.

Asheim, B.T. and M.S. Gertler (2005), 'The geography of innovation: regional innovation systems', in J. Fagerberg, D.C. Mowery and R.R. Nelson (eds), *The Oxford Handbook of Innovation*, Oxford: Oxford University Press, pp. 291–317.

Aslesen, H.W. (2003), 'Knowledge intensive business services and regional innovation. Consultancy in city regions in Norway', paper presented at 'Reinventing Regions in a Global Economy', Regional Studies Association International Conference, Pisa, Italy 12–15 April 2003.

Aslesen, H.W. and A. Isaksen (2001) 'Oslo, in what way a creative city', *European Planning Studies*, Special Issue on 'Creative cities and social costs'.

Aslesen, H.W. and A. Isaksen (2004), 'Knowledge intensive services: vehicle for industrial development in large cities', paper prepared for the XIVth International Conference of RESER – 2004, 'Information and Communication Technologies and Service Relationship in the Global Economy: A Challenge for Europe', 23–24 September, Castres and Toulouse.

Aslesen, H.W. and A. Isaksen (2005), 'Kunnskapsintensive tjenester: motor i storbyenes næringsliv?', in E. Vatne (ed.), *Norske storbyers funksjon i norsk økonomi*, Olso: Norges Forskningsråd, pp. 199–227.

Aslesen, H.W. and O. Langeland (2003), 'Knowledge economy and spatial clustering: the role of knowledge-intensive business services and venture capital firms in the innovation system', paper presented at the DRUID Summer Conference 2003 on Creating, Sharing and Transferring Knowledge, the role of Geography, Institutions and Organizations, Copenhagen 12–14 June.

Aslesen, H.W., L. Stambøl and A. Isaksen (2004), 'KIBS and industrial development of cities. Labour mobility, innovation and client interaction', paper prepared for the 44th Congress of the European Regional Science Association, Porto, 25–30 August.

Audretsch, D.B. and M.P. Feldman (1996), 'R&D spillover and the geography of innovation and production', *American Economic Review*, **86** (3), 630–40.

Bathelt, H., and A. Malmberg and P. Maskell (2004). 'Clusters and knowledge: local buzz, global pipelines and the process of knowledge creation', *Progress in Human Geography*, **28**, 31–56.

Bessant, J. and H. Rush (1995), 'Building bridges for innovation: the role of consultants in technology transfer', *Research Policy*, **24**, 97–114.

Brown, J.S. and P. Duguid (1998), 'Organizing knowledge', *California Management Review*, **40** (3), 90–111.

Bryson, J.R., R.W. Daniels and B. Warf (2004), *Service Worlds. People, Organisations, Technologies*, London & New York: Routledge.

Cantwell, J. and S. Iammarino (2003), *Multinational Corporations and European Regional Systems of Innovation*, London & New York: Routledge.

Cooke, P. (2002), *Knowledge Economies. Clusters, Learning and Cooperative Advantage*, London: Routledge.

Cooke, P. and C. De Laurentis (2002), 'The index of knowledge economies in the European Union: performance rankings of cities and regions', *Regional Industrial Research Report 41*, Centre for Advanced Studies, Cardiff University.

Daniels, P. (1991), *Services and Metropolitan Development: International Perspectives*, London: Chapman and Hall.

Daniels, P.W. and J.R. Bryson (2002), 'Manufacturing services and servicing manufacturing: knowledge-based cities and changing forms of production', *Urban Studies*, **39** (5–6), 977–91.

Dosi, G. (1997), 'Opportunities, incentives and the collective patterns of technological change', *Economic Journal*, **107**, 1530–47.

Edquist, C. (2005), 'System of innovation: perspectives and challenges', in J. Fagerberg, D. Mowery and R. Nelson (eds), *The Oxford Handbook of Innovation*, Oxford: Oxford University Press.

Eurostat (2002), 'Employment in high tech and knowledge intensive sectors in the EU continued to grow in 2001', *Statistics in Focus, Science and Technology*, 04/2002.

Feser, E.J. (2002), 'Tracing the sources of local external economies', *Urban Studies*, **39**, 2485–506.

Fischer, M.M., J. Revilla Diez and F. Snickars (2001), *Metropolitan Innovation Systems*, Heidelberg: Springer.
Frenken, K., F. Govanoort, T. Verburg and R.A. Boschma (2002), 'Variety and regional economic growth in the Netherlands', *Papers in Evolutionary Economic Geography No. 05.02*, Utrecht University.
Gallouj, F. (2002), 'Knowledge-intensive business services: processing knowledge and producing innovation', in J. Gadrey and F. Gallouj (eds), *Productivity, Innovation and Knowledge in Services. New Economic and Socio-Economic Approaches*, Cheltenham, UK and Northampton, MA, USA: Edward Elgar.
Gambardella, A. (1993), 'Innovazioni tecnologiche e accumulazione delle conoscenze: quale modello per le piccole e medie imprese negli anni '90', *Piccola Impresa/Small Business*, **2**, 73–89.
Gertler, M. (2004), *Manufacturing Culture. The Institutional Geography of Industrial Practice*, Oxford: Oxford University Press.
Giuliani, E. (2002), 'Cluster absorptive capability: an evolutionary approach for industrial clusters in developing countries', paper presented at the DRUID Summer Conference, Copenhagen/Elsinor, Denmark.
Giuliani, E. (2005), 'The structure of cluster knowledge networks: uneven and selective, not pervasive and collective', paper presented at DRUID Tenth Anniversary Summer Conference, 2005.
Gottman, J. (1961), *Megalopolis: The Urbanized Northeastern Seaboard of the United States*, Cambridge, MA: MIT Press.
Grabher, G. (1993), *The Embedded Firm. On the Socioeconomics of Industrial Networks*, London & New York: Routledge.
Griliches, Z. (1992), 'The search of R&D spillover', *Scandinavian Journal of Economics*, **94**, 29–47.
Harrison, B., M.R. Kelley and J. Gant (1997), 'Innovative firms behaviour and local milieu: exploring the intersection of agglomeration, firm effects and technological change', *Economic Geography*, **72**, 233–58.
Hauknes, J. (1998), 'Services in innovation. Innovation in services', *SI4S Synthesis paper*, *S1*, Oslo: STEP.
Henderson, J.V. (2003), 'Marshall's scale economies', *Journal of Urban Economics*, **53**, 1–28.
Hertog, P. den (2000), 'Knowledge-intensive business services as co-producers innovation', *International Journal of Innovation Management*, **4** (4), 491–528.
Hertog, P. den, R. Bilderbeek, G. Marklund and I. Miles (1998), 'Services in innovation: Knowledge Intensive Business Services (KIBS) as co-producers of innovation', *SI4S Synthesis Paper No. 3*, Oslo: STEP.
Hoover, E.M. (1948), *The Location of Economic Activity*', New York: McGraw-Hill.
Howells, J. (1988), *Economic, Technological and Locational Trends in European Services*, Aldershot: Gower.
Howells, J. and J. Roberts (2000), 'Global knowledge systems in a service economy', in B. Andersen, J. Howells, B. Hull, I. Miles and J. Roberts (eds), *Knowledge and Innovation in the New Service Economy*, Cheltenham, UK and Northampton, MA, USA: Edward Elgar.
Hudson, R. (1999), 'The learning economy, the learning firm and the learning region: a sympathetic critique of the limits to learning', *European Urban and Regional Studies*', **6** (1), 59–72.

Hyypiä, M. and M. Kautonen (2005), 'Expertise, proximity and KIBS–client relationships – theoretical considerations based on empirical observations on service interactions in knowledge intensive industries in Helsinki, Finland', paper presented at the XV International Conference of RESER, Granada, Spain, 22–24 September.

Isard, W. (1956), *Location and Space-economy. A General Theory Relating to Industrial Location, Market Areas, Land Use, Trade and Urban Structure*, Cambridge, MA: MIT Press.

Jacobs, J. (1969), *The Economy of Cities*, New York: Random House.

Jaffe, A.B. (1986), 'Technological opportunity and spillover of R&D', *American Economic Review*, **76**, 984–1001.

Jaffe, A.B. (1989), 'Real effects of academic research', *American Economic Review*, **79**, 957–70.

Jaffe, A., M. Trajtenberg and R. Henderson (1993), 'Geographic localization of knowledge spill-overs as evidenced by patent citations', *Quarterly Journal of Economics*, **108**, 577–98.

Jakobsen, S.-E. and H.W. Aslesen (2004), 'Location and knowledge interaction between head office and KIBS in city areas', revised version prepared for the XIVth International Conference of RESER – 2004, 'Information and Communication Technologies and Service Relationship in the Global Economy: A Challenge for Europe', 23–24 September, Castres and Toulouse.

Keeble, D. and L. Nachum (2002), 'Why do business service firms cluster? Small consultancies, clustering and decentralization in London and southern England', *Transactions of the Institute of British Geographers*, **27**, 67–90.

Kox, H. (2002), *Growth Challenges for the Dutch Business Services Industry. International Comparison and Policy Issues*, The Hague: CPB Netherlands Bureau for Economic Policy Analysis.

Lambooy, J.G. (2002), 'Knowledge and urban economic development: an evolutionary perspective', *Urban Studies*, **39**, 1019–935.

Lane, P.J. and M. Lubatkin (1998), 'Relative absorptive capacity and interorganizational learning', *Strategic Management Journal*, **19**, 461–77.

Lundvall, B.-Å. (1992), 'User–producer relationships, national systems of innovation and internationalisation', in B.-Å. Lundvall (ed.), *National Systems of Innovation*, London: Pinter.

Lundvall, B.-Å. and B. Johnson (1994), 'The learning economy', *Journal of Industrial Studies*, **1** (2), 23–42.

Marshall, A. (1920), *Principles of Economics*, London: Macmillan.

Marshall, A. (1930), Principles of Economics, 8th edn, London: Macmillan.

Marshall, J.N. (1988), *Services and Uneven Development*, Oxford: Oxford University Press.

Miles, I. (1999), 'Services in national innovation systems: from traditional services to knowledge intensive business services', in G. Schienstock and O. Kuusi (eds), *Transformation Towards a Learning Economy*, Report No. 213, Helsinki: The Finnish National Fund for Research and Development – Sitra.

Miles, I. (2001), 'Taking the pulse of the knowledge-driven economy: the role of KIBS', in M. Toivonen (ed.), *Growth and Significance of Knowledge Intensive Business Services*, Helsinki: Uusimaa T&E Centre's Publications 3.

Miles, I. (2003), 'Knowledge intensive services' suppliers and clients. Ministry of Trade and Industry Finland', *Studies and Reports 15/2003*, Helsinki: Edita Publishing.

Nählinder, J. (2005), 'Innovation and employment in services. The case of knowledge intensive business services in Sweden', PhD dissertation at Linköping Studies in Arts and Science, Linköping University.
Nelson, R.R. (1995), 'Co-evolution of industry structure, technology and supporting institutions and the making of comparative advantage', *International Journal of the Economics of Business*, **2** (2), 171–84.
OECD (2001), *OECD Science, Technology and Industry Scoreboard. Towards a Knowledge-based Economy*, Paris: OECD.
Polanyi, M. (1967), *The Tacit Dimension*, London: Routledge & Kegan Paul.
Rogers, E.M. (1983), *Diffusion of Innovations*, New York: The Free Press.
Sassen, S. (2000), 'Cities in the global economy', in R. Paddison (ed.), *Handbook of Urban Studies*, London, Sage, pp. 256–72.
Sassen, S. (2001), *The Global City*, Princeton, NJ: Princeton University Press (New updated edition, originally published in 1991).
Saviotti, P.P. (1996), *Technological Evolution, Variety and the Economy*, Cheltenham, UK and Brookfield, USA: Edward Elgar.
Scarborough, H. and R. Lannon (1989), 'The management of innovation in financial service sector: a case study', *Journal of Marketing Management*, **5** (1), 51–62.
Scott, A.J. (1998), *Regions in the World Economy*, Oxford: Oxford University Press.
Simmie, J. (ed.) (2001), *Innovative Cities*, London: Spon Press.
Stambøl, L. (2005), 'Arbeidsmarkedsmobilitet og framvekst av kunnskapsbasert tjenesteyting', in E. Vatne (ed.), *Storbyene i kunnskapsøkonomien. Arena for kunnskapsdeling og nyskaping*, Oslo: Scandinavian Academic Press.
Storper, M. (1995), 'The resurgence of regional economies ten years later: the region as a nexus of untraded interdependencies', *European Urban and Regional Studies*, **2**, 191–221.
Storper, M. (1997), *The Regional world. Territorial Development in the Global Economy*, New York & London: Guilford Press.
Storper, M. and R. Salais (1997), *Worlds of Production: The Action Framework of the Economy*, Cambridge, MA: Harvard University Press.
Storper, M. and A.J. Venables (2002), 'Buzz: the economic force of the city', paper presented at the Druid Summer Conference on 'Industrial Dynamics of the New and Old Economy – Who is Embracing Whom?', Copenhagen/Elsinore 6–8 June.
Strambach, S. (2001), 'Innovation processes and the role of knowledge-intensive business services (KIBS)', in K. Koschatzky, M. Kulicke and A. Zenker (eds), *Innovation Networks. Concepts and Challenges in the European Perspective*, Technology, Innovation and Policy, 12, Series of the Fraunhofer Institute for Systems and Innovation Research (ISI), Heidelberg: Physica-Verlag.
Sundbo, J. (1997), 'Management of innovation in services', *The Service Industries Journal*, **17** (3), 432–55.
Swyngedouw, E. (1996), 'Neither global or local "glocalisation" and the politics of scale', in K. Cox (ed.), *Spaces of Globalization: Reasserting the Power of the Local*, New York: Guilford Press, pp. 137–66.
Toivonen, M. (2004), 'Expertise as business. Long-term development and future prospects of knowledge-intensive business services (KIBS)', Helsinki University of Technology, Department of Industrial Engineering and Management, Doctoral dissertation series 2004/2. Espoo.
Toivonen, M. (2006), 'Future prospects of knowledge-intensive business services (KIBS) and implications to regional economies', *ICFAI Journal of Knowledge Management*, **4** (3), 18–39.

Torre, A. and A. Rallet (2005). 'Proximity and localization', *Regional Studies*, **39** (1), 47–59.
Vatne, E. (ed.) (2005), *Storbyene i kunnskapsøkonomien. Arena for kunnskapsdeling og nyskaping*, Oslo: Scandinavian Academic Press.
Werner, R. (2001), 'Knowledge-intensive business services in the Oulu region – business development and geographical linkage', in M. Toivonen (ed.), *Growth and Significance of Knowledge Intensive Business Services*, Helsinki: Uusimaa T&E Centre's Publications 3.
Wood, P. (1996), 'Business services, the management of change and regional development in the UK: a corporate client perspective', *Transactions of the Institute of British Geographers*, **21**, 649–65.
Wood, P. (2002a) (ed.), *Consultancy and Innovation. The Business Service Revolution in Europe*, London & New York: Routledge.
Wood, P. (2002b), 'Services and the "new economy": an elaboration', *Journal of Economic Geography*, **1**, 109–14.
Wood, P. (2005a), 'Urban revival and knowledge intensive services: the case of the English "core cities"', *Working Paper No. 18*, The University of Birmingham School of Geography, Earth and Environmental Sciences.
Wood, P. (2005b), 'A service-informed approach to regional innovation – or adaptation?', *Service Industry Journal*, **25** (4), 429–45.
Yeung, H.W. (1998), 'The social–spatial constitution of business organizations: a geographical perspective', *Organizations*, **5** (1), 101–28.

11. Specialization and heterogeneity in small national economies: the Nordic countries

Åge Mariussen

INTRODUCTION

Specialization is often seen as a cause of path dependency and other misfortunes of inflexibility. As global markets change, one might assume small countries with highly specialized national economies may get locked into declining sectors. To the Nordic countries, the oil crises in the 1970s were a crude wake-up call. Other problems were to come in the late 1980s and early 1990s.

This chapter explains why and how, in different ways, the Nordic countries are constructing unique business systems opening for wide networks of learning, thus enabling highly successful forms of global market adaptation.

GARBAGE-CAN HETEROGENEITY

Based on the the experiences of these crises, Kristensen and Levinsen (1983) discussed the 'small country squeeze'. The characteristics of this squeeze can be described as follows. Small countries have open economies. Unlike the situation for large national economies such as the USA, which can live well for many decades with a negative trade balance, the foreign trade of small countries represents a large part of GDP. Their balance of trade has a direct and forceful impact on a number of core factors in the economy such as the stability of the national currency, rate of interest, the stock market, employment, the well-being of the citizens – in short, the wealth of the nation. What is more, because of its restricted size, a small country is not likely to have a very broad range of globally competitive clusters providing a secure overall net export balance of trade. Instead, small countries – somewhat similar to regions within larger countries – frequently exploit the advantage of being small through specialization in a limited number of globally competitive industries. But when these specialized sectors experience declining markets, small national economies may have nowhere else to go. One solution to this squeeze is increasing knowledge heterogeneity.

This article takes as a point of departure two positions of national business system theory. First, institutions regulating work organization, labour relations, and governance in national systems co-evolve and hence tend to make up nationally specific institutional complementarities stimulating certain forms of business organization, yet discouraging others. Second, when put under external pressure, differences between national systems are likely to be reproduced rather than disappear through convergence towards some global 'best practice' (Whitley, 1999).

These different *national business systems* encourage and discourage different forms of learning and innovation, adapted to different *technological paradigms*. By way of introduction, and in terms of paradigms, let us briefly look at mobile phones and cars.

First Phase: New Paradigm Creation

New paradigms, like the mobile phone, are rare events. They start with radical innovations. In the beginning they may require substantial long-term R&D investment under conditions of extreme uncertainty. Doz and Kosonen (2005) describe how NOKIA created the mobile phone handset paradigm during the period 1990 to 1994. The *founding* of the mobile phone paradigm required '*exceptional foresight and insight, imagination and reframing capabilities*'. Here, strategic agency is at centre stage. The new technology is carefully protected through patents and other strategies of corporate knowledge appropriation. Through internal R&D, the corporate actor accumulates a private knowledge monopoly that protects the technology and enables the super-profits that follow in the second phase.

Second Phase: Exploiting the Paradigm

The growing market rewards the successful entrepreneur with super-profits. New and improved versions of the product are invented and marketed, initial problems removed. The product becomes more accessible and markets continue to grow. As new *producers* enter the market, technological competition is triggered. New competitive models are created through R&D investments. This, in turn, increases the need for efficiency in R&D. Research is increasingly outsourced and shared among several firms. In the mobile phone industry, at an early stage during the 1990s success story, major producers such as Swedish Ericsson and Finnish Nokia developed sophisticated strategies of modularization. At the same time, the producers and their networks accessed external research facilities at universities. The development of ICT-based triple helix clusters in major university cities commenced. Through outsourcing and subcontracting R&D, *markets* were put in place between suppliers of R&D-based products and their users, the mobile phone producers developed (Moen and Lilja, 2001, 2004). Through *competition* between R&D-driven suppliers, and through sharing of R&D development costs, it was possible to start economizing R&D. This enabled improvements in productivity. At the same time, this process undermined knowledge privatization.

Third Phase: Riding the Paradigm

Knowledge externalities, hence competitors, come into play. Fairly similar products emerge in the market and a dominant design is established. Consumers start to compare prices instead of technical novelties. In the face of increasing price competition, the challenge becomes to specialize and to increase efficiency in production. The corporation has to shift from relying purely on technological competition to advantages of scale, price competition and design.

At this point, when the large-scale organization is trying to adapt to increasing pressure to become more efficient, we encounter the problem known as 'cognitive limits' or 'boundaries of rationality' in organizational theory (March and Simon, 1958, p. 170). Specialized forms of knowledge are compartmentalized. Incremental forms of innovation within the established rules of the paradigm come at a lower cost and with less uncertainty. Employees are monitored, work procedures standardized and compartmentalized. These dynamics are likely to lead into trajectories where corporate actors are locked into somewhat more narrow paths.

But it does not stop there. This drive towards cost-cutting also hits internal R&D investments. Organizations do not continue to learn though internal R&D investments, but rather through external market relations. Standardized production, later R&D also, is moved into low-cost countries.

Fourth Phase: Value Chain Disintegration. Learning through the Market

Car manufacturing in Germany is such a mature paradigm, exposed to fierce global price and technology competition. As pointed out by Herrigel (2004) in a recent analysis of this industry, contradictory pressures to cut costs and innovate new products at the same time results in *vertical disintegration*. This leads to role ambiguity where the character of relations between customers and suppliers, in particular the specific role that the customer and supplier will play in their relation, is always ex ante highly uncertain. This uncertainty is a characteristic of garbage-can learning.

Garbage-can Capitalism

Through vertical disintegration, coordination through rigid hierarchies of large corporate actors is melting away. We are instead moving into *non-hierarchical forms of coordination* (Sabel, 2005), which create conditions for what Pierce once referred to as 'abductive' or 'explorative' forms of learning. In organizational theory these forms of learning are sometimes associated with the *garbage-can* model (March and Olsen, 1976). Learning in the garbage can is characterized by a *disconnection* of problems, solutions, actors, and opportunities of making choices (March and Olsen, 1976, pp. 27), which instead are *allowed to move around and produce a random stream of new re-combinations of opportunities, problems and solutions, characteristic of garbage-can learning*.

These random experiments were precisely what managerial compartmentalization was set up to prevent. The garbage-can model may be seen as a problematic state of affairs, since there is no order and capacity for strategic agency. The potential of this form of learning is that of making different, perhaps even inconsistent, experiments with other actors all the time, at least something may succeed, and you may be one of the survivors.

Garbage-can learning also increases the possibility of cross-sector and cross-paradigm knowledge fertilization; in short, it increases heterogeneity. This is what we have referred to as 'Mode 3'[1] in the introduction to this book, and describe in more detail in Chapter 5.

Inside the garbage can, experimental learning is taking place in a random fashion, *through spot market relations*. Certainly, garbage-can learning is not the only alternative. First, external network relations between firms may be more or less institutionalized. Second, the national business system writers have identified several *managerial alternatives* to the garbage can, such as the Anglo-Saxon version of neo-liberal capitalism, and the German forms of organized capitalism. Before going into detail, we briefly look at the contrast between, on the one hand, hierarchies of managerial capitalism, and garbage-can or market learning on the other.

MANAGEMENT OR MARKET?

One indicator of garbage-can learning may be found in the European Working Conditions Survey. In this survey, randomly selected adults in 27 European countries who are employed or self-employed (roughly 1000 were interviewed in each country) were asked: 'Who determines the pace of work?' At this point, most Nordics (73 per cent in Denmark, 78 per cent in Norway) say that *some external actor, such as the customer*, determines their deadlines, as opposed to 68 per cent in the entire EU27. In a somewhat more restricted form of work organization, which may have similarities to a hierarchical organization, one would expect *the boss* (a superior) to be monitoring progress. Many other Europeans also answer that *their boss is monitoring their progress* (35.7 per cent in the EU27, and as many as 47 per cent in the UK). However, the *boss* is obviously absent in the Nordic countries. In Sweden, only 16.4 per cent report that their pace of work is monitored by their boss, and in Finland we are down to 15.5 per cent. In an organization where employees are directly confronted by the demands of customers or other external actors, competent people, able to act autonomously to adapt to diverse demands, are needed. If these autonomous actors are allowed to learn new things and apply new ideas at work, the garbage-can learning model is closer at hand. On the other hand, in work situations where the boss is in control, there is less freedom to apply new ideas, and less incentive to learn new things. This kind of work situation would be closer to some hierarchical model of routine work, locked into a limited managerial rationality. Not surprisingly, we find several related questions that correlate pretty well with this issue of who determines pace of work. On the question of whether you are learning new things as a part of the main job, 90 per cent of the Finns and 89.3 per cent of the Swedes agree (with Danes and Norwegians right behind) as opposed to just 66 per cent of the Germans and 69 per cent of respondents in the EU27.

Learning is also related to local problem-solving. Nordics are allowed to apply their own ideas at work to a greater extent than other Europeans. Here, Sweden and Finland are in the lead, with 73 per cent and 72 per cent, as opposed to only 49.8 per cent of the Germans, 58 per cent in the UK and 58.4 per cent in the EU27. In pulling these questions together, and analysing the scores at the national level, we get the factor shown in Table 11.1, which we will refer to as 'market learning'.

If we compute this factor for 22 European countries, it explains 67.8 per cent of the variance in the five variables included. The highest scores on this indicator are taken by the Nordic countries, Norway, Sweden, Denmark and Finland, and in addition Switzerland and the Netherlands. The lower scores are in southern and eastern European countries, such as Hungary, Turkey, Portugal, the Slovak Republic, Spain and Greece. Among other European countries, the leaders are Ireland, Belgium, Austria and Slovenia.

In Figure 11.1 the relation between the market learning factor (2005) and GDP/capita for 2005 as reported by the OECD is highlighted. The

Table 11.1 Component matrix[a]

Variables included in factor	Factor component
Pace of work determined by customers	.706
Pace of work determined by boss	−.846
Learning new things at work	.967
Applying own ideas at work	.784
Paid training during last 12 months	.793

Notes:
Extraction Method: Principal Component Analysis.
[a] 1 component extracted.

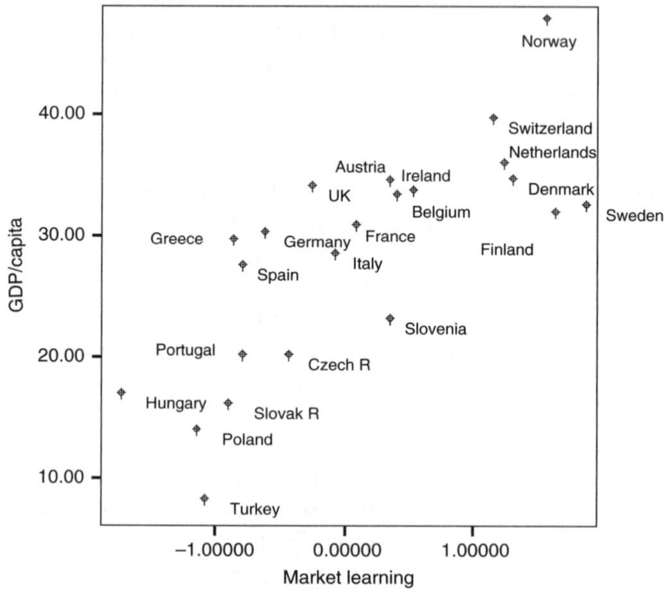

Source: Fourth European Working Conditions Survey, OECD.

Figure 11.1 The relation between market learning and GDP per capita

correlation between the market learning factor and GDP per capita is +0.756. If we take one of the input variables, the question of whether external actors such as customers determine the pace of work, and correlate it with GDP/capita the answer is a correlation of +0.858. Similarly, the correlation between GDP/capita and monitoring of the pace of work by the boss is negative, −0.546. With $N = 23$, these correlations are significant at the 1 per cent level.

In high cost countries with high GDP per capita and low wage differences, such as the Nordic countries, organizing where employees are monitored by a boss is simply a luxury few can afford. Many industries in high cost countries realized this some years ago, and adjusted to it through the *relocation* of standardized work. This relocation results in a concentration of standardized work monitored by bosses in low cost countries with low GDP per capita and fairly low wages, such as Hungary or Poland, thus contributing to the low 'market learning' score in these countries.

Now, what we refer to here as 'market learning' may be seen as independent of the logic of learning usually referred to in the innovation system discourse, that of R&D investments. Market learning may be learning quite simple things, like incremental adaptations to adjust services to customer demands, or other spot market relations based on a restricted knowledge base.

However, from what we have said above, market learning may also include relations between actors in highly sophisticated and mature technological paradigms. In the mature paradigm case, such as the car industry discussed by Herrigel, market learning relies on a high level of shared knowledge among the participants inside the garbage can. In mature co-evolving industrial sectors, this shared knowledge base is accumulated historically through R&D investments undertaken at earlier stages in the evolution of the paradigm. This history may put up invisible or visible walls around the can, with formidable barriers to entry. Market learning may well be supported by public and private investments in R&D. Adaptations to the pressure of the market may well include internal R&D investments – or access to openly available, R&D-based knowledge.

This open-ended relation between 'market learning' and R&D investments as a share of gross domestic product is shown in Figure 11.2. Eastern and southern European countries with a low level of market learning are also investing relatively little in R&D. Some of the large and rich countries with high GDP per capita in the core of Europe, such as Germany and the UK, combine low levels of market learning with high levels of R&D investments. From the national business system literature, these countries are known for different forms of managerial coordination. To make this point, a few stylized facts may be in place.

The German Model

In the German forms of capitalism, coordination is based on cooperation in high-trust relations between managers and unions, engineers and blue-collar workers, enabling learning that combines tacit and codified knowledge. These hierarchies are regulated through institutionalized rules of the

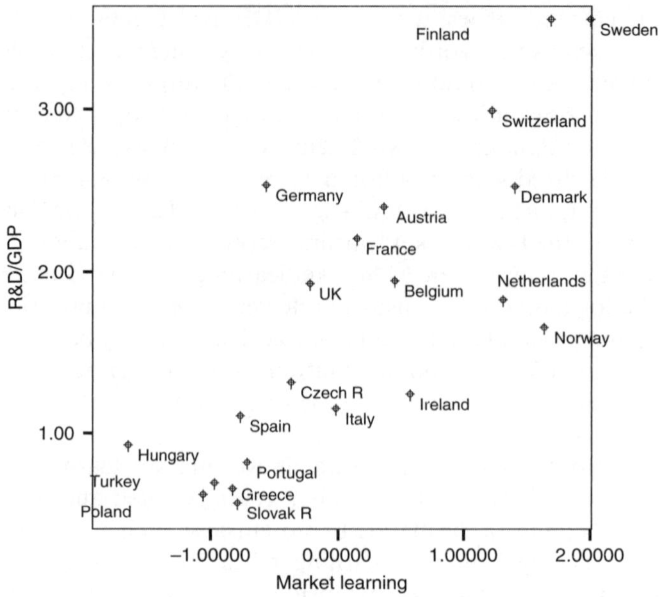

Source: Fourth European Working Conditions Survey, OECD.

Figure 11.2 The relation between market learning and R&D as a share of GDP

games that protect the rights of employees. Labour contracts are expected to be long-lasting. Learning inside firms is enhanced through intramural investments in R&D. Trust relations of highly coordinated market economies may include unions, managers, state level actors, and industrial partners and networks. The existence of these high-trust relations, however, does not remove the boss. On the contrary, the boss may play an important role as guardian of the rules of the game. Given these negotiated rules, German industries have problems in generating enough internal heterogeneity and flexibility to be able to adapt to increasing global market pressure, through outsourcing and competition between internal work and external subcontractors. The story Herrigel tells us from the German car industry is about the transformation of these rules of the games of German industry in the direction of increased market learning. The point of this example is that instead of relocating to a low cost country, work organizations in high cost countries may indeed transform themselves and *become learning*, through *value chain disintegration*. Value chain disintegration increases the capacity for garbage-can learning, and in that way enables industries to survive, despite the disadvantages of expensive labour.

The UK Model

In the Anglo-Saxon model, coordination is top-down and hands-off. Protection of work contracts is low. Workers are kept in place by a stripped-down welfare system with low support for the unemployed. The boss is a crucial actor. The freedom to outsource gives corporate management the option to radically change technologies and create new paradigms. Large Anglo-Saxon corporate actors are constructed in a way that enhances *internal* heterogeneity in order to be able to adapt to changing environments. Large corporations typically include several competing groups of R&D-based technologies, controlling different forms of knowledge related to different technological paradigms. This gives the corporate management a broad repertoire, enabling changes and adaptations in turbulent times, where German counterparts are stuck within their more stable high-trust relations. Corporate management typically coordinates competitive games between different forms of knowledge. The management may also freely choose to outsource specialized functions. This, according to Whitley, is why the US and UK economies are better at radical science-based product innovations than the German and Japanese economies. The ways in which these firms learn are shaped by the corporate management. This top-down coordination, again, relies on the institutionalized assumption that codified knowledge, which is the form of knowledge within which top-down coordination operates, has a superior status in relation to tacit knowledge or craft-based skills, the knowledge of the operators.

Barriers to Learning in Large Countries

Both the German and the UK business systems have built-in barriers to market learning.

- In the German case, the barrier is protection of employees' rights, restricting flexibility, outsourcing and market learning with outsiders.
- In the UK case, the barrier is top-down coordination based on codified knowledge, executed by a managerial layer, which may restrict bottom-up learning and problem solving.

In these diametrically opposing ways, the German and UK institutional complementarities lead to a high score on 'monitoring of the boss' – and a correspondingly low score on our market learning indicator.

Large countries like the UK and Germany may keep up their institutional complementarities in the face of new global market challenges.

Policy leaders in Nordic countries on the other hand, are painfully aware of the small country squeeze. In taking a closer look inside the Nordic garbage cans, we discover adaptations, where the Nordics are modifying their business systems.

CONSTRUCTING NORDIC GARBAGE CANS

In considering the various ecosystems of the Nordic countries, it may be appropriate to start with the co-evolution of technological paradigms and institutional complementarities. When looking at the Nordic countries, it becomes obvious that there is a broad variety in terms of specialization. Differences in specialization, in terms of sectors and related technological paradigms, result in different levels of R&D investment. In mature industries such as food, wood, metals, or petroleum, the ratio of R&D investment in relation to turnover is lower than in 'high-tech' industries characterized by technological competition.

In considering the various ecosystems of the Nordic countries, it is appropriate to start with co-evolution of paradigms and institutional complementarities.

Danish Flexicurity

Industrialization of Denmark commenced with the food industry and agriculture. Agriculture provided craft skills and access to continental consumer markets, combined with institutions of knowledge-sharing and integration of tacit and codified knowledge. Following the democratic Grundvig tradition of popular, high level education combining theory and practice, the food sector and its related education systems shaped the national complementarities. As food industry was modernized, a formidable cluster of related sectors, among them mechanical industries, co-evolved. Even though this cluster has resulted in several other strong export industries, food still is the most important export sector. Other examples today are electronics, machinery, and medical instruments.

The modern version of Danish flexicurity is a garbage can where a high level of distributed skills among autonomous process operators is at the core. Through a high level of labour market mobility, skilled craftsmen move between firms and industries, accumulating and transferring skills and knowledge.

This process is further enhanced through labour market training programmes, often organized via local level union–employer cooperation, and adult education provided by the welfare state. As shown by Peer Hull

Kristensen, the success of the Danish economy is based on the resulting wide-ranging craft-based networks that become powerful instruments for collective learning and knowledge sharing.

Within this context, firms become fairly open systems, forced to compete to access the best skilled manpower around. The boss has a role as facilitator, without slowing down learning or monitoring operator progress. At the same time, Danish firms are not slowed down by German-style regulations of labour markets protecting employee rights. Instead, mobility is enabled by a welfare state that removes the risk of being unemployed through the provision of generous benefits.

Denmark is a close approximation to a 'pure' garbage-can model. The Swedish–Finnish ecologies have a different point of departure. Here, firms are more centre stage, but they have developed institutionalized supporting systems of innovation.

Sweden

The equivalent of agriculture in Denmark is iron ore and forestry in Sweden. Evolving from specialization in the extraction and processing of raw materials, such as iron, Sweden today has a highly diversified export base with nine strong manufacturing sectors with net export surplus. This heterogeneity reflects a technology-driven industrial economy with a high input of R&D in product development and improvement in several important industrial sectors. Like Germany, Sweden combines a high *input* of R&D with a low *output* in terms of new products. This is indicative of well-developed sector innovation systems with high use of R&D in product upgrading. Swedish corporations share many similarities with advanced German firms. They have sophisticated and advanced knowledge bases, highly developed industrial organizations, owners with a deep interest and commitment to technological development, knowledge-driven strategies, and a superb capability in solving complex problems of technological development, often through development coalitions between raw material processors and their technology suppliers. During the 1990s, Sweden demonstrated a high capacity for new path creation in areas of bio-technology and information technology.

Through stronger emphasis on new path creation and intramural R&D investments, Swedish corporations have low turnover. They upgrade the skills of their employees, rather than enhancing labour market mobility. In terms of heterogeneity-creating institutions, the Swedish tradition is a combination of formalized, deep and long-term industrial networks, called 'development blocks' combined with university–industry networking, and more recently, innovation policies coordinated by a specifically dedicated agency (VINNOVA).

The result of this combination of R&D-intensive strategies and market learning is a fairly high level of R&D investments. During the last three decades, the Swedish model has been copied by Finland.

Finland

A characteristic feature of industrial development in Finland is the long-term growth of R&D intensive sectors, in particular electronics. In understanding the institutional preconditions for the 1990 success story, it is necessary to take a look at the geopolitical situation of Finland after World War II, where Finland held a special position in relation to the Soviet Union. Finland was a part of the 'Free World', and could access the planning economy of the East at the same time. As a part of the effort in promoting national independence from the Soviet Union, Finland looked to OECD policies in general, and Swedish policies and institution-building in particular. The major Finnish institutions were copied from Sweden. The Finnish National Fund for Research and Development (SITRA) was established in 1967 to support R&D in private firms. The Science Policy Council (later called the Science *and Technology* Policy Council) was established in 1973. The National Technology Agency (TEKES) was established in 1983, based on a national consensus on the necessity for technological development between politicians, industrialists and trade unions (Lemola, 2005). The long-term focus of these institutions was to *increase investments in R&D* in various parts of the innovation system, and in particular in private industries. In this way, Finland developed electronics industries and other highly innovative sectors supplying the Soviet market, supported by their access to western technologies and western style technology policy. The result was the long-term growth of mechanical and electrical equipment industries.

However, in the 1970s investments in R&D stagnated, and so did the electronics industry. Whereas the smaller Norwegian electronics industry, which had focused almost entirely on the domestic market in the 1970s, got into increasingly deeper problems throughout the 1980s, the larger Finnish electronic corporations could, for a while, rely on the Soviet market access. If we go back to 1981–82, Norway and Finland were equal in terms of the share of GDP invested in R&D. During the early and mid-1980s, the two countries followed each other closely, with growth in private investments in R&D.

The *underlying* industrial structures driving these investments were, however, quite different. The contrast between Finland and Norway in terms of the GERD indicator opened up in 1987. One explanation was that both public and private investments in R&D stagnated, for reasons to be elaborated in the next section. In the latter half of the 1980s, however, Finnish R&D investments continued to grow.

Moen and Lilja (2001) emphasized the significance of *policy-making processes* in this period, where *discursive institutions* successfully mobilized elites on a cross-sector basis. This was the background to the external shock in 1990, as the Soviet market disappeared. At this time, Finnish corporations already had experience of using R&D to recapture market shares.

The NIS innovation policy approach was anticipated by Freeman (1986). Already in 1990, the NIS concept was defined as the basis of the Finnish technology policy. The swift translation of this approach into the heart of the policy-making process in Finland is explained by Moen and Lilja (2005) with the capacity for horizontal coordination and decision-making in Finland. The core of these policies consisted of moving the Science and Technology Policy Council into the heart of policy-making to replace a group of private banks that disappeared during the crises. Nokia used this policy approach, and created the new GSM mobile telephony success story. Finland joined the EU in 1994, at the time when NOKIA was surging ahead, and the Finnish success story came to have a deep influence on EU policies, including the Lisbon process.

The Nokia success in taking the leap into becoming a fully fledged R&D-driven corporation is stunning. Moen and Lilja explain this success through forms of intersector coordination where technology and innovation policy became paradigmatic through discursive institutions, mobilizing cross-sector elites. Such factors helped to override the rigidities of the institutional complementarities of neo-corporatism. Instead, they found that cross-sector ties facilitated the relocation of horizontal coordination in the national system of innovation.

In this way, Finland combines the strengths of the coordinated business system, which dominates in countries like Germany and Sweden, with a strong influence from the US business model. Later, an assembly system using subcontractors evolved through national and global networks.

Accordingly, it should not come as a surprise that the Finnish government simply *called off NIS* in a conference organized by the Pro-Act programme in Tampere in March 2006. Here, the representative from the Minister of Trade and Industry, stated:

> Globalization is one dynamic force that drives the renewal of societies and economies. In order to be in the driver's seat, we must understand this phenomenon. Nations have to analyse what such things as the relocation of manufacturing and internationalization of research and development means for them. What are the national and regional strategies to profit and success?
>
> Without doubt, the role of innovation policy will be important when we answer these questions, and its role and character must also be reconsidered. What

> worked fifteen years ago will not do it again, and we must respond with new ideas and ways of action. (Minister of Trade and Industry, Mauri Pekkarinen)

In 2006, national and regional strategies for profit and success yet again became an open issue, with innovation policy needing to be reconsidered. This reconsideration is now going in the direction of policies promoting global technology platforms. In this way, through the global technology platforms, Finland is again breaking out of the limitations of the small country.

On the surface, this may sound very German. However, inside the Finnish extended enterprise and technology platforms, there is a market driven, Anglo-Saxon approach to business organization, with a project economy characterized by competition and an ever-present customer.

Norway

The equivalent to the farming and food industry for Denmark and metals and timber for Sweden and Finland, is shipping and petroleum for Norway. Today, Norwegians are well known for their ability to rapidly team up and recombine different forms of highly specialized knowledge into new products. The knowledge ecosystem that enables these outcomes is the result of co-evolution of several sectors in the Norwegian economy throughout the latter part of the 20th century. At the core of this process were two dynamic, closely related, and, in the small country perspective of Norway, large industries – shipping and petroleum. Throughout the early 20th century, Norwegian maritime industries formed the basis of dynamic transnational communities of sailors, ship-owners, industries and markets looking for transnational transportation services, harbour operators, shipyards, and maritime supply and equipment industries. Within this global field, the Norwegians developed their own unique way of operating, quite distinct from their main competitor, the UK. This distinction was based on a completely different way of work organization, embedded in different national business systems (Mariussen and Fraas, 2003).

In terms of soft regulations of this field, the name of the game was craft-based authority in a learning organization, operating within – literally speaking – an open horizon. Given what we have just said, the authority of the commander-in-chief is that of the first among equals. This created an atmosphere of frank and open communication and knowledge sharing, drawing on thick layers of overlapping knowledge created by career patterns, combined with well-developed expertise among senior crew members. In this context, UK-style protection of some pieces of codified knowledge as a basis of authority was looked on with deep suspicion.

Most fortunately for the Norwegian shipping industry, American oil companies discovered what was to be one of the world's largest reserves of oil off the coast of Norway, just as the shipping industry, including shipbuilding, was moving to Asia. Similarly, most fortunately for the global oil industry, it met the challenge of exploring for oil and gas in increasingly deeper waters and in an extremely hostile environment when it came to climate, in a country which already had a very well developed innovation system handling maritime technologies.

Oil and shipping are complex industries that combine radical entrepreneurialism in developing new global transportation services, oil provinces and new technologies, with routine cash cow operations, chasing productivity in standardized products in markets characterized by price competition. On the entrepreneurial side of this equation, different knowledge elements must be combined and recombined, often in surprisingly new ways and on an experimental (pragmatic) project-related basis.

These sectors constitute complex wholes that were parsed into parts involving several other sectors in the domestic Norwegian economy, forming a large, flexible and highly heterogeneous set of interrelated, and yet disintegrated, clusters and sectors. This dynamic parsing out may be explained through *value chain disintegration* in these two industries which occured during the latter part of the 20th century. The footprints of these forms of interaction are found in the economy of Norway, through interaction and coordination between industrial actors with different forms of knowledge, visible through patterns of *trade* between sectors. These factors, importantly, are related, with overlapping sectors. This field, bridged by cross-cutting sectors, enables certain truly remarkable entrepreneurial achievements, where entrepreneurs move horizontally across different clusters.

The long line in Norwegian economic policy after World War II is the question of how to be able to direct investments in the direction of more productive industries. In the Norwegian context, enhancing productivity came to imply a focus on raw material-based sectors. A characteristic feature of the Norwegian export base today is the sustained importance of export industries, either directly based on natural resources, such as petroleum and fishing, or *indirectly* depending on natural resources, such as metals and chemicals using cheap domestic supplies of hydroelectric power, the food industry based on fishing and aquaculture, refined petroleum products, and pulp and paper.

This does not mean that the Norwegian economy may be seen as just 'running with the paradigms' of these sectors. Quite the contrary, the Norwegian success story in exploiting natural resources to a large extent depends on a sustained R&D effort where the state has played an active role as 'helping hand'. This has turned Norwegian offshore deep-sea petroleum production and aquaculture into stories where the Norwegians are exploiting natural resources in ways nobody has done before. Seen in relation to the *turnover* of the clusters, R&D investments are relatively moderate as compared to 'high-tech' sectors, such as pharmaceuticals, ICT, medical instruments, and cars, which are more strongly developed in the three other Nordic countries, and where R&D investments are closer to the core of the corporate sector. Most Norwegian corporate actors outsource most of their R&D – and even share knowledge and new cutting-edge processing technologies with other producers in their clusters, while several Swedish and Finnish corporations on the other hand invest heavily in R&D they keep to themselves.

As pointed out by Whitley, in economies characterized by generic corporations, the helping hand of the state is the major actor, with a capacity to generate change (Whitley, 2001). In the post World War II years, the Norwegian state most certainly had a 'helping hand policy', and was actively involved in developing what was then regarded as new high-tech industries such as military industry and metal production. However, these policies were basically promoted by non-formalized networks of social democrats, union leaders, managers, and industrial champions. These networks were referred to as 'mixed administration' (Hernes, 1975). Eventually, mixed administration made mistakes, which brought them into disgrace. For instance, industries owned by the state developed a nasty habit of relying on subsidies from their owner, relaxing their own profit expectations. State-owned industries were privatized and subsidies to unprofitable industries were stopped during the early 1980s (Mariussen et al., 1996). In this way, the traditional Norwegian export industries were put under pressure to restructure and adapt to the requirements of global competition. In 1986, an initiative was taken to pick winners by investing in new, high-tech industries. State support was to be given to certain technology areas that were seen as future industries, such as information technology, biotechnology, oil and gas technology, organizational theory and environmental technologies. This attempt at diversification turned out to be politically unsustainable. At the end of the 1990s, the banking sector was deregulated, and a bubble economy based on rapidly growing domestic consumption started. Both private debt and deficits in state budgets were allowed to accumulate. This combination contributed to rising wages and a rising cost level. This made the situation for the raw material-based export industries outside the oil sector (the mainland economy) even worse. This sector had an old problem with productivity. In the days of mixed administration, it had enjoyed privileges resulting from state subsidies and easy access to cheap resources. The challenge was to promote 'producer friendly' institutional arrangements supporting productive activities, and reducing the options for entrepreneurs to compete for money from the state, or 'grabbing' (Mehlum et al., 2005).

After 1990, the focus was again on the raw material processing industries. Key measures in economic policy were a stable and long-term financial policy, combined with 'a broad structural policy'. In this way, an alliance was made between raw material processors and the regime of economists that later was going to control the oil fund (income to the state based on Norwegian oil production saved in a fund) that started to accumulate in the end of the 1990s. The institution to be responsible for the future oil fund, the Ministry of Finance, had to develop a 'hands-off' policy in relation to all other ministries when it came to budget control. The guard against what came to be seen as populist demands from the high-tech sectors had to be strict and impartial. The result is a policy regime which is enhancing specialization in raw material processing.

CONCLUSION: SPECIALIZATION AND HETEROGENEITY

In comparing the various forms of Nordic capitalism, different distinctions between firms and their environment become obvious (see Table 11.2).

Swedish/Finnish firms are investing in firm-internal learning, through upgrading of employees and intramural R&D investments. At the same time, however, unlike their German counterparts, employees are not protected from the pressure of the market by some soft managerial cushion. This opens up the Swedish/Finnish economies for R&D-intensive new path creation. These learning Swedish/Finnish firms are active in constructing their environment, through dense interfirm and firm–university networks as building blocks for institutionalized innovation systems. In the case of Finland, these attempts are now moving from the level of national innovation systems into global technology platforms. Swedish/Finnish firms and institutionalized networks increasingly may be seen as regulatory mechanisms for competitive project economies, where the customer (the project funder) is monitoring work.

Danish and Norwegian firms are more open-ended and liberal, but in different ways. Danish firms may be seen as coalitions of autonomous craftsmen and other specialists, where management has a function as facilitator. Learning is taking place in wide-ranging networks developing in high mobility regional labour markets. This is a garbage can where autonomous humans are moving around.

Table 11.2 Europe: market learning and wealth, 2005

		Specialization	
		Mature sectors	'High-tech' sectors
Heterogeneity	Firms	Denmark: Firms are open coalitions of experts Norway: Lean processors, outsourcing, knowledge-based suppliers	Sweden and Finland: Accumulation of specialized knowledge, investing in intramural R&D
	Firm environment	Loosely structured distributed knowledge networks	Institutionalized innovation systems as frameworks of competitive project economies
	Policies	Welfare state security Public R&D supporting shared knowledge bases	Innovation system policies, targeting specific networks

The core technologies of the Norwegian specialization – processing of natural resources – require differentiation between engineering and processing skills. Typically, processing industries are lean and stripped down, focusing on core business, and outsourcing learning to a rich ecology of related industries. Here, entrepreneurs are crossing borders between sectors and creating new industries. Several sectors, such as systems engineering, are shared between different raw material-based clusters. Norwegian innovation systems are more weakly institutionalized, and approach the garbage-can model, but they do so in a way that is quite different from Denmark.

NOTE

1. The 'Mode 3 INNOVECO' is in short the nexus or hub of the emerging 21st-century Innovation Ecosystem, where people, culture and technology (forming the essential 'Mode 3 INNOVECO' building block or 'knowledge nugget') meet and interact to catalyze creativity, trigger invention and accelerate innovation.

REFERENCES

Doz, Yves (INSEAD) and Mikko Kosonen (NOKIA) (2005), 'Strategic agility: oxymoron or opportunity?', paper presented at CKIR Workshop 18 August 2005.

Freeman, C. (1986), *The Economics of Industrial Innovation*, 2nd edn, Cambridge, MA: The MIT Press.

Hernes, G. (1975), *Makt og avmarkt Et utgang spunkt for kartlegging av de faktiske maktforhold I det norske samfunn*, Bergen, Oslo and Tromsø: Norwegian University Press.

Herrigel, G. (2004), 'Emerging strategies and forms of governance in high-wage component manufacturing regions', *Industry and Innovation*, **11** (1–2), 45–79.

Kristensen, P.H. and J. Levinsen (1983), *The Small Country Squeeze*, Copenhagen: Forlaget for Samfundsøkonomi og Planlegning.

Lemola, T. (2005), 'Finnish science and technology policy', in G. Schienstock (ed.), *Embracing the Knowledge Economy: The Dynamic Transformation of the Finnish Innovation System*, Cheltenham, UK and Northampton, MA: Edward Elgar, pp. 268–84.

March, James (1998), *The Pursuit of Organizational Intelligence, Malden, MA: Blackwell.*

March, J.G. and J.P. Olsen (1976), *Ambiguity and Choice in Organizations*, Bergen: Universitetsforlaget.

March, J.G. and H.A. Simon (1958), *Organizations*, New York, London and Sydney: John Whitley and Sons.

Mariussen, Å. and M. Fraas (2003), 'Integrating innovation and regional policies incorporating horizontal innovation policy across the board', paper presented at the OECD MONIT Network Workshop, Brussels IWT, 23–24 October.

Mariussen, Å., A. Karlsen and O.J. Andersen (1996), *Omstilling – fra løsriving til ny forankring (Restructuring: From Disembedding to Reembedding)*, Oslo: Norwegian University Press.

Mehlum, H., K. Moene and R. Torvik (2005), 'Crime induced poverty traps', *Journal of Development Economics*, **77**, 325–40.

Moen, E. and K. Lilja (2001), 'Constructing global corporations, contrasting national legacies in the Nordic Forest Industry', in G. Morgan, P.H. Kristensen.

Moen, E. and K. Lilja (2004), *Change in Coordinated Market Economies: The Case of Nokia and Finland*, Helsinki: Helsinki School of Economics Publications.

Moen, E. and K. Lilja (2005), 'Change in coordinated market economies: the case of Nokia and Finland', in G. Morgan, R. Whitley and E. Moen (eds), *Changing Capitalisms: Internationalization, Institutional Change and Systems of Economic Organization*, Oxford: Oxford University Press, pp. 352–79.

Sabel, C.F. (2005), 'Bootstrapping development: rethinking the role of public intervention in promoting growth', paper presented at the Protestant Ethic and Spirit of Capitalism Conference, Cornell University, Ithaca, New York, 8–10 October 2004, this version 14 November 2005.

Whitley, R. (1999), *Divergent Capitalisms: The Social Structuring and Change of Business Systems*, Oxford: Oxford University Press.

Whitley, R. (2001), *Divergent Capitalisms: The Social Structuring and Change of Business Systems*, Oxford: Oxford University Press.

12. Heterogeneity as sectoral specialization: the case of the EU15[1]

Aris Kaloudis

INTRODUCTION

The starting point of this chapter is the observation that the business sector in many EU Member States exhibits clear sectoral specialization patterns, both in terms of R&D and patenting activities and in terms of economic activities. Hence the main question we deal with is whether and how R&D activities in the EU15's business R&D relate to the EU15's technological and economic specialization patterns as compared with the USA and Japan.

In general, if the empirical evidence reveals patterns of congruence between the various dimensions of specialization, one would tend to conclude that the national R&D co-evolves, is well attuned to and supports the economic and structural specialization in the EU15.

Archibugi and Pianta (1992) found that the scientific production of advanced OECD economies shows a much lower degree of specialization compared to the more pronounced patterns of technological specialization. With this work as a starting point, we ask whether national business R&D efforts at a sectoral level within Europe reveal heterogeneous and diverse knowledge production embedded in clear and robust national economic structures.

We analyse OECD and EUROSTAT data at the lowest possible level of sectoral disaggregation, allowing a better understanding (quantitatively and qualitatively) of specialization profiles in the EU15. In brief, the specific tasks comprised in this chapter are the following:

- Collect and compile all data available in OECD or EUROSTAT necessary for the analysis.
- Calculate specialization indices for all EU15 countries across three analytical dimensions: business enterprise R&D expenditure (BERD),

technological activities, more precisely, patenting activity by economic sector and economic activities, that is, value-added, employment and exports at a sectoral level in the respective countries.

- Analyse the specialization patterns based on the computed specialization indices within and between the three analytical dimensions.

We attempted to keep the level of analysis at a disaggregated level similar to that of the OECD STAN database. At this level one may discern and analyse separately and in greater detail activities in low-tech, medium–low-tech, medium–high-tech and high-tech manufacturing sectors as well as in services.

Second, we concentrate on the period 1993 to 2003 as relatively little data are available after 2003. There are two good reasons for commencing in 1993: (1) This provides an 11-year time series for almost all countries in the EU15, and (2) the growth of ICT began to have a significant impact on the global economy in 1993. Third, considerable time and resources have been allocated in the compilation of the business enterprise R&D expenditure (BERD) by sectors. In many studies in the past it was BERD data that proved to be the 'bottleneck', that is, the data source with most missing values and inconsistencies.

The structure of this chapter is as follows: in the first section we concentrate on methodology issues providing necessary information for understanding analytical steps undertaken, and explaining how the specialization indices are calculated. We believe this methodology allows for a relatively simple, sound and robust analysis of specialization and heterogeneity patterns over time at micro (firm level), meso (sectoral or regional level) and macro (national) levels.

The second section presents specialization indices for the EU15 compared with the USA and Japan. We present and briefly comment on the main specialization patterns within each of the five data building blocks. However, emphasis is placed on how the specialization patterns within the five data building blocks relate to each other. Finally, we conclude with a brief summary.

MEASURING SECTORAL SPECIALIZATION – METHODOLOGY AND DATA USED

We start by defining what we mean by specialization. The specialization index (SI) used here is defined as:

$$SI_{kj} = 100 * \tanh \ln[(X_{kj}/\Sigma_j X_{kj})/(\Sigma_k X_{kj}/\Sigma_{kj} X_{kj})]. \quad (12.1)$$

where X_{kj} indicates the number of patent applications, exports, value-added, and so on of country k in sector j.

Positive values point to the fact that the sector has a higher weight in the portfolio of the country than its weight in Europe (EU15). Negative values indicate specializations below the average.

The mathematical expression within the brackets in (12.1) is generally known either as the Balassa index or as the Revealed Technological Advantage Index. The indicator allows the assessment of the *relative* position of a sector in a country beyond any size effects. Neither the size of the sector nor the size of the country has an impact on the outcome of this indicator.

The logarithmic transformation has the effect that the indicator is symmetrically centred around 1, as the relation of the shares before this transformation is not symmetrical (range zero to infinity) and, in particular, it is not linear.

The tangent hyperbolic transformation has the effect that extreme values (of the logarithmic distribution) are truncated to −1 and +1 (or −100 and +100 when multiplied by 100 as in our case). We apply this transformation because the logarithmic function at the tails of the distribution cannot further be interpreted as quasi-linear. Any correlation analysis of logarithmic specialization indices will be highly affected by these (non-linear) outliers. This new distribution has less extreme tails and more 'linear' behaviour.

For the same reasons, note that comparing specialization indices when they are at the extremities of the scale (above 80 or below −80) is not a straightforward exercise. Absolute values above |80| indicate all very high specialization values. By the same token, differences in specialization indicated by index values +10 and +40 are more substantial than differences indicated by values +70 and +100 because of the non-linear nature of the Balassa specialization indicator prior to the logarithmic and the tangent hyperbolic transformation.

Checking the reliability of specialization indices (SI) we also applied another symmetric SI, calculated as:

$$([(X_{kj}/\Sigma_j X_{kj})/(\Sigma_k X_{kj}/\Sigma_{kj} X_{kj})] - 1)/([(X_{kj}/\Sigma_j X_{kj})/(\Sigma_k X_{kj}/\Sigma_{kj} X_{kj})]) + 1 \tag{12.2}$$

This indicator has been proposed by Brusconi and Geuna (2004), among others. The rationale for this transformation was also to provide symmetric values for the non-symmetric Balassa index. Both (12.1) and (12.2) are tested and compared with each other. Simulations and actual calculations of specialization indices based on real data show that the two specialization indices (in (12.1) and (12.2)) provide very similar results. This means our calculations of the specialization indices rely first and foremost on the reliability of the raw data, not on the choice of the symmetric transformation of the Balassa indices that is applied.

We calculate specialization indices for the entire period, 1993–2003. We then present and compare three-year specialization averages; that is, the average of a specialization index for 1993–1995 and 2001–03 if and when data are available. Three-year averages are more robust and reliable and allow the inclusion of more countries for which no data exist for 1993 or 2003. However, a comparison of the results shows that there are no dramatic differences between specialization indices calculated on the base of three-year averages as opposed to single year observations. Note that other alternative techniques for calculating three- or five-year averages could be applied.

As regards the specialization indices for the EU15, in addition to the EU15 the denominator also includes the USA and Japan. This way it is possible to calculate comparative advantage in these three regions relative to each other.

In countries and sectors with missing values, the specialization indices have been calculated by removing the missing sectors from both the nominator and the denominator in (12.1). The same method has been used for the calculation of specialization indices for the EU15.

For all specialization indices excluding exports, current prices in million euros or local currency for the countries outside the monetary union have been used. Where necessary, we used information on exchange rates provided by the IMF and EUROSTAT (see *European Business, Facts and Figures 2004*, p. 25), for conversions from local currency to the euro. Data on exports are in thousand US dollars.

Limitations of Specialization Indices

Some of the most noteworthy limitations related to the specialization indices are the following:

1. Because they are based on relative shares, specialization indices may exaggerate differences since, by definition, positive specialization in some sectors or scientific fields implies negative specialization in all others. In particular, if one country is strongly specialized in just a few sectors (scientific fields, socioeconomic objectives, and so on) it may show a negative specialization in all other sectors. Figure 12.1 shows a simulation example of an extreme, but theoretically possible, specialization profile of a large country.

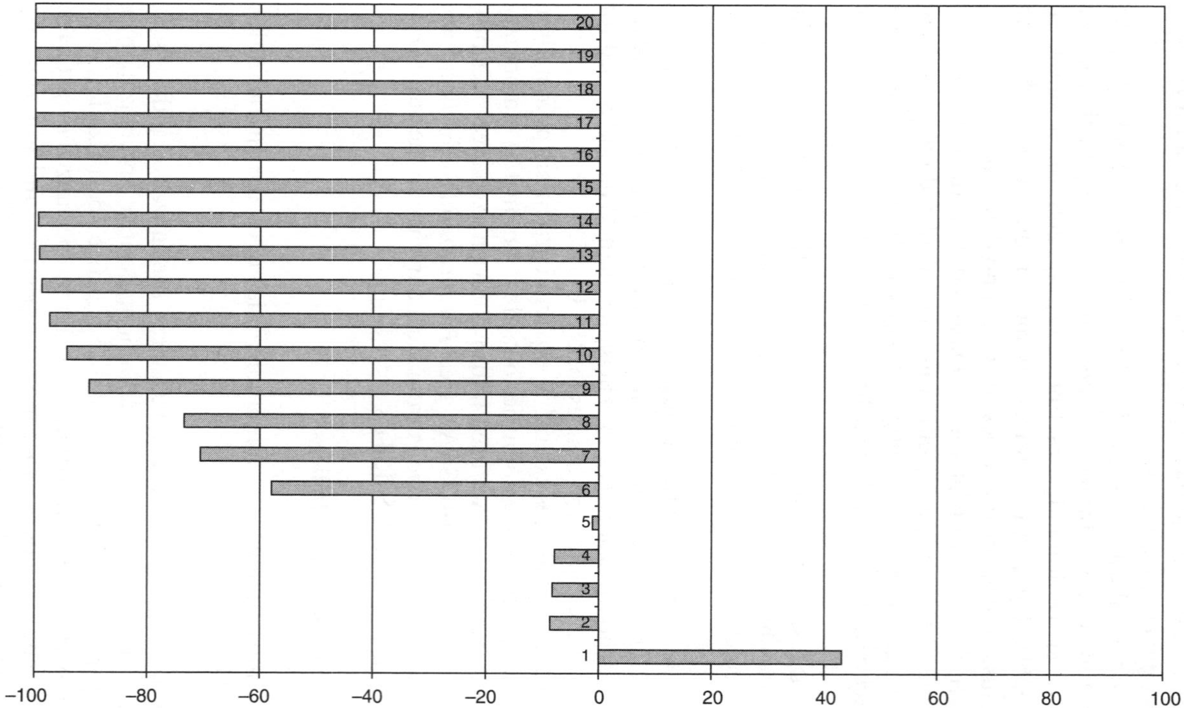

Note: Specialization index with the EU15, the USA and Japan as reference. Maximum specialization +100; Minimum specialization −100.

Figure 12.1 Simulation with specialization indices based on formula (12.1)

2. As a measure of relative advantage, one should also consult the absolute values (raw data) supporting the calculation of indices.

Alternative Methods in Calculating Specialization Indices

As the specialization index in (12.1) supports much of the analysis presented here, some more detail on its construction and properties could be provided. Two aspects are of particular importance and should be mentioned.

1. *Calculating the denominator*: The EU15 (or other) reference shares are based on aggregates of national data and are thus weighted means of national shares. This implies, however, that characteristics of such aggregates are weighted towards those of the larger countries (Germany, France, and so on) so that any 'extreme' data in these large countries feeds through to, and potentially 'unbalances' or 'distorts', the aggregate. Alternatively, one could consider direct comparisons with other countries' shares and calculate unweighted mean shares to use as a denominator. In some cases, this could provide an alternative insight into the country specialization.
2. *Conversion to a common currency in the denominator*: When calculating aggregates, we used information on exchange rates provided by the IMF and EUROSTAT for conversion from local currency to the euro, but we did not use GDP Purchasing Power Parities (PPP) as the *Frascati Manual* (OECD, 2002, Annex 9) recommends. The reason for that is that PPPs are based on consumption patterns, not output patterns (and certainly not R&D patterns), so exchange rates seem to be more appropriate than GDP PPPs.

Data on Business Enterprise R&D Expenditure

Special focus has been given to the compilation of BERD data, where considerable effort was made to provide as complete as possible data for all years, sectors and countries.

Most R&D statistics are obtained through national R&D surveys. The *Frascati Manual* (OECD, 2002) sets out guidelines for the collection of internationally comparable business enterprise research and development (BERD) data, which is generally collected at the enterprise level and allocated to the industrial class of its principal activity. However, certain problems appear in the official Basic Science and Technology Statistics for BERD, known as OFFBERD, statistics. Because the R&D surveys are conducted at the enterprise level, either underestimation or overestimation may occur in some industries where large firms dominate.[2]

Changing the classification of the dominant activity of a firm will also add to the already high volatility of BERD. Despite broad adherence to Frascati definitions, further variations also appear between countries because of differences in the way the R&D surveys are carried out. This is because sampling methodologies or coverage ratios may differ from one country to another.[3] There are also differences in the way that countries classify R&D activity performed by enterprises that are closely associated (not necessarily contractually) with manufacturing firms and R&D activity performed by large public enterprises.[4] Finally, data are often missing for some years or certain industries in a particular country. Many countries in Europe do not conduct annual R&D surveys, creating large discontinuities and limiting international comparability over time. Frequently, for confidentiality or other reasons, statistics for individual industries may not be available separately but suppressed or combined with other data. For these reasons, the OECD recommends that these data should be used with caution.

To improve international comparability and reduce temporal discontinuities and other inconsistencies, the OECD created the ANBERD database which is compatible with the new STAN database (ISIC3) from 1987.[5] Nevertheless, by its nature ANBERD contains many OECD estimates that may differ significantly from the corresponding OFFBERD statistics.[6] Estimates are generally used when borderline institutions and public enterprises are prevalent in the economy, when important deviations from the standard industrial classification are to be found, survey coverage is incomplete, there are discontinuities or breaks in series resulting from a change in industrial classification or survey techniques, and missing years where no survey was carried out.

Since one of the main objectives of this ERAWATCH project was to analyse public R&D as a source of R&D activity in enterprises, it was necessary to use the OFFBERD statistics as our starting point. Statistics are available from the OECD and EUROSTAT that provide BERD by economic activity and source of funds. Both data sources allow for the creation of a dataset that is comparable with data on value-added and employment in the STAN database, but they contain the problems associated with using official R&D statistics. For this reason, time series analysis using these data should be treated with caution. The OFFBERD statistics in the database are given in current national currencies (except the EU total).

ANBERD covers only 15 European countries from which just 11 are among the EU15 (Austria, Luxembourg, Greece, and Portugal are not

Table 12.1 Sources of R&D funding data in OECD

Data	Source	Comments
R&D expenditures by year and by performing sector	OECD Research and Development Statistics 2005	Formerly 'Basic Science and Technology Statistics' Details on national specifications and comparability issues are available in OECD (2004)*
R&D expenditures by performing R&D sector and source of funds	OECD OFFBERD 2005	
BERD by economic sector	OECD OFFBERD 2005 and ANBERD	Gaps in sectors or time series had been filled in using ANBERD. A discussion on national specifications and comparability issues between ANBERD and OFFBERD is available in OECD (2005)
Government funding of BERD	OECD Research and Development Statistics 2005 and Eurostat for a limited number of non-OECD countries	Formerly 'Basic Science and Technology Statistics'

Source: * OECD (2004), Research and Development Statistics, December.

included). OFFBERD provides data for 21 European countries and for 14 countries from the EU15 group.

The gaps in the time series of OFFBERD have been filled using the following methods:

- If the data provided by ANBERD for the specific country and sector are similar to the nearest available years in OFFBERD, then the gaps are filled using ANBERD data. For similarities in methods used for calculations of R&D expenditures in OFFBERD and ANBERD see Table 12.1.
- If no data were available in ANBERD, the missing data were calculated using linear regression analysis (least squares method). In some

exceptional cases, where the results of linear regression were not satisfactory, exponential regression was used.

- If none of the above methods was feasible, the share of the sector to the next available aggregation (e.g. ISIC 351 share to ISIC 35) of the nearest year was used.

Technological Specialization – Correspondence to Sectors

Patent applications, based on the priority year,[7] to the European Patent Office have been used for the computation of technological specialization profiles. The priority year is the year of the first filing of a patent at any patent office worldwide.

Schmoch et al. (2003) provide a methodology for the analysis of the relation between technological knowledge and sector economic performance. The study proposed a concordance matrix linking patent production in the various technical fields to industrial sectors where these patents may have economic relevance. We follow this concordance matrix and calculate technological (patent) specialization by sector. This allows us to compare R&D, technological and economic specializations in a country.

Economic Specialization – Data Used

For the purpose of this project we constructed data on value-added and employment for non-member countries. In the following we provide an overview of the methodological steps used in the construction of these data.

The OECD Structural Analysis (STAN) database was created for the purpose of analysing and comparing industrial performance at a relatively detailed level of activity between countries. It includes annual measures of output, value-added, labour input, investment and international trade at the 2-digit level, plus some further detail above this level on industries that rely relatively more heavily on R&D activity. This permits further exploration into alternative groupings of industries based on R&D intensity, skill intensity or other principle factors. The current version of the STAN database is based on the ISIC3 classification system and has very good coverage of the period between 1993 and 2003.

Other OECD datasets such as the OECD Analytical BERD (ANBERD) database, bilateral trade database, and input–output tables (1990s) are compatible with this breakdown.

THE EU15 VS. THE USA AND JAPAN – SECTORAL SPECIALIZATION

This section discusses the specialization profiles of the EU15, the USA and Japan. The denominator of the mathematical expression (12.1) is the sum of the EU15,[8] the USA and Japan.

Here we will not deal with the numerous and complex data and methodological issues encountered in the analysis of the EU15 profiles; we present only the main results and some hypotheses as to how the EU15 specialization profiles seem to fit together. Hence, this analysis must be considered as only the first step towards a more rigorous investigation in the future.

Total gross R&D expenditure (GERD) in the EU15 Member States increased in absolute values in the last ten years, following the same growth rate as GDP. Thus the R&D intensity remained relatively stable at around 1.9 per cent of GDP (Figure 12.2).

Comparing the financing of business expenditure on R&D by government between the EU15, Japan, the USA and the OECD, two issues are worthwhile mentioning. First, government financing as a share of BERD is steadily declining and, second, they all converge towards 8–10 per cent of BERD. The most dramatic change is observed in the USA where government funding of BERD fell from 18 per cent of total BERD in 1993 to 10 per cent in 2003. The US government tends to finance a higher share of BERD compared to the EU15. We hypothesize that higher shares of government funding of BERD should be related to the clear orientation of the US government budget appropriations or outlays for R&D (GBOARD) towards Defence and Health. The former represents more than 50 per cent of US GBOARD; the latter approximately 25 per cent. For the EU15 and Japan, both socioeconomic objectives represent less than 20 per cent and 10 per cent respectively (see Figure 12.3).

Comparing BERD specialization of the EU15 with that of the USA and Japan (Figure 12.4), the EU15 is specialized in 14 of 22 primary and manufacturing sectors, with high specialization in Mining (only compared to Japan), Textiles (ISIC3 17–19), Aircraft and spacecraft (ISIC3 353), Pharmaceuticals (ISIC3 2423), Electrical machinery (ISIC3 31) and Motor vehicles (ISIC3 34). In Pharmaceuticals, the EU15 exhibits clear specialization (+33) compared with both Japan and the USA. In most of the sectors of European specialization, we also find Japan to be specialized (i.e. Textiles, Motor vehicles, Electrical machinery, Machinery, Non-metallic products, Plastics and Chemicals) while the USA is not. The USA exhibits clear specialization in Wood/Paper products and Printing, and so on (ISIC3 20–22), Office, accounting and computing machinery (ISIC3 30) and in

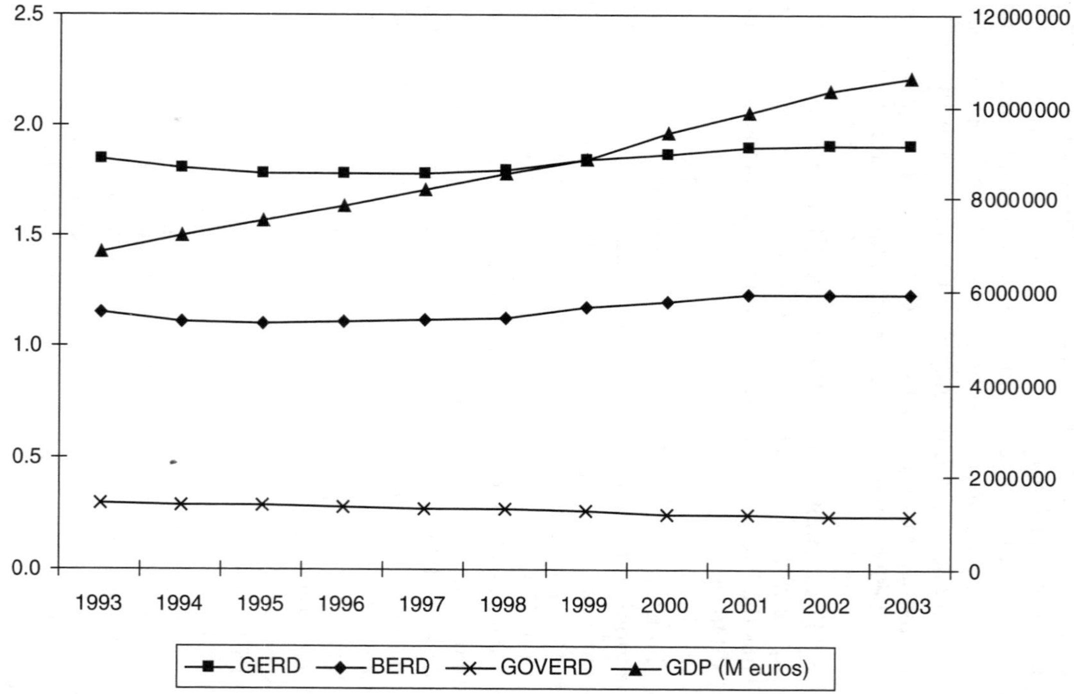

Note: Percentage of GDP in the EU15 (left Y-axis), and GDP in million euros (right axis), 2003.

Source: OECD OFFBERD 2005.

Figure 12.2 ***R&D expenditure by sector of performance***

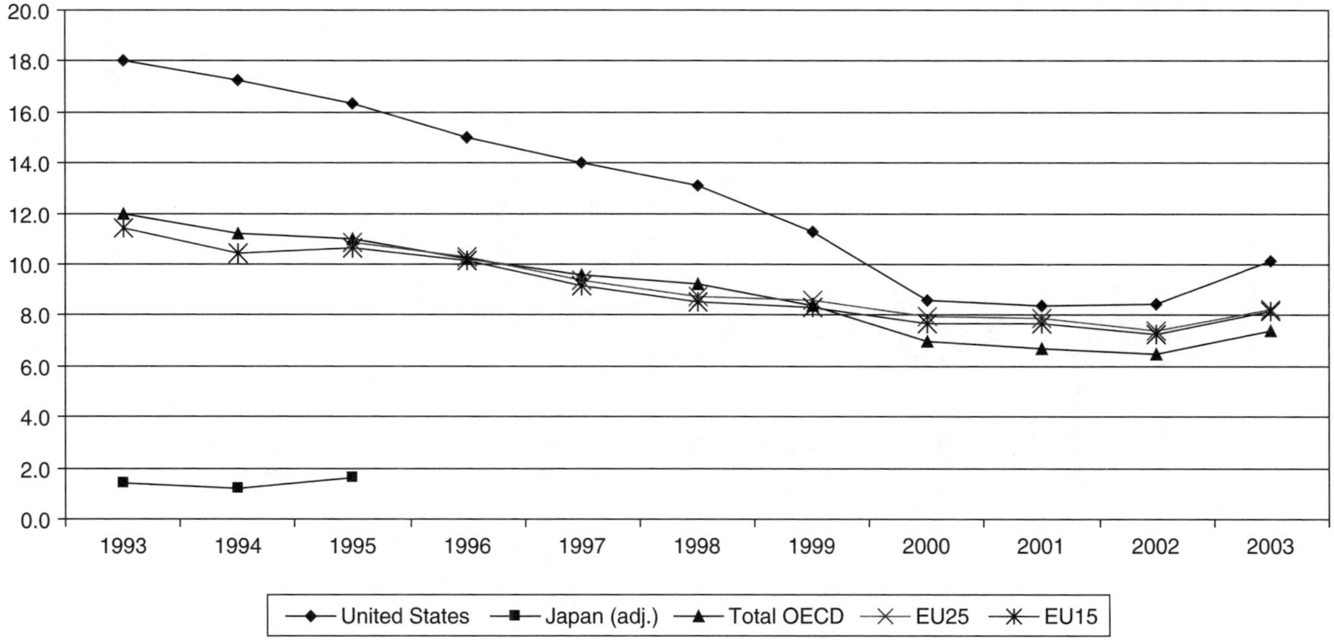

Source: OECD Basic Science and Technology Statistics 2005, own calculations.

Figure 12.3 Percentage of BERD financed by government – comparison of the USA, Japan and the EU

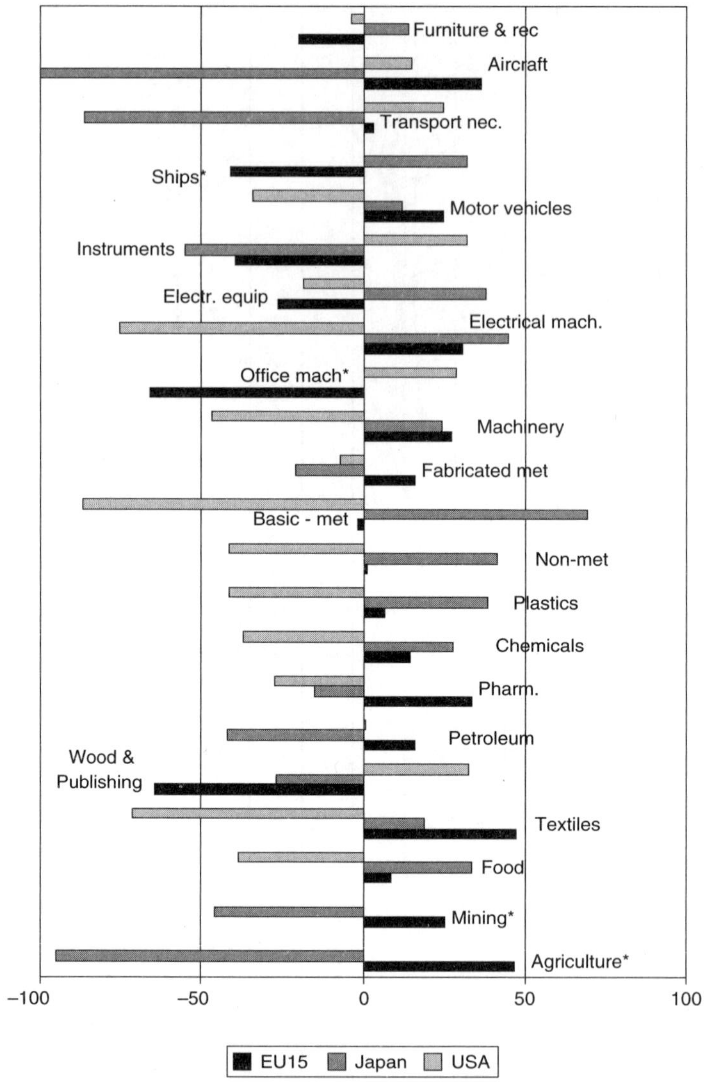

Note: The EU15, the USA and Japan; three-year average of specialization indices, 2001–03. Specialization index with the EU15, Japan and the USA as reference. Maximum specialization: + 100. Minimum specialization: – 100. * Data only for 2 countries.

Source: OECD Basic Science and Technology Statistics 2005, ANBERD 2005, own calculations.

Figure 12.4 BERD specialization in primary sectors of activity and manufacturing

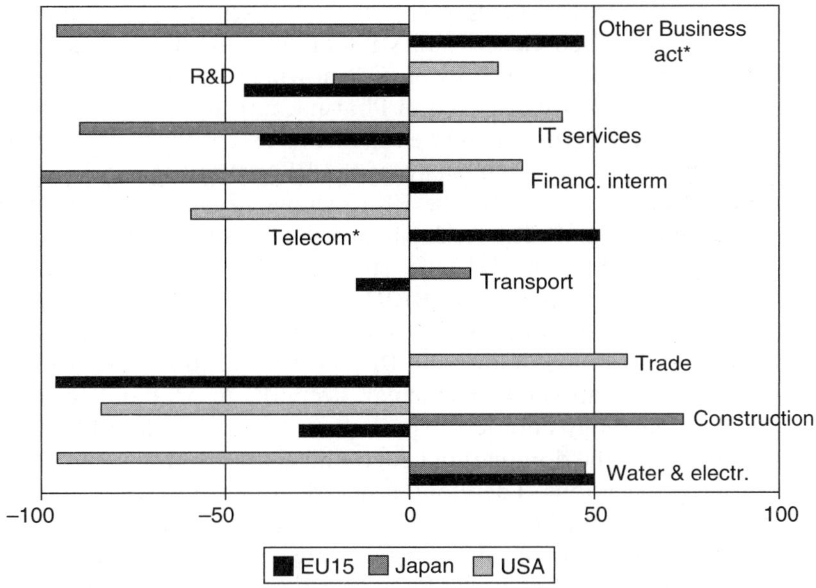

Note: The EU15, the USA and Japan, three-year average of specialization indices, 2001–03. Specialization index with the EU15, Japan and the USA as reference. Max specialization: + 100. Min. specialization: − 100. * Data for 2 countries.

Source: OECD Basic Science and Technology Statistics 2005, ANBERD 2005, own calculations.

Figure 12.5 BERD specialization in services

Medical, precision and optical instruments (ISIC3 33) compared to the EU15.

In Services, the EU15 is R&D specialized in Other business activities (ISIC3 74), Financial intermediation (ISIC3 65–67), Telecommunications[9] (ISIC3 64), Transport and storage (ISIC3 60–63) and Water and Electricity networks (ISIC3 40–41). In water and electricity networks, R&D expenditures are very low in the USA. Conversely, the USA focuses on R&D in IT services (ISIC 72), Financial intermediation and Trade (ISIC3 50–52). Hotels and restaurants are poorly covered in the OECD datasets. Japanese businesses show distinct high R&D activity in Construction (Figure 12.5).

When there are missing values for the USA or Japan, interpretation of the respective specialization indices should be made with caution. As for scientific specialization, the EU15 exhibits poor specialization profiles in many scientific fields, except for Space sciences, Mathematics, Chemistry, Physics, Geosciences and Microbiology. In Space sciences and Mathematics the EU15 is the only region showing positive specialization values. While BERD is specialized in Pharmaceuticals, the EU15 shows no particular strength in related scientific fields except for Microbiology. This possibly explains the high share of the EU pharmaceutical companies doing R&D in the USA. The EU15 shows negative specialization values in Engineering, while both Japan and the USA have positive values in this field. Note that the EU15 was specialized in a number of sectors where engineering is important in BERD (i.e. machinery, textiles, motor vehicles, electrical machinery). On the other hand, specialization in Chemistry and Space are in line with BERD specialization in Aerospace, and Chemicals. Japan shows particular strengths in Material sciences, Physics, Pharmacology and, less expected, in Agriculture and Plant sciences. The USA is the dominating geographical area in terms of volume of scientific publications. Therefore, it acts as a benchmark for the two regions, the EU15 and Japan. Generally speaking, it appears as though the scientific profiles of the three regions are more homogeneous (less extreme values) compared to BERD specialization. Maybe this is the general feature of scientific specialization; that is, all countries are much more specialized in terms of technology and economy than in terms of science.

The EU15 is specialized in patents (technological specialization) (Figure 12.6) in many of the same sectors as in BERD, except for Pharmaceuticals and Chemicals where the USA shows clear signs of technological strength throughout the entire period 1993–2003 – not shown in BERD specialization. Why the USA technological specialization is not shown in the USA BERD specialization is not clear. One plausible hypothesis could be that the USA is strong in R&D but not necessarily in in-house R&D (MNC) in this sector.

Japan shows strong technological specialization only in Office machinery (ISIC3 30), Electrical machinery (ISIC3 31) and in Radio, television and communication equipment (ISIC 32). It is also noteworthy that in Motor vehicles (ISIC3 34) the EU15 is the only economic area showing positive technological specialization; both the USA and Japan exhibit negative technological specialization values in this sector. Also technological specialization shows more homogeneous profiles compared with the profiles in BERD.

The EU15 shows economic specialization, in terms of value-added (Figure 12.7), in Transport nec. (ISIC3 352+359), Ship building (ISIC3 351), Medical, precision and optical instruments (ISIC3 33), Machinery and equipment (ISIC3 29), Fabricated metals (ISIC3 NACE

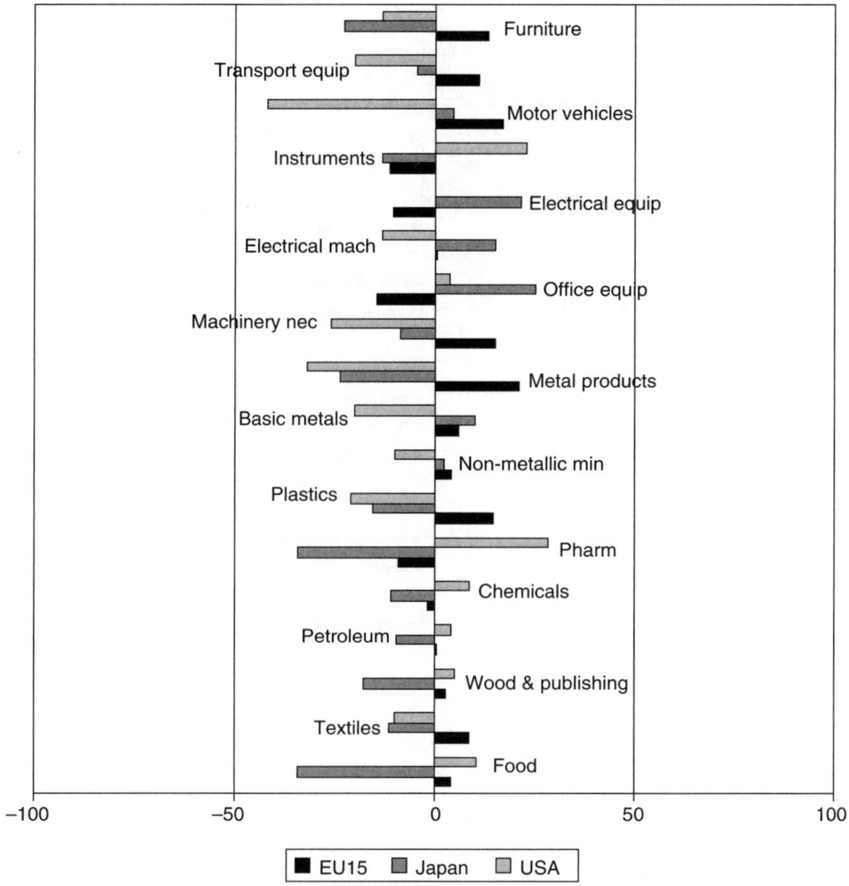

Note: The EU15, the USA and Japan, three-year averages of specialization indices, 2001–03, EPO. Specialization index with the EU15, the USA and Japan as reference. Maximum specialization: + 100. Minimum specialization: − 100.

Source: European Patent Office 2005, own calculations.

Figure 12.6 Technological specialization; manufacturing

28), Plastic products (ISIC3 25), Chemicals (ISIC3 24–2423), Textiles (NACE 17–18) and Agriculture (ISIC 01–05). In services (Figure 12.8), the only sector of EU15 strength seems to be Transport and storage (ISIC3 60–63). The USA shows clear strength in IT services, while Japan dominates in Water and electricity networks (ISIC3 40–41), Construction (ISIC3 45), Hotels and restaurants (ISIC3 55) and Real

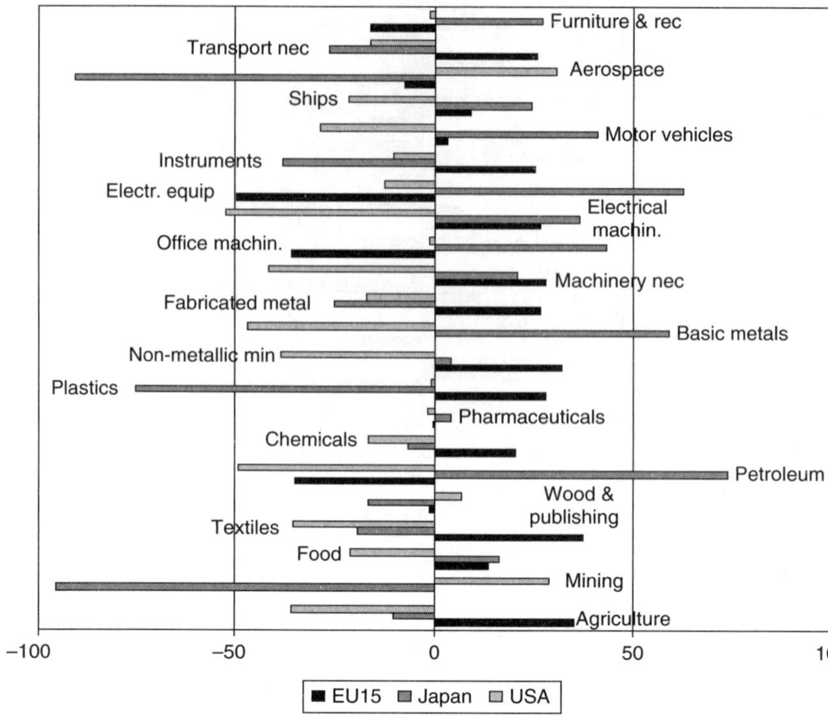

Note: The EU15, the USA and Japan, three-year averages of specialization indices, 2001–03.

Figure 12.7 Economic specialization; value-added in primary sectors of activity and manufacturing

estate and renting (ISIC3 70–71). In Aircraft (ISIC3 353), Pharmaceuticals (ISIC3 2423), and Petroleum products (ISIC3 23) the EU15 is not specialized in value-added, but while it is specialized in BERD (Figure 12.7).

In terms of employment (see Figures 12.9 and 12.10), the EU15 is specialized in the same sectors as in value-added specialization, except for Mining (ISIC3 10–14) and Pharmaceuticals (ISIC3 24–2423).

Export specialization of the EU15 is observed in Pharmaceuticals, Refinery of Petroleum and Plastics, as well as in the less technology-intensive sectors such as Furniture, Textile, Food, Mining, Metal and Basic metal products, Wood and Publishing and Non-metallic products (Figure 12.11). Comparing export specialization with BERD in the case of the EU15, the mismatching is apparent.

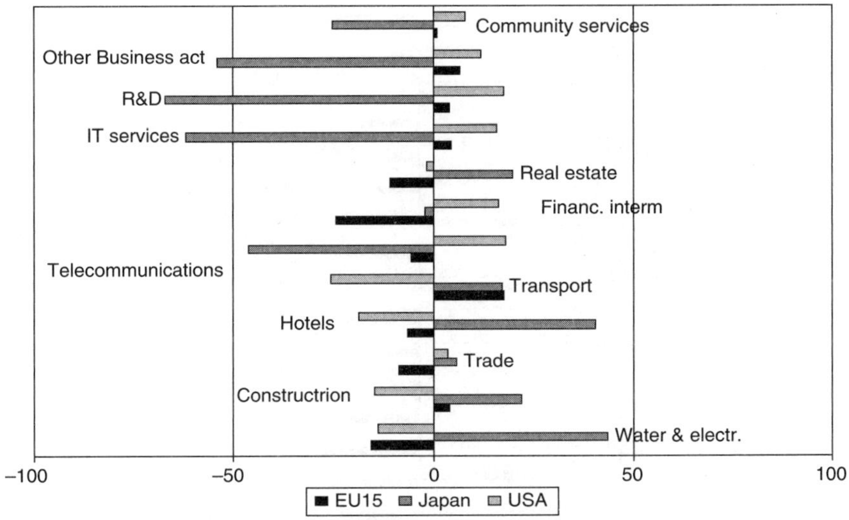

Note: The EU15, the USA and Japan, three-year averages of specialization indices, 2001–03. Specialization index with the EU15, the USA and Japan as reference. Maximum specialization: + 100. Minimum specialization: − 100.

Source: OECD, STAN 2005, own calculations.

Figure 12.8 Economic specialization; value-added in services

SUMMARIZING THE RESULTS FROM SPECIALIZATION ANALYSIS

The correlation analysis in this section will help us to draw some policy conclusions from the specialization patterns of the EU15 as summarized in Table 12.2. The WP1 country reports in the Annex 1 report include also correlation analyses of the specialization patterns for all individual ERAWATCH countries, when data are available.

Correlation Analysis – the EU15 vs. Japan and the USA

Tables 12.2 to 12.4 summarize the results from the correlation analysis of specialization indices in BERD, value-added and employment for the three regions – the EU15, Japan and the USA. Table 12.5 shows how the specialization indices for BERD in the EU15, Japan and the USA correlate in two periods of time (averages 2001–03 and averages 1993–95).

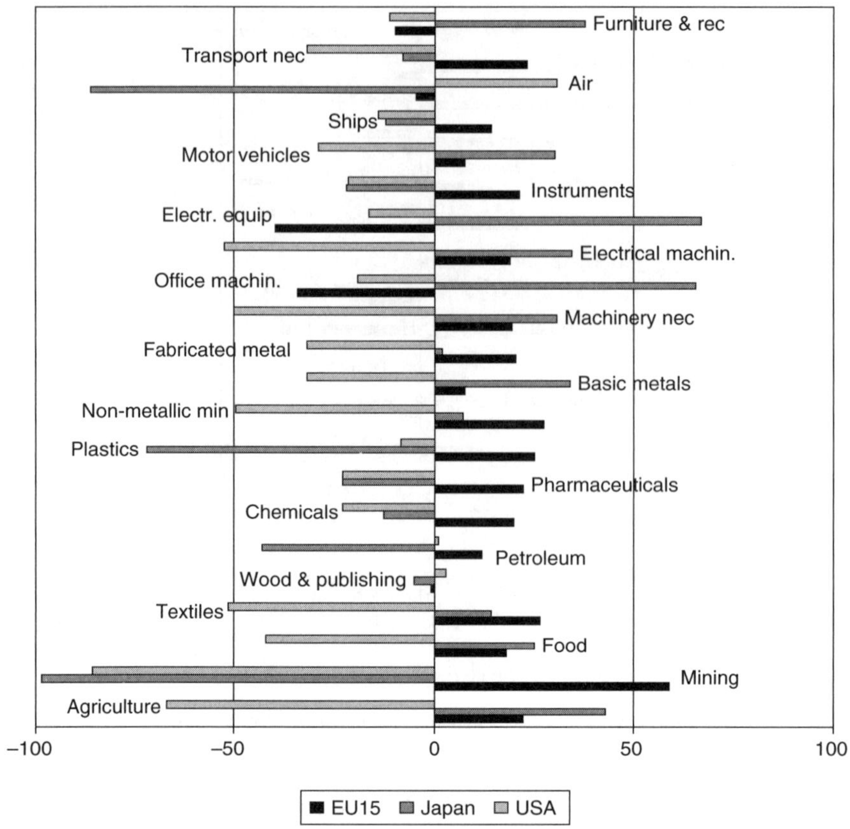

Note: The EU15, the USA and Japan, three-year averages of specialization indices, 2001–03.

Figure 12.9 Economic specialization; employment in primary sectors of activity and manufacturing

The most notable and unexpected finding is that the BERD specialization profile in the EU15 is correlated neither with value-added nor with employment specialization profiles. Conversely, the USA specialization profile is strongly and positively correlated both with the value-added and with the employment specialization profiles, while in the case of Japan we only find a strong correlation between the BERD specialization profile in 2001–03 and the value-added specialization in 2001–03, but not with the employment specialization profile. In terms of BERD specialization, the EU and the USA have drifted further apart (from −0.34 to −0.65), while Japan and the USA have come somewhat closer (from −0.91 to −0.83).

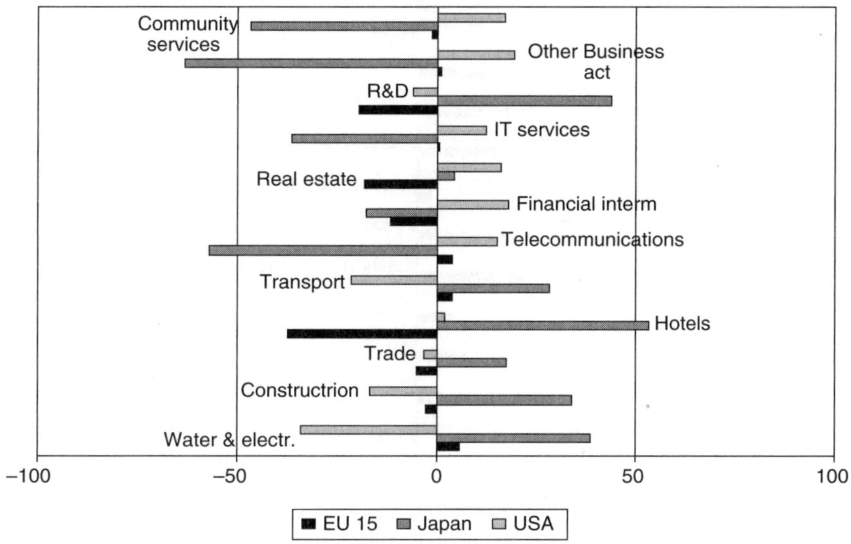

Note: The EU15, the USA and Japan, based on three-year averages, 2001–03. Specialization index with the EU15, the USA and Japan as reference. Maximum specialization: +100. Minimum specialization: −100.

Source: OECD, STAN 2005, own calculations.

Figure 12.10 Economic specialization; employment in services

However, differences between Japanese and the EU15 structures have remained relatively unchanged (−0.10 and −0.11).

Overall, differences between Japan and the USA seem to be broader than between the EU and the USA. Additional points that can be derived from the correlation analysis are:

- Correlations in the BERD over time are strongest in Japan (0.84), and weakest in the EU15 (0.63). This suggests that the biggest inter-sectoral structural change in the BERD is in the EU, and the smallest is in Japan.
- Correlations between employment specialization and value-added specialization are very high over time, and are very similar across all three regions. This suggests that changes over time are relatively less intensive in economic specialization when compared to BERD specialization.
- Correlations between employment specialization and value-added specialization are strongest in Japan and weakest in the USA. This

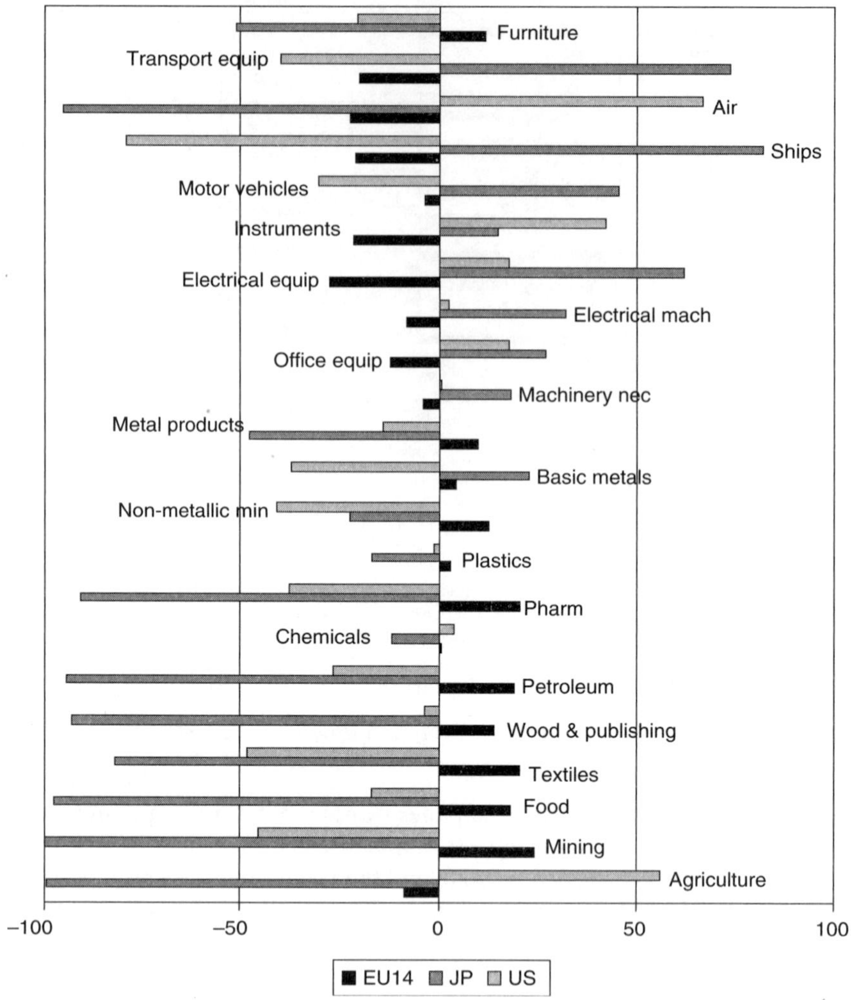

Note: The EU15, the USA and Japan, three-year averages of specialization indices, 2001–03. Inter- and extra-EU15 trade included. Specialization index with the EU15, the USA and Japan as reference. Maximum specialization: + 100. Minimum specialization: − 100.

Source: OECD, COMNTRADE 2005, own calculations.

Figure 12.11 Economic specialization; exports from primary sectors of activity and manufacturing

Table 12.2 Correlation of specialization indices, employment, value-added and BERD for the EU15

Correlations for the EU15	EMP-9395	EMP-0103	VA-9395	VA-0103	BERD-0103	BERD-9395
EMP 1993–95	1					
EMP 2001–03	0.97**	1				
Value-added 1993–95	0.66**	0.68**	1			
Value-added 2001–03	0.61**	0.68**	0.92**	1		
BERD 2001–03	0.43*	0.42*	0.3	0.3	1	
BERD 1993–95	0.41*	0.40*	0.40*	0.21	0.25	1

Note: ** Correlation significant at the 0.01 level (2-tailed). * Correlation significant at the 0.05 level (2-tailed).

Table 12.3 Correlation of specialization indices, employment, value-added and BERD for Japan

Correlations for Japan	EMP-9395	EMP-0103	VA-9395	VA-0103	BERD-0103	BERD-9395
EMP 1993–95	1					
EMP 2001–03	0.98**	1				
Value-added 1993–95	0.72**	0.73**	1			
Value-added 2001–03	0.67**	0.71**	0.97**	1		
BERD 2001–03	0.34	0.33	0.50**	0.48**	1	
BERD 1993–95	0.51**	0.50**	0.47*	0.45*	0.84**	1

Note: ** Correlation significant at the 0.01 level (2-tailed). * Correlation significant at the 0.05 level (2-tailed).

suggests that the largest intersectoral structural change is to be found in the USA, and the smallest in Japan.

If these findings are validated and confirmed by the results of future studies on specialization profiles, policy makers should pay more attention to the reasons why BERD seems to be weakly correlated with economic specialization in the EU15.

Does this relate to activities in the public R&D bases of the EU15 and how they interact with the business sector? Does this have something to do with the increasing internationalization tendencies of demand and supply of R&D services as well as R&D-motivated mergers and acquisitions in Europe? Finally, does the *composition of skills*; that is, high skilled versus low skilled,

Table 12.4 Correlation of specialization indices, employment, value-added and BERD for the USA

Correlations for the USA	EMP-9395	EMP-0103	VA-9395	VA-0103	BERD-0103	BERD-9395
EMP 1993–95	1					
EMP 2001–03	0.98**	1				
Value-added 1993–95	0.56**	0.53**	1			
Value-added 2001–03	0.51**	0.528**	0.93**	1		
BERD 2001–2003	0.55**	0.57**	0.47*	0.53**	1	
BERD 1993–95	0.77**	0.76**	0.71**	0.64**	0.78**	1

Note: ** Correlation significant at the 0.01 level (2-tailed). * Correlation significant at the 0.05 level (2-tailed).

Table 12.5 Correlation of specialization indices for BERD 1993–2005 and BERD 2001–03

Correlations for BERD	EU15-0103	JP-0103	USA-0103	EU15-9395	JP-9395
EU15 2001–03	1				
Japan 2001–03	−0.11	1			
USA 2001–03	20.65**	20.83**	1		
EU 15 1993–95	0.631**	−0.23	−0.23	1	
Japan 1993–95	−0.10	0.836**	−0.75**	−0.26	1
USA 1993–95	−0.27	−0.77**	0.78**	−0.34	**−0.91****

Note: ** Correlation significant at the 0.01 level (2-tailed). * Correlation significant at the 0.05 level (2-tailed), for the EU15, Japan and USA.

within a sector and between sectors, provide any explanations for some of the diverging patterns between R&D and economic specialization in the EU15?

Finally, one of the most complex and intriguing questions to ask is how the sectoral specialization patterns as observed here are related to diversity patterns at micro levels (see Chapters 8 and 9 in this volume) and to innovation policy systems within the EU.

NOTES

1. This chapter is a direct result of the R&D specialization project, a comprehensive data-driven project funded by the EU's JRC IPTS in Seville, as an ERAWATCH activity. Apart

from Aris Kaloudis and Mark Knell from NIFU STEP, other participants in this project were Nikos Maroulis (Logotech, Greece), Pari Patel and Gustavo Crispi (SPRU, UK), Michael Dinges and Martin Berge (Joanneum Research, Austria) and Rainer Fritsch (ISI Fraunhofer, Germany). We thank all these researchers for their contribution to the project and, hence, to this chapter.

2. Not all countries follow a strict enterprise basis for allocating R&D activity to industrial classes. Some countries make a disaggregation of the R&D of their largest, diversified firms into a number of different activities. In other countries, the enterprise approach has been abandoned and data are reported on a product field basis. Countries that report on this basis are more likely to be internationally comparable.

   See also OECD (2005), *Business Enterprise R&D Data by Industry: A Review of ANBERD and Other Issues*, Working Party of National Experts on S&T indicators, Reykjavik, Iceland 15–17 June for more details about different practices related to OFFBERD and ANBERD between countries.
3. In some cases, substantial numbers of small firms are not included in the frame of the major survey. Additional surveys of excluded firms may be undertaken, and appropriate adjustments made, but this is not a universal practice. Treatment of borderline research institutions also varies from country to country. Where government or semi-government agencies and research bodies conduct a substantial amount of industrial R&D, these differences may be important.
4. The *Frascati Manual* states that the R&D carried out by such institutes should be attributed to the 'Research and development' industry (ISIC3 73)
5. An older version of the ANBERD database goes back to 1973 and is compatible with the original STAN database (ISIC2).
6. For this reason, ANBERD is subject to revision because it depends on a number of estimation techniques that are constantly being refined and reviewed.
7. The publication year covers the date of publication at the European Patent Office, whereas the priority year covers the date of the first application, which is much closer to the point in time of the invention. The advantage of using the priority year instead of the publication year is a closer link to other innovation indicators, especially to R&D expenditure.
8. In six sectors where data for BERD in the USA or Japan are missing we are still calculating specialization indices. This is methodologically debatable, as indices for 24 sectors are calculated on the basis of the totals from all three regions, while six sectors are calculated on the basis of just two. Alternatively, one could discard the six sectors with missing information. Though not perfect, we decided to retain (partial) indices on the EU15 specialization in these six sectors sacrificing, perhaps, methodological rigorousness.
9. Only compared with USA; missing data for Japan.

BIBLIOGRAPHY

Archibugi, D. and M. Pianta (1992), *The Technological Specialisation of Advanced Countries: A Report to EEC on International Science and Technology Activities*, Dordrecht: Kluwer Academic Publications.

Brusconi, S., and A. Geuna (2004), 'Specialisation and integration: combining data and publications data to map the "structure" of specialised knowledge', in H.F. Moed, W. Glänzel and U. Schmoch (eds), *Handbook of Quantitative Science and Technology Research: The Use of Publications and Patent Statistics in Studies of S&T Systems*, Dordrecht: Kluwer Academic Publishers.

David, P., B. Hall and A. Toole (2002), 'Is public R&D a complement or substitute for private R&D? A review of the econometric evidence', *Research Policy*, **29** (4–5), 497–529.

Griliches, Z. (1992), 'The search for R&D spillovers', *Scandinavian Journal of Economics*, **94** S29–47.
Guellec, D. and B. van Pottelsberghe de la Potterie (2003), 'The impact of public R&D expenditure on business R&D', *Economics of Innovation and New Technology*, **13** (3), 225–43.
Hatzichronoglou, T. (1997), 'Revision of the high technology sector and product classification', *STI Working Papers*, 1997/2, OECD.
Laursen, K. and A. Salter (2005), 'The fruits of intellectual production: economic and scientific specialisation among OECD countries', *Cambridge Journal of Economics*, **29** (2), 289–308.
McGowan, F., S. Radosevic and N. von Tunzelmann (eds) (2004), *The Emerging Industrial Structure of the Wider Europe*, London: Routledge.
Malerba, F. and L. Orsenigo (1996), 'The dynamics and evolution of industries', *Industrial and Corporate Change*, **5**, 51–87.
Mansfield, E. (1991), 'Academic research and industrial innovation', *Research Policy*, **20** (1), 1–12.
OECD (1996), *Technology and Industrial Performance*, Paris: OECD.
OECD (2001a), *OECD Science, Technology and Industry Scoreboard: Towards a Knowledge-based Economy*, Paris: OECD.
OECD (2001b) *The New Economy. Beyond the Hype. The OECD Growth Project*, Paris: OECD.
OECD (2002), *Frascati Manual: Proposed Standard Practice for Surveys on Research and Experimental Development*, Paris: OECD.
OECD (2004), *Research and Development Expenditure in Industry*, Paris: OECD.
OECD (2005), 'Business enterprise R&D data by industry: a review of ANBERD and other issues', Working Party of National Experts on S&T indicators, Reykjavik, Iceland, 15–17 June.
Papaconstantinou, G. and W. Polt (1997), *Policy Evaluation in Innovation and Technology: An Overview*, volume based on the conference: 'Policy Evaluation in Innovation and Technology', Paris: OECD.
Pavitt, K. (1984), 'Sectoral patterns of technical change: towards a taxonomy and a theory', *Research Policy*, **13**, 343–73.
Pavitt, K. (1991), 'What makes basic research economically useful?', *Research Policy*, **20**, 109–19.
Pavitt, K. (1998), 'The social shaping of the national science base', *Research Policy*, **27**, 793–805.
PREST, CSI – Ecole des Mines, CSIC-SPRITTE, and SISTER (2002), 'A comparative study of public, semi-public and recently-privatised research centres in Europe', CBSTII contract ERBHPV2-CT-200-01, Final project report.
Salter, A.J. and B.R. Martin (2001), 'The economic benefits of publicly funded basic research: a critical review', *Research Policy*, **30** (3), 509–32.
Sandven, T. and K. Smith (1998), 'Understanding R&D intensity indicators', IDEA paper, STEP Group, Oslo.
Sandven T., K. Smith and A. Kaloudis (2005), 'Structural change, growth and innovation: the roles of medium and low tech industries, 1980–2002', contribution to the conference, 'Low-Tech as Misnomer: The Role of Non-Research-Intensive Industries in the Knowledge Economy', Brussels, 29–30 June, available at http://www.pilot-project.org/aboutpilot/about_pilot.html.

Schmoch, U., F. Laville, P. Patel and R. Frietsch (2003), 'Linking technology areas to industrial sectors: Final report to the European Commission', DG Research, Karlsruhe, Paris, Brighton, November 2003.

Smith, K. (2004), 'Measuring innovation', in J. Fagerberg, D. Mowery and R. Nelson (eds), *The Oxford Handbook of Innovation*, Oxford: Oxford University Press, Chapter 6.

UNCTAD (2005), *World Investment Report: Internationalisation of R&D Activities*, Geneva: United Nations.

United Nations (1997), *Structure and Change in European Industry*, Geneva: United Nations.

von Tunzelmann, N. and V. Acha (2004), 'Innovation in "low-tech" industries', in J. Fagerberg, D. Mowery and R. Nelson (eds), *The Oxford Handbook of Innovation*, Oxford: Oxford University Press, Chapter 15.

13. Heterogeneity and international R&D collaboration

Elias G. Carayannis

INTRODUCTION

Knowledge does matter: but the question is when, how, and why. Moreover, with the advancement of economies and societies, knowledge matters even more and in ways that are not always predictable or even controllable.

Moreover, collaborative, team-based research is now a prevalent mode for conducting fundamental scientific research in many fields. Also, collaboration in scientific research is often both transorganizational and transnational in nature; that is, collaborations frequently involve researchers based within different organizational entities and located in different countries (Georghiou, 1998).

Public–private research collaborations are one mechanism by which firms can access or create critical knowledge for use in industrial innovation. Facilitating linkages between public research organizations and firms is viewed as a critical mechanism for increasing the efficiency and outputs of national innovation systems. However, very little is known about how non-domestic firms are involved in collaboration with national public sector research institutions, and the role that these trans-national collaborations play in industrial technology development.

Glocal knowledge is a journey of insight and discovery in the emerging global 'knowledge village'. Perspectives from and about different parts of the world and diverse human, socioeconomic, technological and cultural contexts are presented and interwoven to produce an emerging new worldview on how specialized knowledge that is embedded in a particular sociotechnical context can serve as the unit of reference for stocks and flows of a hybrid, public/private, tacit/codified, tangible/virtual good that represents the building block of the knowledge economy, society and polity.

GloCalizing (global/local) networks (see Carayannis and Campbell, 2005), coupling together different national innovation systems and transnationally linking heterogeneous networks of knowledge producers, knowledge carriers and knowledge users, are thus becoming crucial components of the global, real and virtual knowledge architectures and infrastructures.

Two major purposes of this chapter are:

- To add to the theories and concepts of knowledge further inputs, such as suggesting a linkage between systems theory and the understanding of knowledge, emphasizing multi-level systems of knowledge and innovation, summarized also under the term of 'Mode 3' (Carayannis and Campbell, 2005).
- To leverage this diversified and conceptually pluralized understanding to support practical and application-oriented decision-making and policy-making with regard to the optimization of knowledge creation, diffusion and use systems as well as the leveraging of knowledge for other purposes and objectives that bear directly on economic performance, technological advancement and civic development.

Specifically, this chapter leverages a database consisting of about half a million collaborative R&D projects and containing one form of output from public–private research collaborations between US and European Union organizations: peer-reviewed scientific and technical publications. This is a subset of the ISI Science Citation Index database, containing publications in academic science journals from 1988 to 1997 where at least one author listed an address in the USA and at least one listed an address in the European Union. The database has been supplemented with information on each firm's industries, patent production, and financial performance over the same time period.

This bibliometric analysis addresses the following three key questions:

1. How prevalent are such collaborations in the international science and technology enterprise, and what are their trends in growth and research configurations?
2. How are such collaborations related to various aspects of technological innovation in firms, such as the nature of technology development in particular industries, and corporate processes for acquiring and exploiting new scientific knowledge?
3. How does participation in such collaborations relate to firm-level innovation and corporate performance?

The following method is used to conduct the analysis:

1. Identify different patterns in collaboration over successive overlapping timeframes (1988 to 1991, 1991 to 1994, 1994 to 1998).

2. Using cluster analysis, divide the firms in each industry into different strategic groups based on their patenting output and financial performance.
3. Analyze differences in patterns of participation in collaborations among these groups using a *t*-test of mean values of the dependent variables.

Note that this study does not allow comparisons with firms that do not engage in these collaborations. It also does not allow comparison with different forms of collaboration (e.g. purely domestic collaborations). There is also the problem that co-authorship is an incomplete indicator of actual research collaborations, and that the relationship between collaborative behavior and R&D productivity or firm performance could be spurious.

A significant stream of research in science and technology policy is dedicated to describing and analyzing national innovation systems and the relationship between the features of a nation's innovation system and its global technological competitiveness (Nelson, 1993). According to Metcalfe (1995), a national innovation system is:

> a set of distinct institutions which jointly and individually contribute to the development and diffusion of new technologies and which provide the framework within which governments form and implement policies to influence the innovation process. As such it is a system of interconnected institutions to create, store and transfer the knowledge, skills and artifacts, which define new technologies.

A key assumption of the theory supporting the concept of a national innovation system is that innovation is a boundary-spanning activity, which often involves multiple organizations working collectively, either consciously or unconsciously. This supposition is supported by the finding that much of the basic research that advances scientific knowledge is now conducted as a collaborative activity, rather than as a collection of individual efforts. Studies in the field of scientometrics have found, for example, that co-authored articles constitute a large and growing share of the scientific literature produced each year (Narin and Withlow, 1990; Melin and Persson, 1996). As stated succinctly by Ziman (1994), the organizational units of modern science are not individuals, but *groups*.

The academic, industrial and governmental institutions in an economy each play different roles in the processes of knowledge creation, transfer and application. Universities, for example, typically integrate research, teaching and study to enable both knowledge creation and diffusion

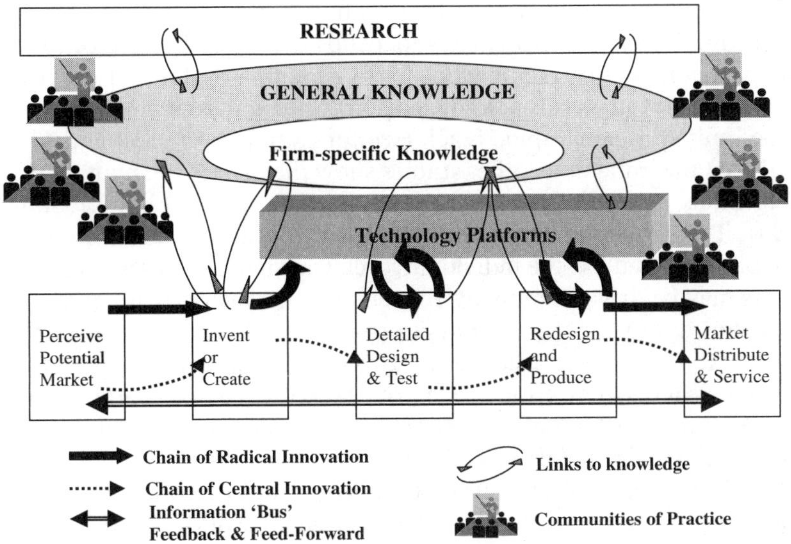

Source: Rosenbloom and Spencer (1996).

Figure 13.1 Total process model of industrial innovation

(see Clark, 1995), and are viewed as the primary force in knowledge creation for many national innovation systems (Godin and Gingras, 2000). In the industrial R&D process, a firm (or group of firms) may draw on knowledge from the body of scientific research and integrate it with firm-specific knowledge to develop and distribute a new technology (see Figure 13.1). Governments often fund research activities (at government-owned facilities as well as external research organizations), and also set important policy and regulatory constraints that determine the direction and conduct of research. Since all three sectors are important to the complete range of innovation in a nation, this observation implies that a nation can improve its capacity to innovate by optimizing the *efficiency* and *effectiveness* with which these national institutions communicate and collaborate with each other. Almost any collaborative activity spanning organizational boundaries involves some kinds of transaction costs, in the form of preparing for collaboration and managing the joint and individual activities of participants. Collaborations can become more efficient as these transaction costs are minimized, and they may become more effective as the improved management of collaborations increases the probability of successful and useful results from collaboration.

The significance of a national innovation system to national economic and technological competitiveness depends in part on the degree to which that particular nation is embedded in transnational systems of innovation. Several studies suggest that knowledge transfer across organizations tends to be limited in geographic reach (see for example Arundel and Geuna, 2001). At the same time, other studies show that a growing number of collaborations in scientific research cross national boundaries (Georghiou, 1998). The motivations for such collaboration are varied, complex and often interrelated. At the individual level, transnational research collaborations may be driven simply by convenience, as one participant may have data that another needs to pursue research of mutual interest (Melin, 2000). At the more formal level, government-sponsored organizations such as CERN (Centre Européen pour la Recherce Nucléaire) in Geneva or the European Union's Framework Programme provide both the settings and the monetary incentives to promote international research (Georghiou, 1998). Viewed more broadly, groups of collaborations indicate the presence of scientific collaboration networks (Newman, 1999), where sets of researchers engage in repeated collaborative activities in different configurations of institutions.

The concept of the national innovation system is both complicated and reinforced by the growth of international research collaborations in both basic and applied scientific fields. In particular, collaborations that extend across institutional sectors (i.e. between public sector organizations and private firms) can be viewed as either increasing or eroding differences in national technological capabilities – generalized as 'techno-globalist' versus 'techno-nationalist' arguments (Ostry and Nelson, 1995). The so-called 'techno-nationalist' perspective views such collaborations as a way for companies to compensate for technological weaknesses in their home countries by tapping overseas research capabilities, thus leading to a convergence in transnational technological capabilities. The 'techno-globalist' perspective views these collaborations as a natural trend towards internationalism in research, where national innovation systems are increasingly integrated into a global research network led by multinational corporations (see Figure 13.2).

One key feature that differentiates specific national innovation systems from their counterparts in other countries is knowledge specialization: the tendency of institutions in a national innovation system to focus on specific types of technological innovation or fields of scientific research rather than others (Brusoni and Geuna, 2001). At both the research level and the product-market level, certain nations (especially smaller nations) may show a bias in their production of knowledge and technologies towards certain areas of innovation and study. This specialization is not

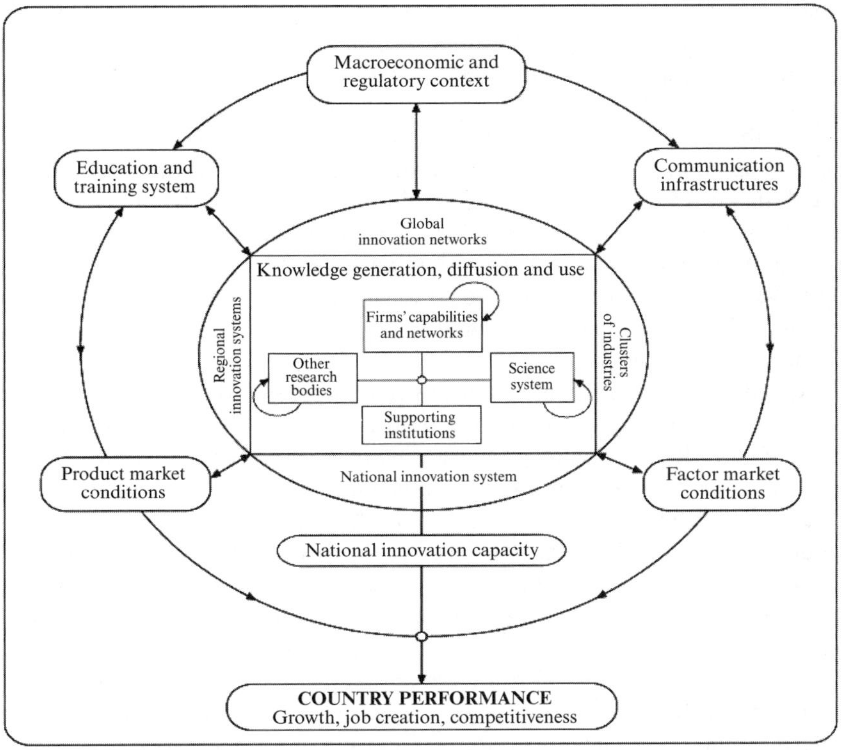

Source: OECD (1999).

Figure 13.2 Relationship among national institutions and global innovation systems

counter-intuitive, as smaller nations need to specialize as a means of managing more strategically the limited resources available to their national innovation systems.

Answering the debate between techno-globalism and techno-nationalism requires further study in the patterns of knowledge specialization and their relationship to international research collaboration. If techno-globalism can support the positive aspects of international research collaboration, it must show that this form of cooperation tends to strengthen the national innovation systems of those countries that participate in such collaborations. Techno-nationalism would argue that this participation would weaken the overall national innovation system, especially by reducing differences in national technological competitiveness. Resolving this dispute is an extremely daunting task and goes far beyond the scope of a chapter such as this. Instead, we intend to examine the relationship between national knowledge specialization and transnational research collaboration by analyzing patterns in cross-sectoral (public–private) research partnerships between the United States and the nations of the European Union.

Collaborative, team-based research is now one of the – if not the – most significant modes of activity in the global scientific community.[1] Groundbreaking advances in research domains are generally the result of the efforts of multiple investigators. Recognition in the field shows this trend; of the 36 Nobel Prizes in physics that were jointly awarded between 1901 and 1999, 21 were awarded in the 30 years since 1970. As stated succinctly by John Ziman, 'the organizational units of modern science are not individuals, but *groups*' (Ziman, 1994, p. 227).

Anecdotal and statistical evidence shows moreover that collaboration in scientific research is increasingly global in nature. That is, the groups of researchers who are involved in scientific progress often span one or more nations in origin, location and/or sponsorship. Georghiou (1998) provides data on the rise of global research cooperation at the personal (researcher), organizational, and national levels. At the researcher level, he provides rough figures on the rise in international co-authorship of scientific articles, particularly among industrialized nations. He goes on to note that co-authorship typically understates actual collaboration, as in at least one case (the International Human Frontier Science Program) a formal program for global scientific cooperation often yielded papers where members of an international team chose to publish separately. The increasing use of the Internet for academic and research communication also provides a new mechanism for scientists to identify potential collaborators and to initiate new joint research projects (Stead and Harrington, 2000).

Another significant trend in recent cases of scientific collaboration is the increase in cross-sectoral cooperation, where researchers in a group are employed by government, private industry, and/or academic and other non-profit institutions. As with global cooperation, there have been a number of formal programs in multiple countries that foster collective government–university–industry (GUI) research collaboration (Carayannis and Alexander, 1999). Informal collaboration across sectors has a longer history, as evidenced again primarily by co-authorship of scientific papers. Even in Japan, co-authorship studies reveal a rich pattern of industry–university collaboration, where no such cooperation was thought to have occurred (Pechter and Kakinuma, 1999; Hicks, 1993).

In this chapter, we review the scale, scope and intensity of cross-national, cross-sectoral research collaboration through the analysis of historical data on co-authorship of scientific publications. The collaborations we study are defined as those where at least one researcher is from the United States and at least one is from a European nation (defined as the 15 nations of the European Union, before expansion), and where at least one researcher is employed at the time of the collaboration by private industry and one by a public institution (government agency or a public or private university). The first part of the chapter reviews existing literature on the analysis of co-authorship data, and discusses the limitations of this form of analysis and typical strategies to mitigate those limitations. The second part describes a preliminary study of cross-national, cross-sectoral scientific collaborations covering the years 1985 through 1997, where we examined the scale (volume of co-authored papers), intensity (co-authored papers versus other kinds of co-authorship), and scope (patterns in co-authorship) of cross-national, cross-sectoral collaborations. The conclusion discusses significant trends and patterns derived from this study, and their implications for further research into these types of collaborations.

ANALYSIS OF RESEARCH COLLABORATION VIA CO-AUTHORSHIP DATA

Previous studies that utilize co-authorship data to explore research collaborations discuss how co-authorship does not provide a comprehensive analysis of all scientific cooperation (Melin and Persson, 1996; Katz and Martin, 1997). In some cases, these limitations are structural and cannot be overcome through data manipulation. Instead, alternate research methods must be used to capture the nature of the collaboration. In other cases, these limitations are a result of problems in data entry or data structure in publication databases, and can be mitigated through systematic analysis.

First, the publishing practices of scientific researchers may distort the reflection of collaboration in bibliometric data. As cited above, researchers may be engaged in a collaboration, but may choose to publish their findings in separate, single-author articles. This is particularly the case in multi-disciplinary collaborations, where the researchers involved come from different scientific fields. If the researchers are employed at universities, each may decide to publish only in journals within his or her specific discipline, as such journals carry more weight in tenure decisions by their respective departments. Co-authorship also varies by field, with co-authored publications playing a major role, for example, in indicators of university–industry collaboration in biotechnology (Zucker et al., 1998).

Second, there are discrepancies in how researchers list themselves as authors that often confound the analysis of collaboration patterns. For example, in some cases a prominent researcher is listed as a co-author on a publication primarily for the prestige of that name, which does not reflect the actual contribution to the work published. Also, a co-author's institutional affiliation on a publication may reflect their employment at the time the work was accepted for publication, but not at the time the research was conducted. For example, a doctoral student's thesis research might be published after that student has joined a firm, but list the student as an industry researcher.

Third, there are simple errors or inconsistencies in how co-authorship data is presented on a publication and how it is entered into the most prominent US database of scientific publications, the Scientific Citations Index from the Institute for Scientific Information. In our review of this database for this study, we found many instances where, for example, an author was listed with multiple affiliations because of ambiguities about the nature of the institution. In one case, a researcher at a university laboratory funded by the French research organization CNRS was listed as affiliated with the university, the CNRS laboratory at the university, or with CNRS, depending on the paper.

Finally, we recognize that co-authorship is only a partial indicator of the complete scale and scope of research collaboration, as pointed out by Katz and Martin (1997). Other indicators include the signing of formal collaborative research agreements between institutions (Oliver and Liebeskind, 1997), joint participation in scientific conferences, and reciprocal citation of publications. Other authors (such as Melin, 2000) argue that co-authorship is a reasonable proxy for the presence and character of research collaboration. We therefore propose to use this indicator as our primary measure of research collaboration, but supplement it with other research methodologies to investigate the topic of cross-national, cross-sectoral research collaborations.

RESEARCH QUESTIONS

For the purposes of our study, we focus on research collaborations defined as activities to investigate both fundamental and applied scientific phenomena, where the research output (a publication or set of publications) includes at least two authors affiliated with different institutions. The institutions can be for-profit (private sector) or not-for-profit (public sector) organizations. Our interest is in research collaborations between researchers in institutions in the European Union and the United States.

Okubo and Sjöberg (2000) used an initial analysis of co-authorship data and case study research to show how the Swedish R&D system is changing as co-authorship between universities and firms, particularly foreign firms, becomes a dominant feature of scientific research in Sweden. Other authors describe the rise in industry–university research in the science systems of several nations, including the USA, the UK and Russia. From these and other studies, some common questions of interest on the nature of research collaborations are:

- Are research collaborations increasing or decreasing over time?
- How do patterns of collaboration differ across scientific disciplines, institutions, and national systems?
- What are the objectives and outcomes of research collaboration?
- Does collaboration improve the productivity of scientific research?
- How does collaboration between different types of institutions affect the motivations and mode of research at those institutions?

As a starting point for our research on cross-national, cross-sectoral research collaboration, we chose to focus on purely descriptive analysis. Using the data set described below, we investigated the following research questions:

1. What is the overall scale of research collaborations, measured by the number of cross-national, cross-sectoral co-authored publications, and what have been the trends in that scale over time?
2. What is the overall intensity of research collaborations, measured by comparing these collaborations to all cross-national co-authored articles, and what have been the trends in intensity over time?
3. What is the overall scope of research collaborations, measured by the scientific field at the focus of each collaboration, and what have been the trends in scope over time?

The descriptive findings are then used to suggest interesting researchable hypotheses for future study.

REVIEW OF PRELIMINARY STUDY

The preliminary study is based on the analysis of a data set for co-authored scientific publications covering the years 1988 to 1997. To simplify the research, as described below, we chose to analyze the interactions among four primary sectors: US private sector institutions (firms) or US public sector institutions (universities and government agencies), and European private sector institutions or public sector institutions.

Data Set

The Institute for Scientific Information (ISI) provided the data. The data used in this study were assembled by ISI at the request of the Washington, DC representative office of the European Commission. These data correspond to scientific papers published in peer-reviewed journals in the physical, natural and social sciences, with publication dates between 1988 and 1997. Further, papers in the data set were only those where at least one author gave an address in a European nation *and* one author gave an address in the United States. This produced a data set of 133 693 individual articles (referred to as *articles* hereafter), 259 353 individual researchers (*names*) and 33 135 entries in the list of organizations.

Closer inspection revealed numerous inconsistencies in the data fields that complicated the analysis of these publications. The most difficult part in preparing the data pertaining to the industrial participation was to identify which organizations were in fact companies. In brief, the following complementary approaches were used. First, all organizations with an address outside the European nations of interest (the EU15) or the United States were eliminated from the analysis. From the remaining organizations, those with an identifying abbreviation commonly used by private firms, such as Inc, Ltd, GmbH, Corp, BV, and so on, were extracted. The names of these organizations without such abbreviations were then searched in the address field to capture all organizations associated with known firms. For example, Philips BV led to Philips Components, Philips Medical Systems, and so on.

Of the remaining organizations, all obvious institutions of a non-industrial nature were extracted to a separate list. Again, this was accomplished primarily by searching for common designations of public sector institutions (Univ, Akad, and so on for universities; Hospital or Hôpital for hospitals, and so on). Searches were conducted via the Internet and private databases to identify which of the remaining organizations were also private firms. As there were still thousands of organizations remaining, a further filter was applied to remove any organization that published fewer than three articles over the 11-year timeframe, under the assumption that such organizations contributed little to research collaboration activities. At the end of this process, the names of organizations were reviewed manually to eliminate minor variations in the name associated with each one, and to account for remaining discrepancies, such as the following:

- Organizations may have merged during the study process. The original organization name was retained, although later analysis takes into account such mergers.

- Organizations are occasionally identified by the name of a department, subsidiary, or other sub-unit. These organizations were standardized under the name of the parent organization (for example, Thomas Watson Research Center was renamed IBM).
- Organizations have changed their names over time, for example because of spin-offs. The names of these organizations were preserved as in the database, but again some of these examples were targeted for special analysis (for example, Bell Labs is associated with AT&T or with Lucent depending on the year).

The result of this process was a consolidated list of 2524 independent companies. Plugging these individual companies into the original database led to a selection of 18 869 articles, to which 68 245 authors' names (researchers) were linked. Furthermore, the list of authors' names were then scanned to eliminate any misspellings that could be detected, and to establish the exact affiliation of each author with a particular institution for a given point in time (keeping in mind that researchers may have moved among institutions during the study period).

A majority of the articles are produced by a mix of industrial, university, national and non-profit laboratories. While the addresses of the industrial partners have been cleaned up, this is not the case for the other laboratories. In the rest of this note all entities that are not part of industry are considered as public institutions.

Most European or American public institutions are located in their country of origin. The same is not true for companies, which may have subsidiaries on both sides of the ocean. The same company can, indeed, perform research in Europe and in the USA. Because the goal is to look at research links across the Atlantic, a company's nationality in any specific document is defined according to its physical location or address, not the address of its headquarters. Thus the US subsidiaries or operations of companies based in Europe are designated as US firms, and vice versa.

Descriptive Methodology

Once the company list had been established, a data file was created that listed every document authored by a researcher from any of those 2524 companies,. Each of these articles was then linked to any co-authors located across the Atlantic. For example, a publication by a researcher at 3M in the United States was linked to any co-authors on that publication from European organizations. The non-industrial institutions found in this new data file were then flagged as public institutions in the USA or Europe. In other words, there were four categories of organizations:

1. US private firms (US_{PR}),
2. US public institutions (US_{PU}),
3. EU private firms (EU_{PR}),
4. EU public institutions (EU_{PU}).

This classification scheme enabled us to categorize every document in the data file as representing one or more configurations of research collaboration, as shown in Figure 13.3.

In summary, we searched the data file to identify the following types of articles:

Group 1: Articles involving only co-authors from one or more US firms and from one or more European public institutions only.
Group 2: Articles involving co-authors from one or more US firms and from one or more European public institutions, and from one or more US public institutions.
Group 3: Articles involving co-authors from one or more US firms and from one or more European firms only.
Group 4: Articles involving co-authors from one or more US firms and from one or more European public institutions, and from one or more European firms.
Group 5: Articles involving co-authors from one or more US firms and from one or more European public institutions, and from one or more European firms and from one or more US public institutions.
Group 6: Articles involving co-authors from one or more US firms and from one or more European firms, and from one or more US public institutions.
Group 7: Articles involving co-authors from one or more US public institutions and from one or more European public institutions, and from one or more European firms.
Group 8: Articles involving co-authors from one or more US public institutions and from one or more European firms only.

To address the research questions at issue in this study, we divided all articles in the database where at least co-author is a private firm into the eight groups listed above. Note that these groups are mutually exclusive. We then analyzed this group of articles along the following measures:

- Performer distribution: number of articles classified into one of the eight groups listed above.

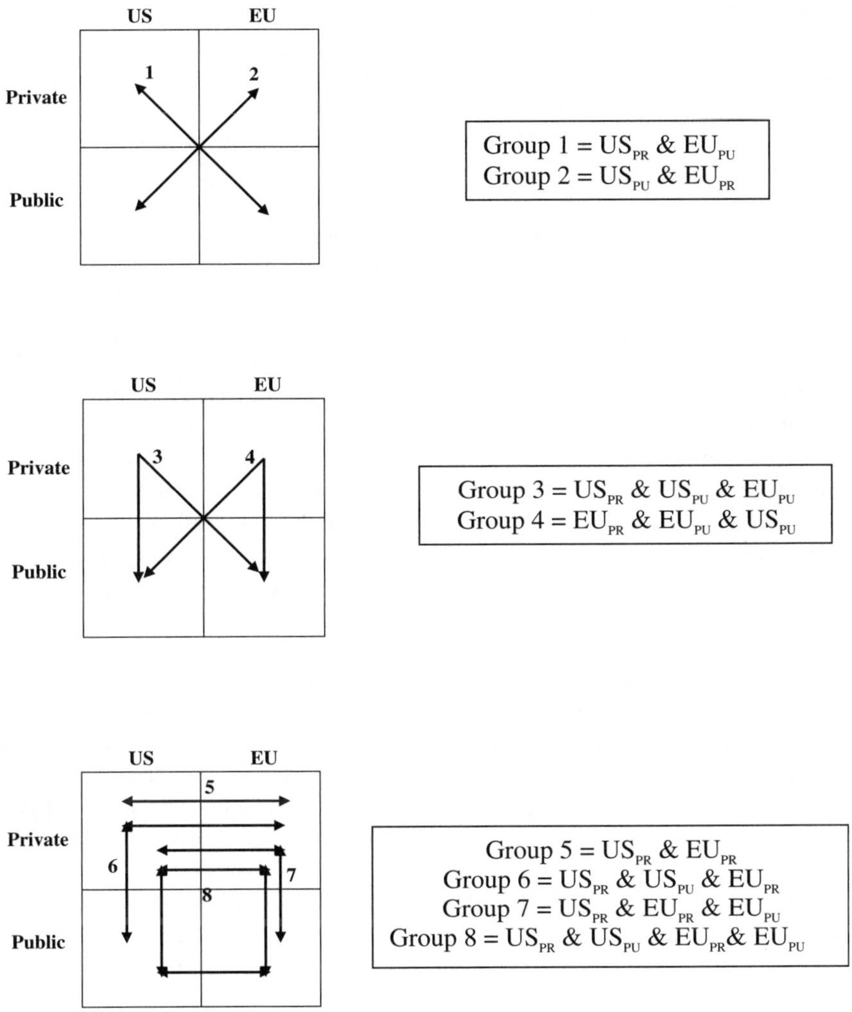

Figure 13.3 Configurations of EU–US research collaboration by type of institution

- Annual trends: number of articles for each year classified by participation of US or European firms.
- Field distribution: number of articles in each scientific field classified by participation of US or European firms.

These measures enable us to address the research questions as follows.

Findings from Descriptive Analysis

As noted above, the initial search for articles with at least one private sector co-author yielded a subset of 18 869 articles over the 11-year study period. The descriptive analysis examines this subset using the measures noted in the previous section.

Distribution of articles by participating organizations

Table 13.1 shows the distribution of all articles identified, categorized by the groupings of participating organizations. This, in turn, reflects the intensity with which each type of partner is engaged in transatlantic collaborations, as measured by the share of all articles in the subset. Clearly the US companies are more active in transatlantic research collaborations than European private companies. US firms, which are present in Groups 1 through 6, appear as co-author organizations in 75 percent of the articles in the subset, as opposed to 39 percent for European firms (which are present in Groups 3 to 8). On the other hand, the European public institutions, which are present in 79 percent of the articles (Groups 1, 2, 4, 5, and 7), are more active than their US counterparts, with 55 percent of the articles (Groups 2 and 5 through 8).

Examining these groups according to the four categories of participants, the combination with the largest output involves European public institutions and US companies (Groups 1, 2, 4 and 5). It is three times bigger than the collaborations involving American public institutions and European companies (groups 5 to 8). Roughly 40 percent of the articles have at least one European and one American public institution. 14 percent have

Table 13.1 Distribution of the articles according to the partners involved

	US_{PR}	US_{PU}	EU_{PU}	EU_{PR}	No. of articles	Share of articles
Group 1	+		+		6 617	35.1%
Group 2	+	+	+		4 858	25.7%
Group 3	+			+	1 243	6.6%
Group 4	+		+	+	730	3.9%
Group 5	+	+	+	+	248	1.3%
Group 6	+	+		+	420	2.2%
Group 7		+	+	+	2 488	13.2%
Group 8		+		+	2 265	12.0%
Total articles	14 116	10 279	14 941	7 394	18 869	100%
Share of articles	74.8%	54.5%	79.1%	39.1%	100%	100%

companies from both the USA and the European Union. For the latter group, however, it is important to note that a substantial share of these collaborations involve two branches of the same company located across the Atlantic from each other.

For collaborations involving participants from three or more of the four sectors identified, the most common configuration involves US firms, public institutions, and European public organizations. The frequency of this grouping is nearly twice that of the comparable grouping of European firms and public institutions working with US public organizations. This is consistent with several observations derived from prior research on collaborations: that US firms are more active in research collaboration with European firms, and that US public organizations (especially universities) are more active in research collaborations with private firms than European public institutions.

Annual trends in research collaborations

Results from Tables 13.2 and 13.3 show the evolution over the years of the industrial participation in transatlantic collaborations. As shown in Table 13.2, the percentage of articles with at least one co-author originating from a private firm grew in every year until 1993, and then remained relatively stable except for a slight decline in 1997 and 1998. This means that for most of the study period, companies involved in transatlantic research collaborations increased their publishing activity at approximately the same pace as public institutions engaged in similar activities.

Table 13.2 Annual trends in S&T cooperation for the entire database and for company participation

Year	Total articles	Articles with firms	Articles published by firms as a share of all articles
1988	7289	889	12.2%
1989	8215	1053	12.8%
1990	8887	1157	13.0%
1991	11196	1617	14.4%
1992	13165	1914	14.5%
1993	13887	2021	14.6%
1994	15802	2278	14.4%
1995	17197	2518	14.6%
1996	18402	2635	14.3%
1997	19655	2787	14.2%
Total	133695	18869	14.1%

A more detailed picture is found in Table 13.3. The different groups correspond to the various configurations of industrial participation in published articles. To summarize, US private company participation in these publications is relatively stable over the period, while European companies' participation grew slightly (albeit at a slower pace during the second half of the study period). Significantly, transatlantic collaborations involving both US and EU private firms increased substantially during the period, with these articles accounting for 10.1 percent of all articles in the data set from 1988 and 16.1 percent from 1997.

A further analysis shows that the two components of this participation, namely collaborations involving units within the same firm and those involving different firms, both increased at roughly the same pace (accounting for about half of such articles in each year). At the same time, the collaborative publications involving American but not European companies decreased from 65 percent to 60 percent, reflecting an apparent increase in relative terms in the participation of EU firms in collaborations with US private firms and EU public institutions. These changes indicate a slow shift during the study period towards a greater degree of industrial participation, particularly by European firms, in transatlantic research collaborations, but no significant increase in transatlantic collaborations involving private firms and public institutions.

Table 13.3 Annual trends in transatlantic collaborative articles with American and/or European firm participation

Year	US_{PR} total (Groups 1 thru 6)	US_{PR} but not EU_{PR} (Groups 1 & 2)	EU_{PR} total (Groups 3 thru 8)	EU_{PR} but not US_{PR} (Groups 7 & 8)	US_{PR} & EU_{PR} (Groups 3 thru 6)	Intra-firm US_{PR} & EU_{PR}	Inter-firm US_{PR} & EU_{PR}
1988	74.92	64.79	35.21	25.08	10.12	4.95	5.17
1989	73.31	62.01	37.99	26.69	11.30	4.84	6.46
1990	76.40	63.87	36.13	23.60	12.53	5.01	7.52
1991	74.83	62.77	37.23	25.17	12.06	4.21	7.85
1992	74.35	60.55	39.45	25.65	13.79	6.06	7.73
1993	73.73	60.76	39.24	26.27	12.96	5.69	7.27
1994	74.67	60.36	39.64	25.33	14.31	6.01	8.30
1995	75.81	61.04	38.96	24.19	14.77	6.91	7.86
1996	74.00	58.06	41.94	26.00	15.94	7.44	8.50
1997	75.74	59.67	40.33	24.26	16.07	7.43	8.65

Trends in distribution of collaborations by field of research

Table 13.4 provides the distribution of articles by research discipline for the entire database of EU–US articles and for the sub-set pertaining only to those articles with company participation. Note that as any article can be classified under multiple fields, the total for the entire table exceeds the actual number of articles in the data set. One notable fact highlighted in Table 13.4 is the concentration of company participation in specific disciplines. As one might expect, firms tend to focus on fields with high industrial relevance: pharmacology, material science, computer science, and so on. This is reflected in the high percentage of articles within these specific fields.

Table 13.4 Field distribution of articles in the entire database and in the sub-set corresponding to company participation

Fields	Totals for all articles	Totals for articles involving firms	Articles with firms as a share of all articles (%)	Index of industrial participation in field
Agricultural Sciences	1043	160	15.3	1.07
Astrophysics	7880	506	6.4	0.45
Biology & Biochemistry	14012	1831	13.1	0.91
Chemistry	11075	2212	20.0	1.39
Clinical Medicine	27007	3812	14.1	0.99
Computer Science	1090	265	24.3	1.70
Ecology/Environment	1715	164	9.6	0.67
Economics & Business	1457	39	2.7	0.19
Engineering	7366	1476	20.0	1.40
Geosciences	4849	480	9.9	0.69
Immunology	4234	994	23.5	1.64
Materials Science	2265	735	32.5	2.27
Mathematics	3894	135	3.5	0.24
Microbiology	3974	649	16.3	1.14
Molecular Biology & Genetics	8216	835	10.2	0.71
Multidisciplinary	5095	723	14.2	0.99
Neuroscience	7107	841	11.8	0.83
Pharmacology	2724	936	34.4	2.40
Physics	23389	3542	15.1	1.06
Plant & Animal Science	4872	410	8.4	0.59
Psychology/Psychiatry	2190	83	3.8	0.26
Total	145454	20828	14.3	1.00

The index in the last column is obtained by dividing the percentage of articles by field with industrial participation (fourth column) by the percentage of all articles related to that field (total of fourth column, i.e. 14.3). This reflects the intensity of the industrial participation in each field. This index is simply another way to show the relative degree of industrial participation across fields, as those fields with a higher index score have greater industrial participation. However, the index is especially useful in the further analyses shown in Tables 6 and 7, as the index indicates the relative participation by the type of institution (EU or US, and public or private).

The results from Table 13.5 indicate that European and US companies do not have the same priorities. The articles from the company database were first split into two overlapping groups, those with US firm participation and those with EU firm participation (note that some articles involve both types of firms). The articles in the first case have an American company in the address field; the articles in the second case have a European company in the address field. The next step was to express, for each field, the percentage of articles from the company database that involve firms from either the USA, the EU, or both. The values in the first two columns of Table 13.5 are index scores derived from these percentages, as done in Table 13.4.

Clearly, the European companies tend to focus on life sciences and on engineering. This does not mean, however, that the absolute number of articles in pharmacology or in agriculture is the highest. It only means that, within pharmacology or agriculture, the European firms' output is higher than the average. Their US counterparts have more interest in computers, materials science, chemistry, and other physical sciences. US firms also have a high index in basic sciences such as astrophysics, mathematics and physics. Further analysis showed that in physics and mathematics, there are a small number of active US firms that participate in these collaborations. For physics in particular, a significant number of articles are linked to the AT&T or IBM basic research laboratories. Each of these two companies has a share of about 8 percent of the articles in the data sub-set and more than half of their activity is in physics.

Table 13.5 also shows the index scores for articles with companies from both sides (strong industrial involvement) and for articles with academia from both sides (strong university involvement). The data in these last two columns allow for further selection between fields that are more academic and those that have a high level of industrial cooperation. For example, agriculture has a higher proportion of collaboration between firms, but astrophysics is the realm of universities. Computer science, which has a high index in Table 13.4 (i.e. a priority for industry) is below average in

Table 13.5 Field distribution for all articles with an American company, all articles with a European company, all articles with both types of company, and all articles with American and European public institutions

Fields	Index of US_{PR} participation	Index of EU_{PR} participation	Index of participation by both US_{PR} and EU_{PR}	Index of participation by both US_{PU} and EU_{PU}
Agricultural Sciences	0.88	1.61	2.06	0.89
Astrophysics	1.30	0.12	0.11	1.97
Biology & Biochemistry	0.92	1.10	0.85	0.99
Chemistry	1.08	0.98	1.34	0.72
Clinical Medicine	0.93	1.12	0.97	1.12
Computer Science	1.15	0.69	0.95	0.97
Ecology/Environment	0.83	1.35	1.11	1.17
Economics & Business	0.93	0.84	0.18	1.66
Engineering	0.96	1.20	1.33	0.88
Geosciences	1.04	0.92	0.98	1.48
Immunology	1.01	0.95	0.92	0.96
Materials Science	1.09	0.96	1.35	0.76
Mathematics	1.13	0.47	0.21	0.89
Microbiology	0.95	1.22	1.34	0.93
Molecular Biology & Genetics	0.98	0.85	0.48	1.25
Multidisciplinary	1.05	0.77	0.64	1.27
Neuroscience	0.74	1.47	0.95	1.02
Pharmacology	0.91	1.60	2.18	0.71
Physics	1.12	0.63	0.61	0.93
Plant & Animal Science	0.93	1.32	1.51	0.89
Psychology/Psychiatry	0.71	1.49	0.85	1.38
Total	1.00	1.00	1.00	1.00

both cases, which means that the bilateral relation is the preferred mode of cooperation. On the other hand, environment, which has a low index in Table 13.4, has a high level of cooperation between firms. When the collaboration takes place in that field, companies tend to work together. Since the index for academic relations is also high, it means that the level of multilateral cooperation is high.

Tables 13.6 and 13.7 go one step further. They look at how the links between American firms or European firms with other partners affect the

Table 13.6 Fields distribution for articles with US firms according to various partners

Fields	EU_{PR} participation	No EU_{PR} participation	EU_{PU} participation	EU_{PU} and US_{PU} participation
Agricultural Sciences	2.34	0.68	0.71	0.64
Astrophysics	0.09	1.22	0.45	2.27
Biology & Biochemistry	0.93	1.02	1.08	0.94
Chemistry	1.24	0.94	1.13	0.69
Clinical Medicine	1.04	0.99	0.92	1.09
Computer Science	0.83	1.04	1.00	1.09
Ecology/Environment	1.33	0.92	0.63	1.32
Economics & Business	0.19	1.19	0.71	1.84
Engineering	1.39	0.91	0.89	0.93
Geosciences	0.94	1.01	0.53	1.68
Immunology	0.91	1.02	1.13	0.87
Materials Science	1.24	0.94	1.10	0.73
Mathematics	0.18	1.19	1.28	1.08
Microbiology	1.42	0.90	0.95	0.83
Molecular Biology & Genetics	0.49	1.12	1.02	1.26
Multidisciplinary	0.61	1.09	0.89	1.36
Neuroscience	1.29	0.93	0.96	0.90
Pharmacology	2.41	0.67	0.79	0.50
Physics	0.55	1.11	1.20	0.98
Plant & Animal Science	1.62	0.85	0.91	0.77
Psychology/Psychiatry	1.19	0.96	0.68	1.33
Total	1.00	1.00	1.00	1.00

field distribution. In Table 13.6 the primary key for the selection of articles is the presence of an American company (groups 1 to 6 in table 13.1). All indexes in the table reflect the relative repartition of articles with at least one US firm involved, according to different criteria. The secondary key is whether or not a European company is present. The group *without* EU firm participation is further divided into articles with public institutions from Europe only and articles with public institutions from Europe and the US. While these last two groups are homogeneous (they correspond respectively to the two first groups in Table 13.1), the group involving EU firms includes groups 3 to 6 from Table 13.1.

There are differences of patterns between the various groups in Table 13.6. For example, the preferred partnership in agriculture or

Table 13.7 Field distribution for articles with European firms according to various partners

Fields	US_{PR} participation	No US_{PR} participation	US_{PU} participation	Both EU_{PU} and US_{PU} participation
Agricultural Sciences	1.28	0.84	1.06	0.65
Astrophysics	0.93	1.04	0.41	1.62
Biology & Biochemistry	0.77	1.13	1.11	1.14
Chemistry	1.37	0.79	0.88	0.71
Clinical Medicine	0.87	1.07	0.99	1.15
Computer Science	1.37	0.79	0.98	0.61
Ecology/Environment	0.82	1.10	1.19	1.02
Economics & Business	0.21	1.44	0.75	2.07
Engineering	1.11	0.94	1.17	0.73
Geosciences	1.07	0.96	0.76	1.15
Immunology	0.97	1.02	0.89	1.14
Materials Science	1.40	0.77	0.80	0.75
Mathematics	0.45	1.31	1.96	0.72
Microbiology	1.10	0.95	0.95	0.94
Molecular Biology & Genetics	0.56	1.25	0.91	1.55
Multidisciplinary	0.83	1.10	0.98	1.20
Neuroscience	0.65	1.20	1.26	1.15
Pharmacology	1.36	0.80	0.88	0.72
Physics	0.97	1.02	1.00	1.03
Plant & Animal Science	1.14	0.92	0.99	0.85
Psychology/Psychiatry	0.57	1.24	1.07	1.40
Total	1.00	1.00	1.00	1.00

pharmacology is industrial, while in molecular biology or physics the most common partner is a public organization. Also, relations with public institutions are focused on European partners in chemistry or immunology and involve both US and European partners in clinical medicine or computer science.

In Table 13.7, the primary key for selection of the articles is the presence of a European company. Articles are first split depending on the presence or absence of an American company. The latter group is subsequently divided into those with US public institutions, and those with both EU and US public institutions. Once again, there are clear variations in the patterns of the different groups.

INTERPRETATION OF RESULTS

To summarize the major findings of interest from this descriptive study:

1. The most common configuration for research collaborations across the entire study period is those with European public institutions and US companies.
2. Transatlantic collaborations involving private firms based in either the USA or Europe are much more common than collaborations involving at least one US private firm and at least one European private firm.
3. For collaborations involving participants from three or more of the four sectors identified, the most common configuration involves US firms, US public institutions, and European public organizations.
4. US firms are more commonly involved in transatlantic research collaborations than EU firms, but EU firms increased their participation rate over the study period.
5. European firms engaged in transatlantic research collaborations focus on life sciences and engineering research, while US firms focus on computers, mathematics and physical sciences research in their collaborations.
6. US firms partnered most frequently with European firms for life sciences and chemical research during the study period, while research with European public institutions only focused on physics, math and biology.
7. European firms partnered with US firms primarily in materials science, computer science, and some life sciences fields, and relied more on US public institutions in mathematics, environmental sciences, and some life sciences.

In view of the differences in the behavior of US and EU firms in both their choice of partners and disciplines of focus for transatlantic research collaborations, we can see that firm involvement in these collaborations is national in nature (US firms utilize these collaborations differently than European firms). However, the research process in industry is becoming global, as US and European firms are clearly accessing foreign pools of human and intellectual capital through their research activities. The national differences appear to be based on the relative strengths of the national innovation systems for each firm's home country. For example, US firms have a relatively higher share of global output for computer technologies than for biomedical technologies, and tend to engage in transatlantic collaborations that are related to that strength. At the same time, European firms have a significant share of total US imports in biotechnology life science technologies, and tend to engage in transatlantic research collaborations that reflect that particular competence (National Science Foundation, 2000).

One other interesting feature of transatlantic research collaborations is the relative concentration of multi-sector, transatlantic partnerships in three fields: astrophysics, molecular biology, and psychology/psychiatry. These fields have different contextual factors, which seems to make them natural areas for collaboration. In astrophysics, international collaboration is often motivated by the need to access large research facilities (such as observatories) that are only available in a few nations. Also, international collaboration may be required to share national datasets in this field. In molecular biology, the International Human Genome Project and the globally distributed nature of biological research would tend to promote multilateral research collaboration. Psychological research also is often enriched through comparative cross-national studies. Note, however, that astrophysics and psychology are also fields with relatively low industrial participation in research.

Future research on this topic is needed to understand more fully the motivations for organizations to engage in transatlantic research collaborations, and in particular what factors tend to promote specific configurations of public–private partnerships across national boundaries. Specifically, we are continuing our data collection, analysis and interpretation to assess the validity, reliability and generalizability of the following preliminary observations stated in the form of propositions and derived from our research so far:

- *Proposition 1A:* US and European firms will collaborate with their transatlantic private sector counterparts primarily in scientific and technical fields where both partners have equivalent but complementary national competences (*complementary collaboration*).
- *Proposition 1B:* US and European firms will collaborate with their transatlantic private sector counterparts primarily in scientific and technical fields where one partner has national competences greater than the other (*outsourcing-based collaboration*).
- *Proposition 2A:* US and European firms will collaborate with their transatlantic public sector counterparts primarily in scientific and technical fields where the firms have strong national competences (*complementary collaboration*).
- *Proposition 2B:* US and European firms will collaborate with their transatlantic public sector counterparts primarily in scientific and technical fields where the firms have weak national competences (*outsourcing-based collaboration*).

- *Proposition 3A:* US and European firms will exhibit increasing frequency of collaboration over time with their transatlantic private sector counterparts primarily in scientific and technical fields where they have strong national competences (*complementary collaboration*).
- *Proposition 3B:* US and European firms will exhibit increasing frequency of collaboration over time with their transatlantic private sector counterparts primarily in scientific and technical fields where they have weak national competences (*outsourcing-based collaboration*).
- *Proposition 4A:* US and European firms will exhibit increasing frequency of collaboration over time with their transatlantic public sector counterparts primarily in scientific and technical fields where they have strong national competences (*complementary collaboration*).
- *Proposition 4B:* US and European firms will exhibit increasing frequency of collaboration over time with their transatlantic public sector counterparts primarily in scientific and technical fields where they have weak national competences (*outsourcing-based collaboration*).

Our study is currently focused on analyzing 'clusters' of transatlantic research collaboration, where specific organizations show a pattern of continuing collaboration over time. Through statistical studies of these clusters, supplemented by case studies on specific clusters with interesting features, it will be possible to create a more comprehensive and detailed framework identifying key factors leading to transatlantic partnerships in scientific research.

NOTE

1. A striking example is the current race to map the SARS genome and develop a diagnostic test and vaccine for it – this effort involves public and private R&D facilities around the world engaged in active collaborative and distributed R&D (*Wall Street Journal*, 30 April 2003).

REFERENCES

Arundel, A. and A. Geuna (2001), 'Does proximity matter for knowledge transfer from public institutes and universities to firms?', *SPRU Electronic Working Paper Series*, Paper No. 73, University of Sussex, Brighton, UK.

Brusoni, S. and A. Geuna (2001), 'The key characteristics of sectoral knowledge bases: an international comparison', *SPRU Electronic Working Paper Series*, Paper No. 69, University of Sussex, Brighton, UK.
Carayannis, E. and J. Alexander (1999), 'Winning by co-opeting in knowledge-driven, complex environments: the formation of strategic technology government–university–industry (GUI) partnerships', *Journal of Technology Transfer*, **24** (2/3), 197–210.
Carayannis, E. and D. Campbell (eds) (2005), *Mode 3: Knowledge Creation, Diffusion and Use in Innovation Networks and Knowledge Clusters: A Comparative Systems Approach Across the US, Europe and Asia*, Westport, CT: Quorum Books/Greenwood Press.
Clark, B. (1995), *Places of Inquiry: Research and Advanced Education in Modern Universities*, Berkeley: University of California Press.
Georghiou, L. (1998), 'Global cooperation in research', *Research Policy*, **27**, 611–26.
Godin, B. and Y. Gingras (2000), 'The place of universities in the system of knowledge production', *Research Policy*, **29**, 273–8.
Hicks, D. (1993), 'University–industry research links in Japan', *Policy Sciences*, **26**, 361–95.
Katz, J.S. and B.R. Martin (1997), 'What is research collaboration?', *Research Policy*, **26** (1), 1–18.
Melin, G. (2000), 'Pragmatism and self-organization: research collaboration on the individual level', *Research Policy*, **29**, 31–40.
Melin, G. and O. Persson (1996), 'Studying research collaborations using co-authorships', *Scientometrics*, **36** (3), 363–77.
Metcalfe, S. (1995), 'The economic foundations of technology policy: equilibrium and evolutionary perspectives', in P. Stoneman (ed.), *Handbook of the Economics of Innovation and Technical Change*, London: Blackwell, pp. 409–512.
Narin, F. and E.S. Withlow (1990), *Measurement of Scientific Cooperation and Coauthorship in CEC-related Areas of Science*, Volume 1, Brussels: Commission of the European Union (EUR 12900 EN).
National Science Foundation (2000), 'Industry, technology and the global marketplace', in *National Science Foundation, Science & Engineering Indicators – 2000*, Washington, DC: National Science Foundation.
Newman, M.E.J. (1999), 'The structure of scientific collaboration networks', *Working Paper*, The Santa Fe Institute, Santa Fe, New Mexico, USA.
Okubo, Y. and C. Sjöberg, (2000), 'The changing pattern of industrial scientific research collaboration in Sweden', *Research Policy*, **29**, 81–98.
Oliver, A.L. and J.P. Liebeskind, (1997), 'Three levels of networking for sourcing intellectual capital in biotechnology', *International Studies of Management and Organization*, **27** (4), 76–103.
Organization for Economic Cooperation and Development (OECD) (1999), *Managing National Innovation Systems*, Paris: OECD.
Ostry, S. and R. Nelson (1995), *Techno-Nationalism and Techno-Globalism: Conflict and Cooperation*, Washington, DC: Brookings Institution.
Pechter, K. and S. Kakinuma (1999), 'Co-authorship linkages between university research and Japanese industry', in L. Branscomb, R. Florida and F. Kodama (eds), *Industrializing Knowledge: University–Industry Linkages in Japan and the United States*, Cambridge, MA: MIT Press, pp. 102–27.
Rosenbloom, R. and W. Spencer (1996), *Engines of Innovation*, Boston, MA: Harvard Business School Press.

Stead, G. and T.F. Harrington, (2000), 'A process perspective of international research collaboration', *Career Development Quarterly*, **48** (June), 323–31.
Ziman, J. (1994), *Prometheus Bound: Science in a Dynamic Steady State*, Cambridge: Cambridge University Press.
Zucker, L.G., M.R. Darby and J. Armstrong (1998), 'Geographically localized knowledge: spillovers or markets?', *Economic Inquiry*, **36**, 65–86.

14. Conclusion

Elias G. Carayannis, Aris Kaloudis and Åge Mariussen

This book explores issues of diversity and heterogeneity in national knowledge systems. In Figure 1.4 in the Introduction, we acknowledge the presence and interactions of input, process, and output factors in the knowledge society and economy manifested via co-existence, co-opetition, co-evolution, and co-specialization processes. We have further studied and discussed the ways and means in which diversity and heterogeneity influence how knowledge is created, diffused and used.

Our discussion of knowledge systems is open-ended. We have attempted to provide an emerging conceptual framework to serve as the 'intellectual sandbox' and 'creative whiteboard space' of the mind's eyes of 'knowledge weavers' (*Wissensweber*)[1] across disciplines and sectors as they strive to tackle the 21st-century challenges and opportunities for socioeconomic prosperity and cultural renaissance based on knowledge and innovation.

As a result of the glocalized nature and dynamics of state-of-the-art, specialized knowledge one needs to cope with and leverage two mutually reinforcing and complementary trends:

a. *Micro–macro*. The symbiosis and co-evolution of top-down national and multi-national science, technology and innovation public policies, and bottom-up technology development and knowledge acquisition private initiatives; and
b. *Multi-level*. The leveling of the competitive field across regions of the world via technology diffusion and adoption accompanied and complemented by the formation and exacerbation of multi-dimensional, multi-lateral, multi-modal and multi-nodal divides (cultural, technological, socioeconomic, and so on).

NOTE

1. The term constitutes the brainchild or *conceptual branding* of the authors as part of this journey of discovery and ideation.

Index